ADAPTIVE AND ~~RE~~ PRO ACTIVE

S.D.L.C. PROJECT MANAGEMENT

AGILE MEETS PMBOK(R)

MEETS PM YOU

AUTHOR: Joshua Boyde

DATE: 2015

A WORK IN PROGRESS

Dedication

This book is dedicated to all those projects that did not work out as was diligently planned; irrespective of how hard the Project Team strived to make it a success.

Firstly, I wrote this book purely for the enjoyment of writing, and

secondly, to repay all those who have taught me a thing or two during my career.

Hence, I now "pay it forward".

I hope you enjoy my eBook, and consider purchasing it in the physical form.

If you do happen to find this book useful to your understanding of System / Software Development Life Cycle (SDLC) project management then please pass it onto others.

JoshuaBoyde.
@Hotmail com

"I'm pleased that your words reflect so many of the things that I have experienced, and that you are passing on good examples of how to deal with these issues. I think this is turning into a better, more useful day to day tool for new or even experienced project managers.

[Update] ... I'm seeing this as a very useful 'how-to' guide; because, although your book documents what to me is common sense, I continue to encounter large and small organizations where this is evidently not the case. I would be tempted to have a hard copy on my desk just to say to people – here – do this."

Michael McAuliffe
Program Director

Foreword

"NO PLAN SURVIVES THE FIRST CONTACT INTACT".

> Paraphrasing Field Marshal Helmuth von Moltke (the Elder), 1800-1891.
> And, as a statement now considered as one of Murphy's Laws of Combat.

This book has been crafted for both the project management novice who is ready to confront their first real project, through to the seasoned veteran with several project battle campaigns under their belt.

Described within is the practical application of field-tested project management techniques to actual situations and prevailing circumstances where commercial realities have to be given serious consideration.

This book is based on many years of "real-world" System Development Life Cycle (SDLC) project management, the adaptation of the Project Management Body Of Knowledge (PMBOK®), the blending of Agile techniques, elements from other practices & principles, and the incorporation of the past experiences & lessons learnt from the various industrial backgrounds of those persons who graciously contributed to this book's creation.

Notes

This book is not intended to be like your usual textbook; hence, the use of language, colloquialisms, 3rd person comments, and a turn of phrase that best articulates the situations and prevailing circumstances being described in the various example cases.

That is, this book has been crafted as a practical guide for managing real-world SDLC projects, rather than as a purely academic or theoretical reference.

Please forgive the abundant use of ampersands "&" as a binder of interlinked thoughts and concepts; *e.g. "needs & wants", "monitoring & control", and "processes & procedures".*

Thank You

Thank you for opening my book, "**Adaptive & Proactive – S.D.L.C. Project Management – Agile meets PMBOK®, meets PM you**".

I would also like to thank the following contributors.

SME – Subject Matter Expert

Critique & "jack of all trades" SME: Steven Lipke

Project & Program Management SME: Michael McAuliffe

Earned Value Performance Measures & Budgeting SME: Allan Pedersen

References, Credits & Copyrights

This book is based on the practical knowledge and real-world experiences of the author, the reviewers, and those persons who happened to chip in their thoughts, opinions, and the details of their issues of concern. ... As well as, those work colleagues who took the opportunity to express their frustration with project management.

This book has used my other project management textbook, "**A Down-To-Earth Guide To SDLC Project Management**" (2nd Edition) as the primary source ... remoulding & evolving its contents to better suit the current situation & prevailing circumstances of the projects at hand, ... also, the PMBOK® Guide, public domain materials, thank you Wikipedia, and specific textbooks where referenced.

Every effort has been made to give credit where credit is due and to provide references where possible.

If you decide to use any of the contents of this book in your own works or studies then please give credit to whom credit is due.

Preface

Project management is so much more than knowing all of the theory and using the correct jargon words.

At the end of the day; the success or failure of your project, and subsequently the success or failure of you as a project manager will greatly depend on:

- ☐ Your ability to "adapt" to the project's current situation & prevailing circumstances confronting your employer and your customer. ... And,

- ☐ Your ability to "proactively" be ahead of the game so as to be in the right place at the right time, with the right people & resources at your command.

...

Well, I would like to add my 2 cents worth, that in my opinion, project success is also dependent on;

\# Your ability to "improvise" to maximize the utilization of that which is available to you, at that time and in that particular place. ... As well as,

\# Your ability to "communicate" sufficiently to be understood and to interpret the real situation that is at hand. ... And,

\# Some good old fashioned, "luck".

Table of Contents

1.	A PLACE TO BEGIN	13
1.1.	PRELUDE	13
1.2.	RECYCLING & REUSE OF A PREVIOUS WORK	15
2.	THEORY … A LITTLE BIT	19
2.1.	INTRODUCTION	19
2.2.	OVERVIEW	20
2.3.	PROJECT LIFE CYCLES	21
2.3.1.	LINEAR-WATERFALL PROJECT LIFE CYCLE	21
2.3.2.	ITERATIVE PROJECT LIFE CYCLE	22
2.3.3.	AGILE PROJECT LIFE CYCLE	25
2.3.4.	HYBRID PROJECT LIFE CYCLE	29
2.4.	PLAN – DO – CHECK – ACT	33
2.4.1.	PROBLEM SOLVING CYCLE	33
2.4.2.	CONTROL CYCLE	34
2.5.	PROJECT RESCUE & RECOVERY	36
2.5.1.	PROJECT RESCUE	36
2.5.2.	PROJECT RECOVERY	38
2.6.	R.I.S.C. MANAGEMENT	39
2.6.1.	RISK MANAGEMENT	41
2.6.2.	STAKEHOLDER MANAGEMENT	42
2.6.3.	CHANGE MANAGEMENT	44
2.6.4.	RULES TO R.I.S.C. MANAGEMENT	45
2.7.	PROJECT VARIABLE STAR	46
2.8.	PROJECT MANAGEMENT PROCESS	49
2.8.1.	PMBOK® AS A PROCESS MODEL	49
2.8.2.	"SATISFACTION" VERSUS "PERFORMANCE"	52
3.	THEORY … PMBOK®	59
3.1.	INTRODUCTION	59
3.2.	OVERVIEW	60

3.2.1.	DOCUMENT HIERARCHY	62
3.2.2.	GATING & GOVERNANCE	68
3.3.	**INITIATING PHASE**	**69**
3.3.1.	PURPOSE	69
3.3.2.	WHAT HAS TO BE ANSWERED?	69
3.3.3.	QUESTION: WHAT PROBLEM TO RESOLVE?	70
3.3.4.	QUESTION: WHAT DO THEY WANT?	74
3.3.5.	QUESTION: WHAT WILL WE GIVE & NOT GIVE THEM?	76
3.3.6.	QUESTION: HOW LONG IS IT EXPECTED TO TAKE, AND WHAT WILL IT COST US?	77
3.3.7.	QUESTION: WHAT WILL WE GET IN RETURN?	78
3.3.8.	QUESTION: WHAT IS AT STAKE FOR US?	81
3.3.9.	WHAT TO DO WITH THESE ANSWERS? … GATE	81
3.4.	**PLANNING PHASE**	**93**
3.4.1.	PURPOSE	93
3.4.2.	WHAT HAS TO BE ANSWERED?	93
3.4.3.	QUESTION: WE NEED >WHAT< , WHEN, AND HOW MUCH?	94
3.4.4.	QUESTION: WE NEED WHAT, >WHEN< , AND HOW MUCH?	98
3.4.5.	QUESTION: WE NEED WHAT, WHEN, AND >HOW MUCH< ?	109
3.4.6.	WHAT TO DO WITH THESE ANSWERS?	110
3.4.7.	PROJECT VARIABLES VERSUS THE BASELINES	113
3.4.8.	QUESTION: HOW WILL WE KNOW WE GOT IT RIGHT?	114
3.4.9.	QUESTION: WHAT CAN GO WRONG?	118
3.4.10.	PLANNING PHASE COMPLETION	119
3.5.	**EXECUTING PHASE**	**125**
3.5.1.	PURPOSE	125
3.5.2.	WHAT HAS TO BE ANSWERED?	127
3.5.3.	QUESTION: HOW EXACTLY WILL THE TEAM DO IT?	128
3.5.4.	QUESTION: IS THE TEAM DOING IT?	130
3.5.5.	QUESTION: IS THE TEAM CHECKING IT?	133
3.5.6.	QUESTION: IS THE TEAM DONE YET?	136
3.5.7.	WHAT TO DO WITH THESE ANSWERS?	140
3.5.8.	EXECUTING PHASE COMPLETION	141

3.6.	**MONITORING & CONTROL PHASE**	145
3.6.1.	PURPOSE	145
3.6.2.	A CONTINUAL ITERATIVE PROCESS	147
3.6.3.	[SCOPE] MONITORING & CONTROL	149
3.6.4.	[TIME] MONITORING & CONTROL	151
3.6.5.	[COST] MONITORING & CONTROL	153
3.6.6.	[QUALITY] MONITORING & CONTROL	170
3.6.7.	CHANGE MANAGEMENT	176
3.6.8.	RISK MANAGEMENT	181
3.6.9.	STAKEHOLDER MANAGEMENT	185
3.6.10.	[PEOPLE] MONITORING & CONTROL	194
3.6.11.	[RESOURCES] MONITORING & CONTROL	197
3.7.	**CLOSING PHASE**	199
3.7.1.	PURPOSE	199
3.7.2.	WHAT HAS TO BE ANSWERED?	200
3.7.3.	QUESTION: DID WE GET IT RIGHT?	201
3.7.4.	QUESTION: DOES EVERYONE AGREE IT IS ALL DONE?	204
3.7.5.	QUESTION: WHAT DID WE LEARN FROM THIS?	208
3.7.6.	QUESTION: CAN EVERYONE SAY GOOD BYE AS FRIENDS?	212
3.7.7.	CLOSING PHASE COMPLETION	213
4.	**YOUR PM TECHNIQUES**	217
4.1.	INTRODUCTION	217
4.2.	A PEOPLE FIRST, FOUNDATION	218
4.2.1.	THE SCENARIO	218
4.2.2.	BUILDING A GOOD FIRST IMPRESSION	218
4.2.3.	RECRUITING THE PROJECT TEAM	223
4.3.	CONSTRUCTING A TEAM	244
4.3.1.	TEAM DYNAMICS & TEAM RELATIONSHIPS	244
4.3.2.	TEAM WORK AND TEAM COHESION	247
4.4.	A PEOPLE FIRST MANAGER	258
4.4.1.	BEING A "HUMANE" PROJECT MANAGER	258
4.4.2.	BEING A "LEADER" PROJECT MANAGER	284

4.4.3.	YOUR PERSONAL PM CREDO	291
5.	**PM YOU … FIELD TEST 1**	**295**
5.1.	OVERVIEW	295
5.2.	INTRODUCTION	296
5.2.1.	THE SCENARIO	296
5.2.2.	BUILDING THAT FIRST IMPRESSION	297
5.3.	AND SO, IT BEGINS	301
5.3.1.	ANSWERS TO THE OPENING QUESTIONS	301
5.3.2.	ADAPT & IMPROVISE ON THE PROJECT'S DOCUMENTATION	304
5.3.3.	ADAPT & IMPROVISE ON THE PROJECT'S DURATION & BUDGET	312
5.4.	GETTING UNDERWAY	313
5.4.1.	OVERVIEW SCHEDULE AS AN ATTACK PLAN	313
5.4.2.	COSTING THE PROJECT TO SUCCESS	315
5.4.3.	PRICING THE PROJECT TO A CERTAIN DEATH	321
5.5.	ALL AHEAD, FULL … "ICEBERG !!"	327
5.5.1.	PROJECT RESCUE … GETTING OUT OF TROUBLE	327
5.5.2.	PROJECT RECOVERY … REDEFINING THE PLAN	331
5.5.3.	THE SCENARIO … FAILURE TO GAIN ALTITUDE	337
5.6.	COPING WITH … "FAILURE"	341
5.6.1.	SELF-DOUBT AND QUESTIONING WHY	341
5.6.2.	DIFFERENCES IN UNDERSTANDING OF SDLC	342
5.6.3.	WORK RELATED STRESS	346
5.7.	"FAILURE" A POSSIBILITY OF FACT	354
5.7.1.	"FAILURE" IS OPINIONATED	355
5.7.2.	PROJECT AUTOPSY	356
6.	**PM YOU … FIELD TEST 2**	**359**
6.1.	OVERVIEW	359
6.2.	INTRODUCTION	360
6.2.1.	THE SCENARIO	360
6.2.2.	BEGINNING FROM A CLEAN SLATE	361
6.3.	AND SO, IT BEGINS	363
6.3.1.	WHAT DO THEY WANT?	363

6.3.2.	WHAT'LL WE GIVE & NOT GIVE TO THEM?	364
6.3.3.	HOW LONG IS IT EXPECTED TO TAKE?	368
6.3.4.	WHAT WILL IT COST?	417

7. PM YOU ... FIELD TEST 3 427

7.1.	OVERVIEW	427
7.2.	INTRODUCTION	428
7.2.1.	THE SCENARIO	428
7.2.2.	START AT THE BEGINNING	428
7.2.3.	MONITORING & CONTROL, TODAY'S CONCERN	430
7.3.	MONITORING & CONTROL	435
7.3.1.	THE PROJECT SCHEDULE	435
7.3.2.	LEVEL OF EFFORT (S-CURVE) & HEAD COUNT	436
7.3.3.	SCHEDULE PERFORMANCE MEASURES	437
7.3.4.	AGILE PERFORMANCE MEASURES	440
7.3.5.	EARNED VALUE PERFORMANCE MEASURES	445
7.3.6.	KEY PERFORMANCE INDICATORS AND PROGRAM REPORTING	455

8. PM YOU ... FIELD TEST 4 469

8.1.	OVERVIEW	469
8.2.	INTRODUCTION	469
8.2.1.	THE SCENARIO	469
8.3.	R.I.S.C. MANAGEMENT	470
8.3.1.	RISK MONITORING & CONTROL	472
8.3.2.	STAKEHOLDER MONITORING & CONTROL	481
8.3.3.	CHANGE MONITORING & CONTROL	502
8.4.	ACTIONS & REACTIONS	513
8.4.1.	HANDLING STAKEHOLDER REQUESTS	513
8.4.2.	DELEGATING DOWNWARDS AND MANAGING UPWARDS	521
8.4.3.	A PROJECT MANAGER'S DECORUM	525

9. PM YOU ... FIELD TEST 5 529

9.1.	OVERVIEW	529
9.2.	INTRODUCTION	529
9.2.1.	THE SCENARIO	529

9.3.	PROJECT CLOSURE	530
9.4.	BUSINESS CHANGE MANAGEMENT	535
9.4.1.	THE EFFECT OF CHANGE ON THE BUSINESS	535
9.4.2.	BUSINESS CHANGE MANAGEMENT PROCESS	539
9.5.	BUSINESS AS USUAL OPERATIONS	544
9.6.	BENEFITS REALIZATION	547
10.	THE END	555
10.1.	ADAPT AND IMPROVISE, ELSE	555
10.2.	PM YOU DON'TS	558
10.3.	OFFICE POLITICS	563
10.4.	IS BEING A PM RIGHT FOR YOU?	571
11.	REFERENCES & INDEX	577

1. A Place To Begin

1.1. Prelude

The objective of this book is to provide you (the reader) with the knowledge, understanding, and field-tested techniques that are necessary to guide your **System / Software Development Life Cycle (SDLC)** project to a successful conclusion.

And, hopefully during your journey of reading this book, the ideas & topics within will serve to aid in the evolution of your own "customized" project management techniques that are tailored specifically to your leadership style and to your personality traits. ... As well as, giving due consideration to your employer's (i.e. the performing organization's) processes & procedures, and also taking into consideration your customer's expectations of, *"what a professional project manager should do"*.

> And, this brings us to the zeroth rule of "Adaptive & Proactive" SDLC Project Management.
>
> **RULE 0: DO NOT expect your project to be deemed a success, just because you precisely followed some prescribed project management methodology; rather, success is dependent on you continually adapting to the current situation & prevailing circumstances.**

I would recommend that, you start by taking some recognized project management methodology which you feel comfortable with, and use this as the foundation on which to build your own project management techniques. ... And then, **continually adapting** "your techniques" **to deal with the current situation & prevailing circumstances**, by emphasizing those aspects which add value while deemphasizing other aspects that won't contribute to achieving your current project's successful outcome.

TERMINOLOGY ... *Two sides to every story.* ... *The situation & circumstances.*

- **Performing Organization** ... the organization that is executing | implementing the project.

- **Customer Organization** ... the organization that has "sponsored" (is paying for) this project to be undertaken on their behalf. Noting that, the customer could be either external OR internal to the performing organization.

- **Current Situation** ... is the project's relationship to the performing organization's present state of being, when given the condition | position that it is now in.

- **Prevailing Circumstances** ... is the project's relationship to the performing organization's (and customer organization's) surrounding socio-economic conditions, which will influence the project's future situation and the subsequent activities that will & won't be undertaken.

THEORY ... *What is a project?*

- A project is a **limited duration unique endeavour** that produces a **one-off set of deliverables** that are not brought about by continually ongoing repetitive operations; i.e. a project is **NOT a "Business As Usual"** (BAU) activity.

- A project has either a **definitive beginning** (Start Date) **and/or a definitive end** (End Date) by when a **specific collection of objectives** will have been **achieved to the satisfaction of the project's stakeholders within pre-agreed performance measures**. ... OR, it is decided that these objectives can no longer be achieved effectively, or there are such significant changes to the requirements that the project has to be re-evaluated (and re-scoped), or these objectives are no longer applicable and thus the project is not required anymore.

In essence, the act of writing and publishing this textbook forms a kind of project.

1.2. Recycling & Reuse of a Previous Work

There is copious amounts of information on SDLC Project Management, and I have previously published a (telephone directory thick) textbook, "**A Down-To-Earth Guide to SDLC Project Management**". ... And, that book did cover a lot of theory.

> "ALL APPLICABLE KNOWLEDGE CAN BE DISTILLED DOWN TO THE ESSENTIAL INFORMATION THAT IS NECESSARY TO UNDERSTAND THE CURRENT SITUATION AND TO COPE WITH THE PREVAILING CIRCUMSTANCES."

Thus, the [Scope - definition] of this current textbook (that you are now reading) is to; primarily concentrate on field-tested project management techniques (which are based on solid theory) that can be applied to real-world situations and prevailing circumstances where the realities of commercial necessities have to be given serious consideration.

> *For this current textbook, many of the diagrams and the notable theory points will be recycled & reused (aka "adapted") from my previously mentioned textbook.*

If the creation of this current (new) textbook was to be considered as **a project**, then such **recycling & reuse would substantially reduce the [People] effort required** to produce many of the diagrams and the determination of what are the important theory points; i.e. to start with **satisfactory [Quality]** of **[Resource]** materials. Thereby, **noticeably reducing the [Time]** required to complete this project. ... And, if the [People] doing this work were being paid on an hourly basis then this recycling & reuse would **significantly reduce the [Cost]** of paying for this work to be done.

And, this brings us to the first rule of "Adaptive & Proactive" SDLC Project Management.

RULE 1: Recycling & Reuse is GOOD.

THEORY ... *Variables and Constraints on a project's success.*

What are the determinates ("parameters") of a project's success or failure?

(1) **SCOPE** ... does the project produce the expected results and does the resultant **deliverables contain the agreed features & functionality?**

(2) **TIME** ... was the project's outputs delivered to the customer (i.e. the persons requesting the product, service, or goods) when it was agreed to be delivered? That is, **did the project deliver to its milestone dates?**

(3) **COST** ... does the cost of the project not exceed the budget that was allocated for undertaking the project? That is, **was the project a profitable endeavour?**

(4) **QUALITY** ... does the resultant **deliverables meet the agreed Acceptance Criteria?**

Additionally,

(5) **PEOPLE** ... did the project **effectively & efficiently utilize those persons** that were **assigned** to do work on the project?

(6) **RESOURCES** ... did the project **effectively & efficiently utilize those material & services resources** that were **allocated** to be used by the project?

There are a couple of important things to note from the above list:

- ❖ Firstly, the existence of **six project variables / constraints** (i.e. "project parameters") that determine whether the project will succeed or fail. These project variables / constraints **of [SCOPE], [TIME], [COST], [QUALITY], [PEOPLE], and [RESOURCES]** will **need to be continually monitored, adjusted,** and when necessary, **adapted** so as **to maintain balance,** else <RISK> losing control over the project.

- ❖ Secondly, the inclusion of the words "agreed", "expected", and "allocated"; as, the **determination of whether a project is deemed "a success" or "a failure" is based entirely on the expectations & perspectives of the people involved with the project**; i.e. **"SUCCESS IS OPINIONATED".** ...And, you can't influence their "opinions" unless you engage in meaningful bidirectional communications with them.

THEORY ... *Risks to a project's success.* ... *Balancing project constraints and variables.*

What is the difference between a project variable and a project constraint?

❖ <u>**Project Variable**</u> ... is an **aspect** (parameter) of the project **that can be ADJUSTED** in response to changes to the project's current situation & prevailing circumstances.

❖ <u>**Project Constraint**</u> ... is a project variable that **has been FIXED to a specific value** (aka a "**Baseline**") that **will not be changed in the foreseeable future without prior authorization** (via some formalized **Baseline Change Control** process).

A baseline is some agreed constant against which the project's current progress is measured & compared, and by which the project's Success | Failure is judged.

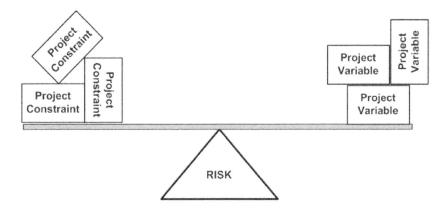

Figure 1: Balance relationship of <Risks> to project variables and project constraints.

NOTE: If **too many** of these project variables become **project constraints then** there may **not be enough flexibility** in the remaining unconstrained project variables **to manoeuvre when problems & issues confront the project.** Subsequently, with the majority of the project variables stipulated as being project constraints then **as pressure is exerted on the project then the <Risk> is that the project will lose stability and some of these project constraints would uncontrollably revert to being project variables,** ... irrespective of those mandated as being constraints.

NOTE: The following chapter is an overview introduction to SDLC Project Management theories & concepts, and these starting points will be expanded upon in later chapters.

2. THEORY ... a little bit

2.1. Introduction

As the reader of this textbook, you are in effect the customer who will decide whether this "textbook project" is a success or a failure. ... Based on your expectation, perspective, and your opinion of the resultant deliverables.

However, your current level of understanding about project management (encompassed by your previous training and past experiences) will greatly influence your decision. Hence, before delving deeply into the application of practical SDLC Project Management, a common foundation to project management will need to be established.

Don't worry, this chapter is a very condensed summation of project management theory, composed of explanation diagrams and notable points adapted from my textbook, "**A Down-To-Earth Guide To SDLC Project Management**" (2nd Edition).

Which brings us to the next rule of "Adaptive & Proactive" SDLC Project Management.

RULE 2: If you want a simple solution, then what would a lazy person do?

Umm, a lazy person's simple solution would be to "recycle & reuse" extracts from that previous textbook.

...

"Though, you shouldn't just cut & paste one-liners from other sources, just because some search engine found that line based on your entered key words. What is also important is the context of where that particular line appeared in the paragraphs of other words, and how does it fit in the overall scheme of things."

2.2. Overview

The collective diagrams in [Figure 2] below are some of the primary theories that will be summarized in this chapter.

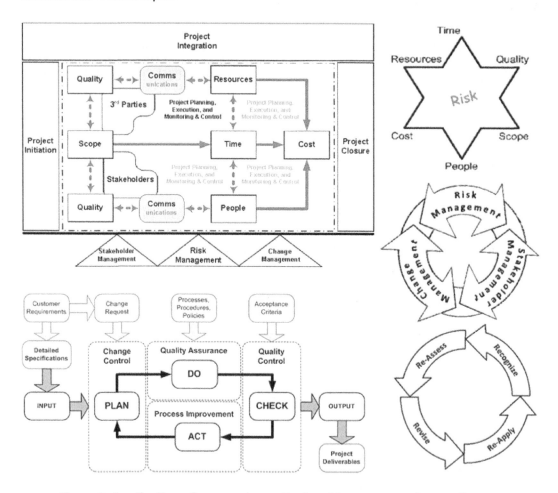

Figure 2: A collection of some primary Project Management theory diagrams.

These theories depicted above, are the PMBOK® inspired "Project Management Process" model, the "Project Variable Star" model, the "PLAN – DO – CHECK – ACT" model, the "R.I.S.C. Management" model, and the "Four Re's" model of Project Rescue & Recovery.

Though, let's start with the basic models which underlie all Software / System Development Life Cycle (SDLC) projects.

2.3. Project Life Cycles

2.3.1. Linear-Waterfall Project Life Cycle

In the beginning, ... when projects started to be considered as logical endeavours rather than as ad-hoc journeys, ... a project was envisioned as a **linear "waterfall" arrangement where each phase sequentially followed the completion of the previous phase**; like water flowing down a stream and toppling over waterfalls along the way, see [Figure 3] below.

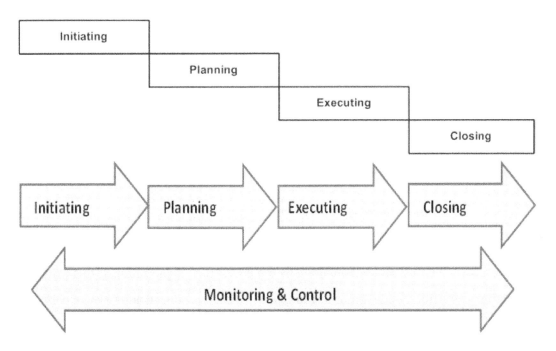

Figure 3: Project Life Cycle phases outlined as a Linear-Waterfall model.

This Linear-Waterfall Project Life Cycle clearly possessed **a systematic way to get from the START | "Initiating"** of the project, through **"Planning"**, **"Executing"**, to finally arrive at the **"Closing" | END**. ... And hopefully, via diligent **"Monitoring & Control"**, the project would be completed in a predetermined [Time] duration, and/or within a prescribed [Cost] budget, and that the pre-agreed [Scope] was delivered in accordance with the pre-agreed Acceptance Criteria so that the pre-agreed level of [Quality] was achieved.

This Linear-**Waterfall Life Cycle model** was "*all fine & dandy*" for producing things **when**;

☐ what the **customer's wanted & needed was clearly understood** by the representatives from both the customer & performing organizations,

☐ there was a **clearly defined body of work with obvious achievement points**,

☐ there was a **dedicated team** to be assigned to work on a **siloed part** of the project, and

☐ the customer was **prepared to wait for the entire project to be completed** before being able **to receive the agreed deliverables**.

Which brings us to the next rule of "Adaptive & Proactive" SDLC Project Management.

RULE 3: To understand a complex thing, then start with a simplified explanation.

2.3.2. Iterative Project Life Cycle

But, what if for the project being undertaken;

(a) the **[Scope] features & functionality to be delivered are not clearly understood** by the Customer's Representatives let alone by the performing organization as this would be better understood once something tangible has been produced and "*played with*",

(b) the customer requires that **some of the [Scope] functionality is to be delivered** "**As Soon As Possible**" and the rest can wait until later on, and/or

(c) the customer only has a **limited [Cost] budget and/or available [Time]** and hence is looking to obtain the best "*bang for their buck*" (i.e. the best **Return On Investment**)?

The simple solution to these above listed problems is to modify ("*adapt*") the Linear-Waterfall Project Life Cycle model so that only a limited number of features & functionality (of the total collection) are completed during each cycle through the project, so that each cycle builds upon the previous cycle, see [Figure 4].

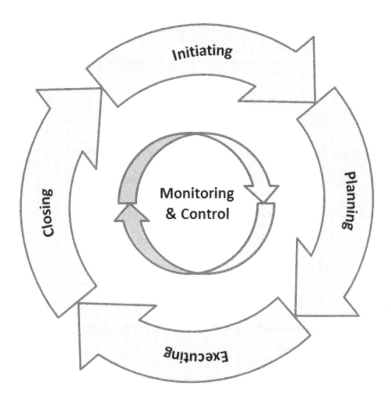

Figure 4: Project Life Cycle outlined as an iterative model.

Hence, **at semi-regular intervals the project's life cycle would return to the beginning,** where the customer & performing organizations' representatives could **re-evaluate** the project's **current situation & prevailing circumstances**; i.e. be **"Change Driven"**.

And, subsequently at the **start of the next cycle**'s Initiating Phase, the project's **deliverables could be refined or completely redefined, and re-prioritized**; i.e. **"Progressive Elaboration"**.

Which brings us to the next rule of "Adaptive & Proactive" SDLC Project Management.

RULE 4: Don't be afraid to take a solution that works for some situations and adapt it to better suit a different situation.

Excellent, ... this Iterative Project Life Cycle model now provides:

(1) The capability to **obtain something usable a lot earlier** than with the full Waterfall Project Life Cycle, and thereby **learn quicker from the outcomes & mistakes made during that previous cycle.**

(2) The earlier partial delivery would enable the project to be **re-evaluated sooner and more often**, and thus by progressive elaboration would provide:

- Earlier **customer feedback** that improves the chances of the project deliverables meeting the customer's expectations, and enables the **earlier detection of missing or misunderstood features & functionality**.

- **More flexibility to change the project's direction** when the current situation & prevailing circumstances deem it necessary to do so.

"Though, there is the potential for the [Costs] to escalate as changing requirements are thrown in with these iterative cycles."

(3) This earlier partial delivery improves the customer's and the other primary stakeholders **belief in** the project's potential for **success because** they have **physical "proof of progress"**.

Where **each iterative cycle could be thought of as a mini waterfall project**; see [Figure 5].

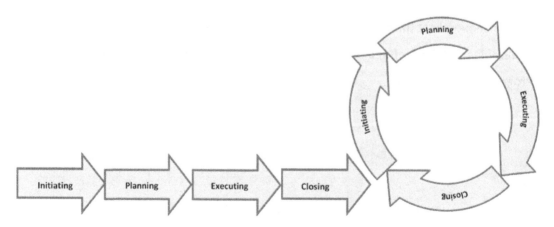

Figure 5: An Iterative Life Cycle as a mini waterfall project.

And, the conclusion of each iterative cycle could be declared a milestone. Therefore, the project would consist of a sequence of iterative milestones; see [Figure 6].

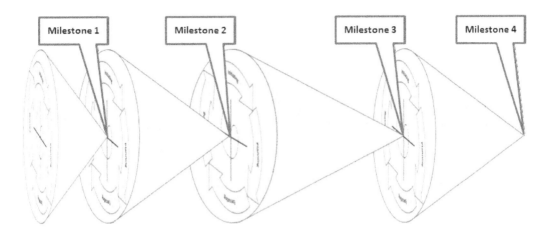

Figure 6: Iterative Project Life Cycle delivered as linear milestones.

However, this **Iterative Life Cycle does not define or limit the duration of each cyclic pass**; hence, "meaningful chunks" of features & functionality would be handed over to the customer at **varying durations** depending on what has to be delivered for each milestone. Consequently, an iterative cycle could be days, weeks, months, even years in duration.

Which brings us to an addendum to "Adaptive & Proactive" SDLC Project Management.

RULE 4^A: Don't be afraid to keep adapting an evolving solution to better suit a changing situation.

2.3.3. Agile Project Life Cycle

But, what if instead of having such variable durations between each of these milestones, there were **consistent "time boxed" periods of only 2, 3, or 4 weeks**; so that the project would be "**sprint**ing" along at **a constant beat, rapidly turning out useful components** that could be **quickly evaluated to verify** that those **completed deliverables** were going to **meet a selected portion of the agreed Acceptance Criteria**. See [Figure 7].

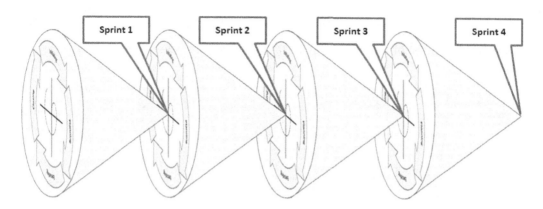

Figure 7: Agile cycles delivered as fixed duration sprints.

It would be best that **these** *"regular as clockwork"* **iterative cycles** should be organized so that **only a limited number** (aka 'Sprint Backlog') **of those remaining priority ordered features & functionality** (i.e. 'To Do Items') **be selected "to be done"** for each 'Sprint' cycle. This *"agile"* process is often represented via a Task Board, see [Figure 8] below.

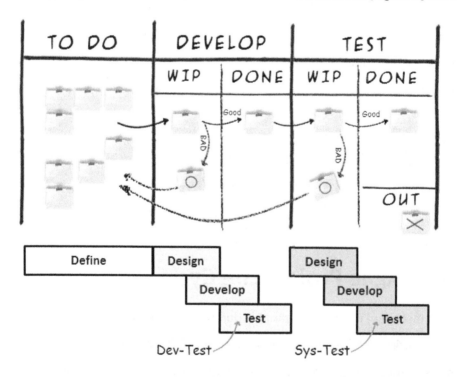

Figure 8: Agile Task Board motion with its own little waterfall sequences for both Development and Testing.

Where, **the motion across the Task Board is** similar to that of waterfall except on an individual 'To Do Item' basis. That is, **akin to "Lean Development" where one feature** (in its simplest form) is "**Defined, Designed, Developed,** and **Tested**".

> *"Though, the scope of the 'To Do Item' needs to be defined as the customer won't necessarily understand what is being provided."*

If for any reasons, a **selected 'To Do Item' has not been completed** (i.e. is **not ready to be deployed** or to be **demonstrated**) **by the end date of the current Sprint** then this incomplete functionality would be **removed from the current Sprint Backlog**, and subsequently **put back into the** remaining accumulation of unselected features & functionality (aka '**Product Backlog**'), so as to be implemented in some later cycle.

> *"With Agile methodologies there would not necessarily be a distinct demarcation between those who develop and those who test (as shown in the previous illustration) as the Project Team is supposed to be cross-functional and self-organizing. However, such a clear separation of development and (Quality Control) testing may be sufficient to satisfy the concerns of those project stakeholders who are more comfortable with waterfall siloing."*

Given the relatively short period of time between each sprint cycle, this would subsequently **limit the window of opportunity for changes to be requested for each cycle's chosen functionality**. Hence, limiting the possibility to slip in additional functionality & features into the current sprint cycle (i.e. limiting "Scope Creep").

Excellent, ... this Agile Project Life Cycle model now provides:

(1) The **high priority features are delivered in the early stages** of the project's life, and therefore these features can be utilized via a usable product before the project is even completed. Thus, **providing a significant "Return On Investment"** for both the customer & performing organizations.

(2) With the **low priority features, being delivered towards the end** of the project's life then there is the **possibility to drop these low priority features** as these comparatively provide **minimal Return On Investment**.

(3) The **regular periodic delivery of usable functionality** (with relatively short durations between deliveries) means that the **customer can be more involved with the evaluation** of the deliverables and therefore **request changes & corrections sooner**; rather than later, as can be the case with the Waterfall and Iterative Life Cycles.

(4) With usable deliverables being provided sooner and more often, the **customer could decide** at say 60-80% of the way through the project **that the deliverables** thus far are **sufficient to be used to generate a revenue stream** for themselves and thereby recoup some of the [Costs] of undertaking the project.

> *"Agile techniques are highly adaptive to rapidly changing situations and prevailing circumstances that were not necessarily known when the project was planned out, let alone when the project was conceived. Also, Agile is especially useful when the project's (Scope) of requirements are not clearly defined; as the project's stakeholders will understand it better once something is presented. Or, when the requirements (Scope) of the overall system is so large that to define it all up front in advance of the implementation commencing would take ages. So with Agile, the project would be undertaken as Rolling-Waves of Define – Design – Develop – Test of specific Use Cases or as evolutions of the whole."*

Which brings us to an addendum to "Adaptive & Proactive" SDLC Project Management.

RULE 3^A: A complex thing, once understood seems to be such a simple thing.

RULE 3^B: A complex thing, is still a complex thing to those who don't yet know or understand that thing.

2.3.4. Hybrid Project Life Cycle

While that was rather simple; however, different levels of the performing organization and the customer organization (i.e. a large portion of the project's stakeholders) would prefer (based on their particular perspectives, expectations, opinions, and past experiences) that the project should be undertaken via either a Waterfall, Iterative, or Agile Life Cycle.

Project Management Hierarchy and Governance Structure

Consider the three different levels of the project management & governance hierarchy.

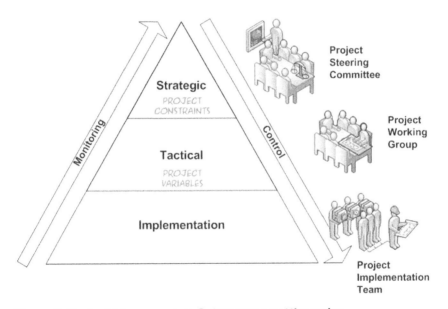

Figure 9: Project Management & Governance Hierarchy.

(1) **Project Steering Committee** … are **drawn from the ranks of senior management** at the performing organization (and maybe from the customer organization). This committee is **concerned with the** "*big picture*" of how the project fits in with the organization's **other projects, programs, and business strategies**. Therefore, the Project Steering Committee would **prefer the Waterfall** Life Cycle (with its logically sequential nature of an approximately known duration and resource utilization) because it is **better suited for laying out the** "**Roadmap**" **of** the organization's **future endeavours and thereby allocating budgets & personnel to such endeavours**.

(2) **Project Working Group** … are **generally drawn from the ranks of middle management** at the performing organization (and possibly from the customer organization). This working group **provides tactical directions for the implementation** of the "*individual pictures*" composing the project; i.e. coordinating & arranging project activities.

(3) **Project Implementation Team** … is composed of those persons (internal and external to the performing organization) that **do the "hands-on" execution of the project**'s activities. Therefore, the Project Implementation Team with their focus solely on executing the project's tasks would **prefer** the "break-it-down" of **the Agile Life Cycle**.

Hybridized Merging of Project Life Cycles

The **Project Working Group acts as the "middleman" between the Project Steering Committee and the Project Implementation Team, interfacing & translating between these two parties.** For the **Project Working Group, the Iterative life cycle is** the **optimum** fit because it can be manipulated to provide the information necessary for the Waterfall life cycle and the control essential for the Agile life cycle not to become chaotic (or rather, be perceived as chaotic by the Project Steering Committee); see [Figure 10] & [Figure 11].

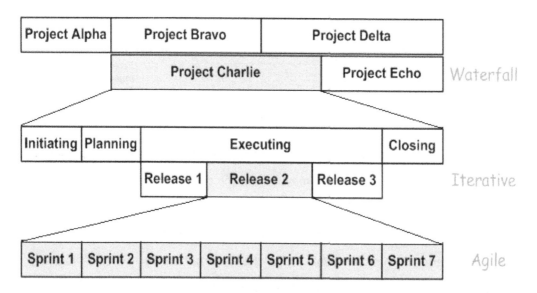

Figure 10: Hybrid relationship between Waterfall, Iterative, & Agile project life cycles.

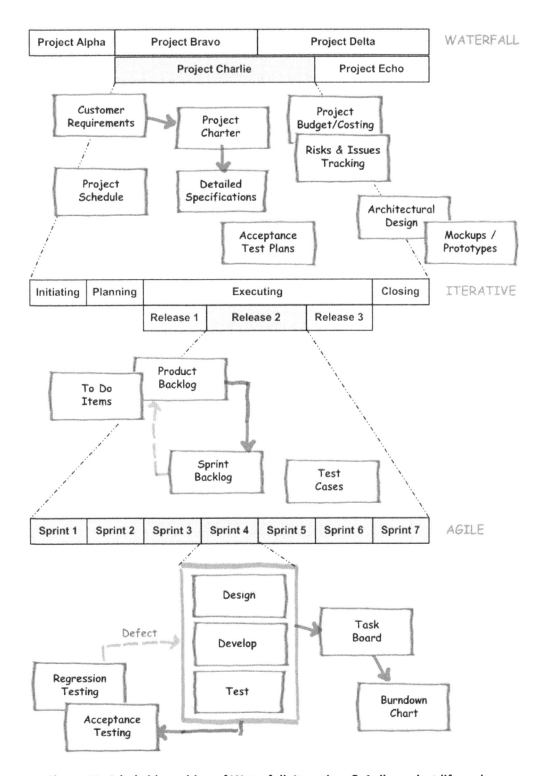

Figure 11: A hybrid mashing of Waterfall, Iterative, & Agile project life cycles.

COFFEE BREAK DISCUSSION ... *What is the optimum duration for an agile sprint?*

Agile is described as having timeboxed Sprint durations of one (1) week to one month duration, with two (2) weeks being typical. ... But, I'll argue that **one calendar month is the ideal** because.   ... *"You better get a cup of coffee or tea for this."*

(1) With each Sprint starting on the first working day of the calendar month and the Sprint ending on the last working day of that same calendar month. People generally know when a new calendar month starts, when it is about to end, and especially when they can expect to be routinely paid in between. Therefore, a pre-existing synchronized time clock resides in each team member's head. Thus, each team member will feel that at the conclusion of each week then they should be another quarter way towards finishing the work to be done for this current Sprint.

(2) Conceptually the calendar month Sprint can be broken up into Week 1 "combined define & design", Weeks 2-3 "combined develop", and Week 4 "combined testing". And, a calendar month is typically 4.33 weeks in duration, which gives a couple of days buffering if required. Whereas with 2 weeks sprints, The Team members can start to feel that they are constantly *"under the gun"* of design – develop – test, and *"don't get a breather"* to consider *"the big picture"* of where the project is heading.

(3) With a 2 week Sprint, that is, 10 working days in duration. The loss of a day due to sick leave, annual leave, diversions onto other work, equates to 10% of the total Sprint duration lost per project team member affected. Whereas with a calendar month Sprint, that is on average 22 working days in duration, the loss of a day per Sprint equates to under 5% per effected team member, which if needs be equates to only 15-30 minutes extra work per day for the remaining work days to catch up.

(4) With calendar month Sprints, that is potentially 3 customer releases per quarter year. A lot of progress can be made in a month, therefore the customer will notice significant advances per release which makes their sign off on payment milestones more justifiable which makes for better expectations on receiving regular cash flow.

2.4. PLAN – DO – CHECK – ACT

2.4.1. Problem Solving Cycle

While hybridizing the application of project life cycles, let's take that Iterative model in [Figure 4], and overlay a problem solving process like a "**PLAN – DO – CHECK – ACT**" cycle (P.D.C.A.) [ref: Deming / Shewhart], as shown below in [Figure 12].

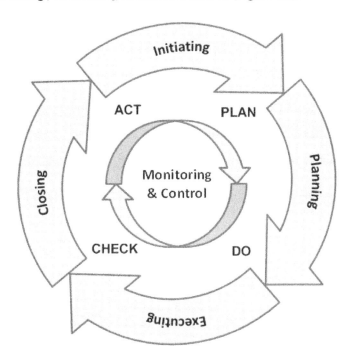

Figure 12: 'PLAN – DO – CHECK – ACT' relationship with an Iterative life cycle.

1.. <u>**PLAN**</u> … **decide what** has **to be do**ne, and **what** are the **expected** results.

2.. <u>**DO**</u> … **implement** what was decided to be done.

3.. <u>**CHECK**</u> … **compare** the **actual** results of what was done, **with** the **expected** results.

4.. <u>**ACT**</u> … **determine** the **cause of** any **differences** between the actual results and the expected results, decide whether **improvements** have to be made, and decide whether it is necessary to **go around** the cycle **again**, and again, and again, and again.

2.4.2. Control Cycle

Let's add a bit of logical structure to this 'Iterative PLAN – DO – CHECK – ACT' model and come up with something that looks a lot more like a **feedback loop process flow**, with its clearly defined INPUTs and OUTPUTs.

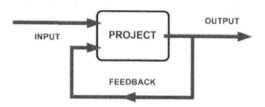

The "**INPUT**" into the project will be easy to define, as this would be the **Customer Requirements**. ... But, the customer organization and the performing organization will **probably negotiate over what is "Wanted" and** that which **is "Needed"**, so let's consider this input to be an agreed **Detailed Specifications**.

As for the "**OUTPUT**" from the **project**, well this of course would be the **Deliverables**; however, the customer & performing organizations will have to come to a **clear understanding of what this is**, so some **Acceptance Criteria** will have to be used.

And of course, once this process is operating then someone is likely to **Request Changes**.

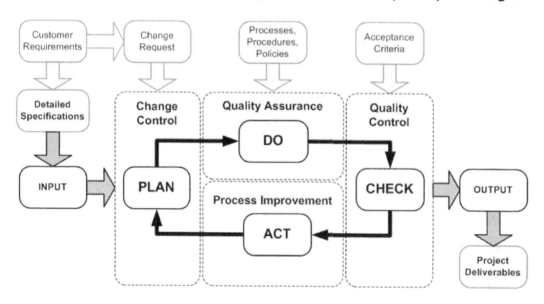

Figure 13: An 'Iterative PLAN – DO – CHECK – ACT' process with respect to Change Control, Quality Assurance, Quality Control, and Process Improvement.

Subsequently, the resultant project process flow would look something like [Figure 13].

Which brings us to the next rule of "Adaptive & Proactive" SDLC Project Management.

RULE 5: A simple thing, can be expanded upon until it becomes a complex thing.

Okay, now that we have constructed this complex looking thing, lets run it through.

1.. **PLAN** The INPUT into this process flow is the agreed Detailed Specifications, and with that a decision has to be made as to **what will need to be done, what are the expected outcomes** of churning through this process, and to then record these expectations in the **Acceptance Criteria**.

2.. **DO** **Implement** what was decided to be done while abiding by the performing organization's **Quality Assurance processes, procedures, and policies** so that the resultant Project **Deliverables will conform** to the definitions & descriptions in the agreed **Detailed Specifications (and any approved & authorized Change Requests)**.

3.. **CHECK** The performing organization's **Quality Control** people will have to **evaluate and verify** that the resultant OUTPUT Project **Deliverables do conform** to the **expected results** as was described in the **Acceptance Criteria.**

4.. **ACT** Determine the **cause of** any **differences** between the actual results and the expected results, decide on whether corrections will be necessary, should **Process Improvements** be made to processes and/or procedures due to inefficiencies & ineffectiveness, and decide on whether it is necessary to **go around** this process cycle **again** and again; I.e. a "**Continuous Improvement**" FEEDBACK loop.

5.. **PLAN** Are there any remaining agreed Detailed Specifications left to do, are there **Change Request**s to that which has already been done, are there Improvements & corrections from the last cycle that need to be implemented the next time round.

Which brings us to an addendum to "Adaptive & Proactive" SDLC Project Management.

RULE 3^C: A complex thing, will have detailed ways of working.

2.5. Project Rescue & Recovery

The 'Iterative PLAN – DO – CHECK – ACT' model in [Section 2.4] is good for those routine project course corrections and agreed adjustments & changes; but, what about when the project (colloquially) *"skids off the road"*, … so what then?

Well, with this colloquialism scenario, there are two distinct parts to handling a project when it suddenly goes wrong. The first part, **Project Rescue are those emergency actions undertaken to save the project from potential failure due to abrupt changes to the current situation and/or to the prevailing circumstances.** Whereas the second part, **Project Recovery are those planned operations undertaken to mitigate a pending project failure after stabilizing the project from a loss of control.**

2.5.1. Project Rescue

It is during the Executing Phase (i.e. the "Implementation Stage") **of the project when things are most likely to go wrong**, and hence when some form of Project Rescue actions will be required. Analogy wise, **Project Rescue** could be thought of as the companion gear that silently **meshes with the project's routine 'PLAN – DO – CHECK – ACT' cycle**, so as to **bring things back under control** if *"things get too out of whack"*.

Project Rescue consists of the following steps:

1.. <u>Recognize</u>   **Realize that a problem exists** and that the problem **is occurring** (is immediately about to occur, or has just occurred & could escalate to being worse).

2.. <u>Reassess</u>   **Evaluate what happened,** and **determine the cause** of what's gone wrong.

3.. <u>Revise</u>   Decide whether it is necessary to do something about it **straight away or can it wait** for a more opportune moment, decide **what exactly has to be done**, decide **what is the expected outcomes**.

4.. <u>Reapply</u>   Implement that which was **decided to** be **do**ne.

5.. **Repeat** from step 1, **until** the situation is **completely back under stabilized control**.

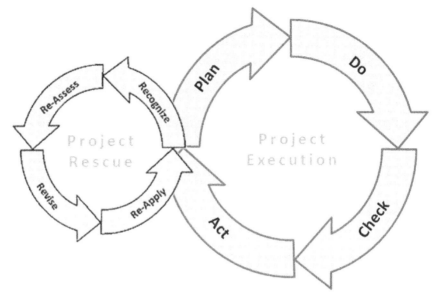

Figure 14: Relationship of Project Rescue to the 'PLAN – DO – CHECK – ACT' Cycle.

While the steps of the four Re's "**Recognize – Reassess – Revise – Reapply**" can be viewed simply as a re-sequencing of 'PLAN – DO – CHECK – ACT' as "**CHECK – ACT – PLAN – DO**"; the difference being that the four Re's is "<u>reactive</u>", rather than **PDCA being** "<u>proactive</u>".

> *"Though, many a project is run in what seems to be a constant state of reactive rescue, rather than as a logical endeavour."*

Hence, given the current situation & prevailing circumstances, **the project's management needs to be able to both proactively & reactively manage the project**.

Which brings us to the next rule of "Adaptive & Proactive" SDLC Project Management.

RULE 6: BAD things do happen.

RULE 6[A]: The occurrence of a BAD thing is not necessarily anyone's (direct or indirect) fault. ... So, it is better to just deal with that BAD thing now. And, not look for someone to blame.

2.5.2. Project Recovery

Once the project has colloquially been *"dug out of the ditch"*, *"peeled off the wall"*, or *"hoisted back up onto stable land"*, then some form of Project Recovery | project salvage will have to be undertaken.

Analogy wise, **Project Recovery** could be thought of as **fixing and/or replacing** those **busted** component **bits** of the current **'PLAN – DO – CHECK – ACT' cycle**, so as to ensure that everything will **stay under control** (before continuing on with the project's journey).

> *"Though, part of the challenge with Project Rescue to Project Recovery is not to end up in a perpetual state of firefighting, as this is not conducive to a timely project delivery."*

...

CONTENTS MOVED to [Section 5.5.2].

NOTE: Project Recovery is too complex a topic to be detailed in an introductory chapter on project management theory. That is, given the significant amount of practical & rational project management knowledge and techniques that are required to successfully recover a failing project; this topic deserves more than a few explanatory pages in this current chapter.

As an aside, take it as observed that, the relocated section on Project Recovery does contain a number of "skull & crossbones" warnings, several mentions of "change", and the active involvement of the project's primary & secondary stakeholders.

Which brings us to an addendum to "Adaptive & Proactive" SDLC Project Management.

RULE 6ᴮ: Just because you fixed up after a BAD thing doesn't mean that same BAD thing can't happen again.

2.6. R.I.S.C. Management

Having now read [Section 2.5] on Project Rescue & Recovery; and, especially given Project Recovery, in [Section 5.5.2], with all of those "skull & crossbones" warnings, the number of times that "change" was mentioned, and how it involves both the customer & performing organizations (inclusive of the Project Steering Committee, the Project Working Group, and the Project Implementation Team). There would appear to be **an underlying intertwined relationship between Risk Management, Stakeholder Management, and Change Management**. That is, you **cannot deal with one of these management matters without** (at the same time) **having to deal with the consequential effects of and the effect on, these other two types of management matters**.

Figure 15: the cycle of interaction between the management of Risk, Change, and Stakeholders.

Let's term an acronym to represent this mutually binding relationship between Risk Management, Stakeholder Management, and Change Management.

R.I.S.C.   [R]isk [I]ssues [S]takeholders [C]hange

Consider that Risks & Issues, and the response to such Risks & Issues, could necessitate that Changes be made to the project parameters [Baselines], and these Risks & Issues could also affect the project's Stakeholders' expectations, perspectives, and opinions on the project.

Conversely, Changes to any of the project parameters of [Scope], [Time], [Cost], [Quality], [People], and [Resources], … see the theory in [Section 1.2] and [Section 2.7], … will introduce <Risks> to the project being completed successfully (as was agreed to), and these Changes will also simultaneously affect the project's Stakeholders' expectations, perspectives, and opinions on the project.

Likewise, the project's Stakeholders could require that Changes be made to the project parameters [Baselines, most often to Scope], and each of these Stakeholder's individual "needs", "wants", "concerns", "expectations", and "perspectives" will influence the <Risk> of these Stakeholders subsequently deeming the project to be a Success | Failure.

Thus, **the purpose of R.I.S.C. Management is to orchestrate those activities, interactions, and relationships that will prevent, mitigate, avoid, or eliminate "BAD things" happening to the project**, in both quantifiable factualities and qualifiable statements.

"R.I.S.C. MANAGEMENT IS DEALING WITH WHAT HAPPENS WHEN EVERYONE WAS PLANNING (AND COUNTING) ON SOMETHING ELSE OCCURRING, BUT IT AIN'T GONNA HAPPEN THAT WAY NOW."

> **NOTE:** The following is a short introduction to each of these components of R.I.S.C. Management; for more in-depth information on the monitoring & control of these components then please refer to the following Sections.
>
> - **Risk Management** — refer to [Section 3.6.8], and [Section 8.3.1].
> - **Stakeholder Management** — refer to [Section 3.6.9], and [Section 8.3.2].
> - **Change Management** — refer to [Section 3.6.7], and [Section 8.3.3].

2.6.1. Risk Management

Risk Management is about managing those <Risks> related to problems that are either:

(1) **Inherent Risk** … is due to the (technical) nature of the project being conducted.

(2) **Management & Control Risk** … is directly related to the management and control of the project constraints & project variables.

(3) **Confrontational Risk** … is situational & circumstantial problems that simply come into being and confront the project.

(4) **Execution vs. Operational Risk** … is where the execution of the project (or as a result of the project's deliverables) affects the organization's routine operational activities.

Though, <Risks> can be categorized into other groupings based on the particulars of the type of project that is being conducted, the industry, and the involved organization's perspectives. … And, the involved stakeholders own **Risk Tolerance** (i.e. "Risk Appetite") **will influence how much of a "risk-taker" they are compared to them being a "risk-avoider"**, and hence, how they will prioritize the importance of dealing with such Risks.

Therefore, **when documenting the project's Risks & Issues, then select Risk Categories** (and grades of Risk Priority) **that make the most sense to the project's stakeholders, and that contribute best to the management of the project** and these associated Risks.

❖ **Risk** … is an uncertain **event or situation (in the future) that could affect the project** and thereby endanger the achievement of the project's desired outcomes.

❖ **Issue** … is a **Risk that has happened (but is yet to be resolved)**, or is now currently **commencing**, or is **definitely about to happen** real soon from now.

A risk becomes an issue when it actually prevents the project from delivering the agreed [Scope], by the agreed [Time], within the agreed [Cost] budget, with the agreed level of [Quality], and/or necessitates the assignment of unexpected additional [People], necessitates the allocation of unexpected additional [Resources].

2.6.2. Stakeholder Management

Stakeholder Management is about managing those <Risks> associated specifically with the stakeholders, based on those stakeholders' "Needs", "Wants", "Concerns", "Expectations", and "Perspectives". … And, subsequently the effect on their "Opinions".

> ❖ **Needs** … a **minimum** of what someone **"reasonably requires"** to be satisfied.
>
> ❖ **Wants** … a **maximum** of what someone **"reasonably desires"** to be satisfied.
>
> ❖ **Concerns** … an **interest in** the **effect on them** or on others.
>
> ❖ **Expectation** … a **belief** about what is **to occur**, and what is **to eventuate**.
>
> ❖ **Perspective** … the way they see the situation, and how they **interpret & understand**.
>
> Those who have a positive expectation / perspective will either try to help the project along, or at least not hinder its progress. Whereas, those who have a negative expectation / perspective will try to either actively impede the project's progress, or they will passively resist.

As stated previously in [Section 1.2], for a project to be deemed a Success | Failure is very much dependent on the expectations & perspectives (and the subsequent opinions) of those stakeholders who have an interest in the project. Where, "interest" means that these people are either positively or negatively **affected by the project**, or these people have a positive or negative **affect on the project's outcomes**.

Which brings us to the next rule of "Adaptive & Proactive" SDLC Project Management.

RULE 7: Stakeholders' expectations, perspectives, concerns, needs and wants must be carefully managed, for the project to be a success.

"IMHO, the only way your gonna find these out is by talking with them."

But first, you have to understand that there are different ways that stakeholders can be categorized or grouped; hence, there are different levels of interaction between them and the project, and thus differences in their associated potential for <Risk> to the project.

A two categories model, see [Figure 16], [ref: Cleland].

(1) **Primary** ... Stakeholders who have direct effects on the project's outcome,

(2) **Secondary** ... Stakeholders who influence the project's outcome and/or are affected by its outcome.

Alternatively, a three groupings model, see [Figure 17], [ref: Clarkson].

(1) **Core** ... Stakeholders that are essential to the project's survival.

(2) **Strategic** ... Stakeholders that are vital to the project at a particular time.

(3) **Environmental** ... Stakeholders that form the backdrop and surrounding environment to the project.

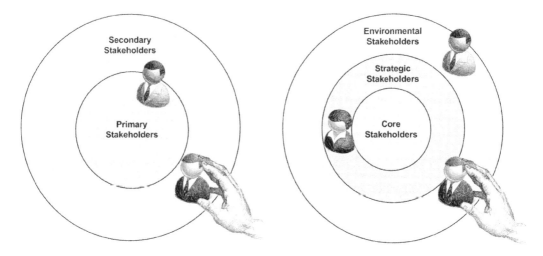

Figure 16: Stakeholder primary-secondary categories model.

Figure 17: Stakeholder core-strategic-environmental groups model.

Though, **during the life of the project, a stakeholder could change** between core, strategic, and environmental. ... And, even appear and disappear.

2.6.3. Change Management

Change Management is about managing those <Risks> related to changes made to the project's [Baselines]. ... Often related to [Scope] changes made with or without consent.

A change will be initiated by one of three means:

(1) **Change Request** ... where some feature or functionality [Scope] as defined by the approved Detailed Specifications will not result in the desired outcomes, or the agreed [Quality] Acceptance Criteria is found to be inappropriate, the agreed [Time] schedule needs new milestone dates, the agreed [Cost] budget **needs to be adjusted**. ... And/or, the authorized allocations of [People] / [Resources] **needs to be amended**.

That is, a **Change Request desires a modification be made to the definition of an existing project baseline.**

(2) **Defect Report** ... where some feature or functionality [Scope] of the project's deliverables does not conform to the agreed [Quality] Acceptance Criteria (nor to the approved Detailed Specifications) and hence this defect **needs to be rectified**.

That is, a **Defect Report desires a modification be made to the implementation so as to conform to the definition of an existing project baseline.**

(3) **Ad-hoc** ... where some feature or functionality [Scope] is **unofficially added** (i.e. "Scope Creep") or **subtracted** (i.e. "Scope Shrinkage") from the project **without** having obtained **prior formal** (written agreement and) **authorization** to do so. Or, the agreed [Quality] | [Time] | [Cost] | [People] | [Resources] are just. ... ... *"Don't worry about it, she'll be right, we'll make it up, as we go"*.

> Which brings us to the next rule of "Adaptive & Proactive" SDLC Project Management.
>
> **RULE 8: Change is inevitable during the life of the project; to expect otherwise is short-sighted and foolhardy.**

2.6.4. Rules to R.I.S.C. Management

Having now touched on Risk Management, Stakeholder Management, and Change Management; there are some general rules for R.I.S.C. Management that should be applied at all times:

(1) **DO NOT Ignore the RISC** … any individual RISC should not be ignored, but rather its existence has to be acknowledged. This is because **ignored RISCs do not usually just fade away**; rather, the negative aspects of an ignored RISC will continue to grow until it has "*snowballed*" **into a major problem that can no longer be overlooked**, nor "*swept under the carpet*" without some major denial & falsification of its existence.

(2) **DO NOT hide the existence of the RISC** … from those who should know; instead, as soon as is feasibly possible **communicate the details of the RISC and the status of the resolution of that RISC** to the relevant project stakeholders.

(3) **DO NOT be overly pessimistic or optimistic** … be realistic about the significance of the RISC, and **represent it accordingly** so as to **result in an appropriate level of action, response, and concern** by the relevant project stakeholders.

(4) **Assign responsibility & ownership** to a relevant project stakeholder … because **without** responsibility & ownership for a specific RISC having been assigned to a certain person or the leader of a group then **everyone involved with that RISC could assume that it's someone else's problem** to be dealt with. … And subsequently, no one actively takes care of that RISC, until it escalates into a major problem.

(5) **Not capable of resolving every single RISC (by yourself)** … because the reality is, the Project Manager is not always capable (and not necessarily skills-competent) of personally resolving every single RISC; hence, other persons (more appropriate) will have to be involved with doing something about the RISC.

2.7. Project Variable Star

Back in [Section 1.2] the concept of project variables and project constraints was briefly explained. ... And, it was stated that **the project's management are responsible for the continuous monitoring, adjusting, and balancing of** these project parameters of **[Scope], [Time], [Cost], [Quality], [People], and [Resources]**, else <Risk> it all coming unstuck.

> NOTE: Indifference to what some project management philosophies may state, IMHO, **<RISK> IS NOT A PROJECT VARIABLE, NOR IS <RISK> A PROJECT CONSTRAINT.**
>
> This is because; the project's management don't deliberately adjust <Risk> in order to control the project. Rather, the **project's management adjust the project variables & constraints so as to react & adapt to the Risks & Issues confronting the project**. However, some of these <Risks> will have resulted from previous adjustments to the project variables and project constraints.

While mentioning project management philosophies, a common model you may encounter is the project's **"Triple Constraints"** which will be represented by a triangular diagram with most often [Scope], [Time], and [Cost] at the points of the triangle. Though, sometimes these points are swapped with either [Quality], [Resources], or [People].

Figure 18: Project Variables / Constraints.

Let's take these variations of the project's Triple Constraints model and stick them all together into one model, a "**Project Variable Star**" per se.

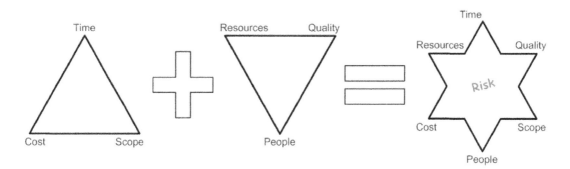

Figure 19: Project Variable Star.

Now that these project parameters have been joined together, what happens **when one or more of these points is deliberately or consequently changed**? The answer is, **there is a subsequent ripple-effect change to one or more of the other project parameters**. ... But, what would be the impact if some of these parameters were fixed as project constraints?

For example;

- *the project's delivery date [Time] is "set-in-stone", ... "it must be delivered by this date or ELSE !!!"*
- *the project has been signed off for a certain price and the [Cost] budget is now set, ... "there ain't no more money that can be spent than this."*
- *the mandatory features & functionality [Scope] has been defined, ... "deliver these things or ELSE !!!", and*
- *these features & functionality must conform to this acceptance criteria [Quality], ... "they ain't gonna accept it if it ain't exactly like this."*
- *All of the available [People] have been assigned to the project, ... "cause we ain't got anyone else left."*
- *The required [Resources] have been procured, ... "we ain't got the budget nor the time available to get any more."*

As the current situation & prevailing circumstances change then there will be subsequent pressure placed on the project parameters; however, **if a project parameter is constrained then the remaining unrestrained project variables will have to be continually adjusted so as to stably manoeuvre the project towards the desired outcomes.**

But, **as more and more project variables are mandated as being project constraints**, as additional pressure is exerted onto the project, **then there is less and less flexibility to maintain control over the project.** If excessive pressure is sustained then eventually one or more of the project variables "explodes" or "collapses". *Where an explosion or collapse can take the form of; a budget overrun [Cost], a schedule blowout [Time], features forgotten or incorrectly implemented [Scope], lack of [Quality] such as high defect counts, fall in morale or [People] resigning, shortfalls in material [Resources] or equipment failures.*

Think of the Project Variable Star as being a child's push-me pull-me toy, when one of the points is pushed in then one or more of the other points pop out. But if some of the points were restrained from popping out and other points were all pushed in at the same time then the remaining point would pop-out, possibly with such force that it could potentially break the toy. Hence, the aim of the game is to stop any of the points from completely dislodging; where the <Risk> is, that one or more will end up scattered across the floor.

Or to put it into academic sounding terminology, … **the project's management needs to maintain control over the balance between the project variables and project constraints, so that as one project variable changes then the other project variables (and project constraints) do not change catastrophically.**

> **NOTE:** This section on project variables & constraints can be interpreted as, one is either "adjustable" OR "fixed"; however, during the project's Executing Phase those project constraints are in-effect "transiently flexible". Provided that when the project is coming to a delivery milestone or is about to finish, then each transitioning project constraint should converge to conform with its defined project constraint, i.e. homes in on the approved baseline. See [Section 3.4.7].

2.8. Project Management Process

2.8.1. PMBOK® As A Process Model

That previous section on the "Project Variable Star" may seem a bit peculiar, but it does directly relate to the **Project Management Body Of Knowledge (PMBOK®)** as outlined in the PMBOK® Guide, and its subdivision into ten knowledge areas of:

(1) **Project SCOPE Management** ... purpose is to ensure that only the agreed work is undertaken, and that only the mutually agreed functionality & artefacts are delivered. Hence, the delivery of nothing more or less than what was agreed to (and signed off) by the representatives from both the customer & performing organizations.

(2) **Project TIME Management** ... purpose is to ensure that the project is completed within the agreed timeframe, and that each of the agreed milestone dates are achieved.

(3) **Project COST Management** ... purpose is to ensure that the project is completed within the agreed & approved budget.

(4) **Project QUALITY Management** ... purpose is to ensure that the project's deliverables conform to the agreed Acceptance Criteria, and to ensure that the Project Team members are following the relevant quality processes & procedures.

(5) **Project Procurement Management** ... purpose is to ensure that the inanimate RESOURCES aspects of the project are handled appropriately, and to ensure that these services & material resources are utilized effectively & efficiently.

(6) **Project Human Resource Management** ... purpose is to ensure that the PEOPLE aspects of the project are handled in a humane manner, and to ensure that the Project Team members are utilized effectively & efficiently.

(7) **Project Communications Management** ... purpose is to ensure the timely and relevant bi-directional exchange of project information, and to ensure that records of such stakeholder interactions are kept for future reference.

(8) **Project Stakeholder Management** ... purpose is to ensure that the project stakeholders' needs, wants, expectations, perceptions and concerns are handled appropriately. Including those 3rd Party vendors & suppliers (both external & internal to the performing organization) who provide services & material resources.

(9) **Project RISK Management** ... purpose is to ensure that the project is not derailed by risks & issues that confront the project.

(10) **Project Integration Management** ... relates to the coordinated integration of all of those previously listed knowledge areas of project management.

Now, if you took the project life cycle models, the 'PLAN – DO – CHECK – ACT' model, the Project Rescue & Recovery model, the Project RISC model, the Project Variable Star model, and grabbed the PMBOK® Guide textbook, put them all into a blender and pushed the MIX button, you'd end up with something like the 'Project Management Process' model below.

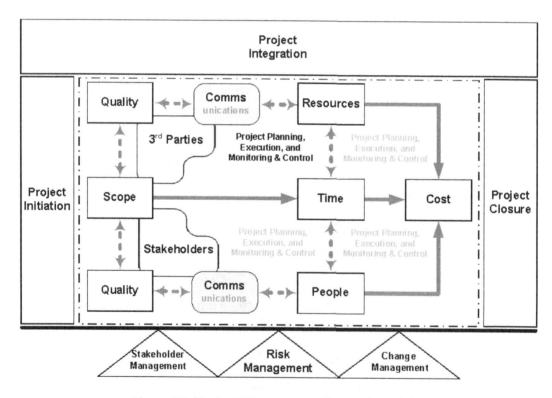

Figure 20: 'Project Management Process' model.

Which brings us to the next rule of "Adaptive & Proactive" SDLC Project Management.

RULE 9: **Don't be afraid to take others theories & concepts, and mix & match these, while combining in your own ideas & philosophies to produce some hybrid thing (that evidentially works for you and for the Project Team).**

RULE 9^A: **Be prepared to explain what you did, why you did it that way, and then stand behind the resultant outcomes.**

Okay, so how to explain this hybrid mixing of; the ten PMBOK® knowledge areas, the six project variables / constraints in [Section 2.7], the five phases of a project's life cycle in [Section 2.3], combine in the three different facets to project RISC Management in [Section 2.6], then work-in the 'PLAN – DO – CHECK – ACT' model in [Section 2.4]?

…

Nah, this ain't gonna happen, when I wrote a telephone directory thick textbook on that; so how about, doing it justice and give it a [Chapter 3] of its own.

Which brings us to the next rule of "Adaptive & Proactive" SDLC Project Management.

RULE 10: **Don't be stupid, if something is going to be a big job then at least give it an appropriate amount of time & effort to be completed properly.**

RULE 10^A: **In the long-run, no one appreciates a half-arsed job, no matter how cheap and quickly it was done.**

2.8.2. "Satisfaction" versus "Performance"

As an aside to this 'Project Management Process' model.

Consider a laptop / tablet / mobile-phone / television / vehicle which you personally paid your hard-earned money to buy; i.e. you (with your own money) bought an asset that year after year will depreciate in financial value while you use it to perform some specific purpose.

Initially, prior to the act of purchasing that particular product, you would have been very interested in the features & functionality [Scope] and the [Quality] of that product when compared to the other competing offerings. Though, as you moved closer to making the final purchasing decision, then the purchase-price [Cost] and the [Time] availability of that product weighed heavily into the equation.

Once you had completed the purchase transaction, ... *and, after the shocked "you paid how much!!!" questioning from your spouse / partner when they saw the resultant credit card bill,* ... in a relatively short period you forgot about exactly how much you had paid and how long it took to arrive. Rather, over the usable life of that product, you either lament or appreciate the "working value" which that particular product gives / gave you.

...

Now consider the above (personal purchase) example case as being the produced deliverable of the project. For some period after the project has finished, the senior management of the performing organization may bemoan how over budget [Cost] and/or how late [Time] the project was delivered. However, it is the customer organization (or more specifically the customer's end-users) who will be continuously exposed to the features & functionality [Scope] that was & was not delivered, and what was the "fitness-for-use" [Quality] of those deliverables.

The combination of [Scope] & [Quality] can be thought of as **"Customer Satisfaction"**; whereas, **the combination of [Time] & [Cost]** is the project's **"Performance Measures"**.

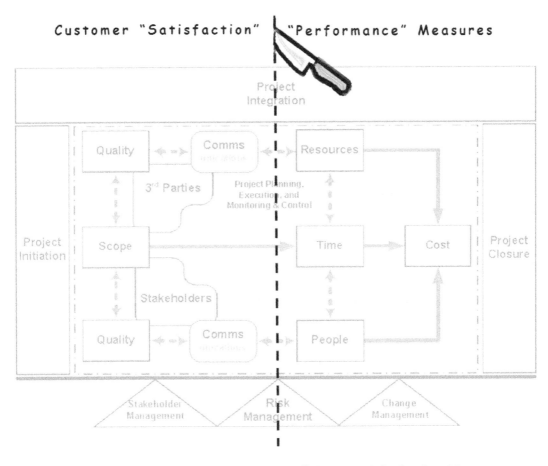

Figure 21: "Satisfaction" versus "Performance" division of the 'Project Management Process'.

Which brings us to the next rule of "Adaptive & Proactive" SDLC Project Management.

RULE 11: Project success is opinionated, based on the "satisfaction" of [Scope] & [Quality] versus the relative "performance" of [Time] & [Cost]

COFFEE BREAK DISCUSSION … *Agile, is just a dirty word.*

Well, with a heading like that, talk about trying to win friends and influence others; a bit irreverent and possibly down right offensive to some. … *"Ya, betta getta cuppa."*

In the real-world of commercial projects, just how often do performing organizations describe themselves as using *"aj-jile"*, yet what they are actually doing is far removed from any truly recognizable agile methodology. They are instead doing a kind of hybrid ad-hoc mix of a very compressed waterfall – iterative project life cycle with the jargon from agile thrown into the blend.

Consider an agile methodology such as Scrum, and list its notable characteristics:

- ☐ **Standup Meeting** … is there a daily (15 minute max.) "scrum" standup meeting held with the Project Implementation Team to go over; what they did in the last day, what blocking issues they face, and what they intend to do for the coming day? … OR, are project team meetings only held once a week (at best), more likely fortnightly, (at worst) every month to just update the project's schedule and update The Team on what's going on with the project and in the performing organization?

- ☐ **To Do Items** … are there such things (*as "user stories", "use cases", "requirements info"*) used to compose the **Product Backlog**, and does the Project Implementation Team choose which to do and not to do for the next Sprint and put this into a clearly defined **Sprint Backlog**? … OR, is there only a Detailed Specifications that the project's management solely choose what & who is to do specifically assigned tasks in the next "tight iterative" cycle? … OR, is the 'To Do Items' just a bullet-point list of the customer's requirements with little to no detail outlining what is intended let alone what the acceptance criteria are?

- ☐ **Task Board** … is the "to do – work in progress – done" motion of each 'To Do Item' represented (either physically or virtually) on a Task Board? … OR, is a task's completion progress represented solely via percentages in the project's schedule?

- ☐ **Burndown Chart** ... is the summation progress through the current set of selected Scope in the Sprint Backlog presented (publically) on a Burndown Chart? ... OR, on a weekly basis, does the project's management come along with a detailed project schedule and (privately) ask each implementer (or The Team leader) for the percentage completes for their assigned tasks?

So with a drum-roll. ... ... Of those *"aj-jile"* projects that you (the reader) have worked on in the real-world, just how many of those notable agile characteristics were true? ... OR, were the listed (non-agile) "OR" cases predominant instead?

I wouldn't be surprised to find that you are nodding your head for the latter "OR" cases; hence, that project is not actually being run using an agile methodology. Yet, they will insist on calling it, *"aj-jile"*.

So, what makes them believe that theirs is truly an agile based project ??

I surmise that this is because they are using "tight iterative" cycles of timeboxed 2-4 weeks durations, instead of "traditional iterative" variable duration cycles of a few months. Essentially, what they have done is super-compress an iterative project into micromanaged chunks of days with the ad-hoc inclusion of Defect Repairs and Change Requests.

And/or, they waterfall plan out the entire project as "tight iterative" releases, and let some of the implementers choose what tasks they want to do, but assign specific skilled tasks that certain implementers must do (irrespective of The Team's individual desires).

Alternatively, they "rolling wave" plan the project as micromanaged bursts of work, that aim to get to the next (arbitrarily) selected milestone marker. As though, they are down in a trench, digging away day after day, to only occasionally pop one's head-up to figure out where they currently are, and only then deciding on what general direction they should be digging towards next.

So, what is your thoughts and opinions on this topic ??

"Agile methodology does not mean taking a waterfall structured project (of upfront defining all of the requirements to the nth degree, prior to the all-in implementation and sequential testing), then along the way, expecting to lob in additional requirements and redefining existing requirements, with 'aj-jile' morphing the project to accommodate such Scope changes. As though, by saying that the project is 'aj-jile' then this somehow immunises the project from the impact of these changes on the agreed Time & Cost of delivery and the effect on the assigned People & Resources. That is, the incorrect belief that 'aj-jile' magically circumvents the need for Baseline Change Request Management, and that 'aj-jile' legitimizes the ad-hoc running & reporting of the project."

"Agile is a perspective change to waterfall, in that agile is taking what you know up to now and building the critical bits first, then building on that, and when necessary reworking (and even trashing some of what has already been done) so as to achieve an optimal solution that will keep the project moving forwards. That is, having never constructed this type of building before (and thus not knowing what the outcome will look like), the waterfaller will insist on firstly completing all of the detailed floor plans and architectural drawings for this monolithic castle; instead of starting with a guardhouse, so as to ensure that the underlying construction principles & techniques are going to work. But alas, the waterfaller will see that proof-of-concept construction as a waste of time, wanting instead to lay the foundations for the entire castle in one go, and expecting to succeed in only one-pass."

Which brings us to an addendum to "Adaptive & Proactive" SDLC Project Management.

RULE 11[A]: Everyone has a valid opinion, the perspectives on that opinion are influenced by the knowledge and experiences of those discussing it.

3. THEORY ... PMBOK®

3.1. Introduction

Okay, that last [Chapter 2] on "a little bit" of project management theory was almost 40 pages long. ... And, unfortunately this chapter on PMBOK® based project management theory is going to be a lot bigger.

However, **to successfully manage that SDLC project** *"across the finish line"*; by the agreed [Time], within the agreed [Cost] budget, with all of the agreed [Scope] of deliverables, with the agreed level of [Quality], while effectively & efficiently utilizing those assigned [People] and allocated [Resources], ... then, **this can't be an ad-hoc** *"make-it-up as you go, adventure"*. Rather, this project has to be a well thought-out and skills trained endeavour.

To use an analogy; if you were going to step into a boxing ring, a mixed martial-arts octagon, or fight in a war zone, then, wouldn't you want to have rote learnt as much of the basic theories and studied those proven field-tested combat techniques that would give you a realistic chance of surviving the encounter. ... OR, are you suicidal?

...

For both the novice & experienced project manager, it is good to occasionally go over the basics. ... And, incorporate | "adapt" your own ideas & lessons learnt experiences; so as to, better your chances of achieving success on your current project, and then on the next.

> **NOTE:** For some, who deal predominately with true agile projects, you may consider that this chapter on PMBOK® is not relevant to you; however, your agile project is often contained within a "Program" of multiple projects which are managed using PMBOK® based methodologies (or similar "structured" methodologies).
>
> *"Better to know the face of thy enemy, than to fight on blindly."*

3.2. Overview

The 'Project Management Process' model introduced in [Section 2.8], and reproduced below in [Figure 20], is the culmination of the walk-through of the PMBOK® derived methodologies that will be summarized in this chapter.

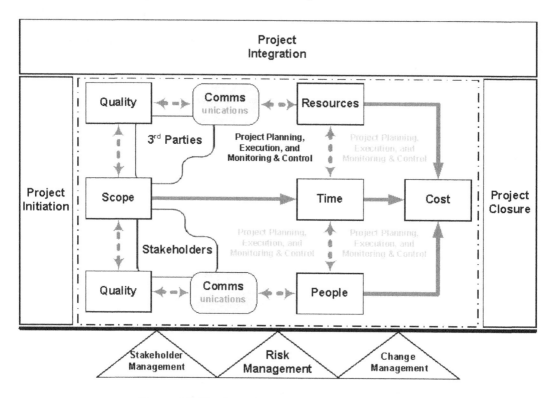

Figure 20: 'Project Management Process' model.

Which brings us to the next rule of "Adaptive & Proactive" SDLC Project Management.

RULE 12: Start at the beginning, and step it through from there, one step at a time until you arrive at the Finish.

RULE 12[A]: Consider where that next step is intended to go. ... Will it lead, down the wrong path?

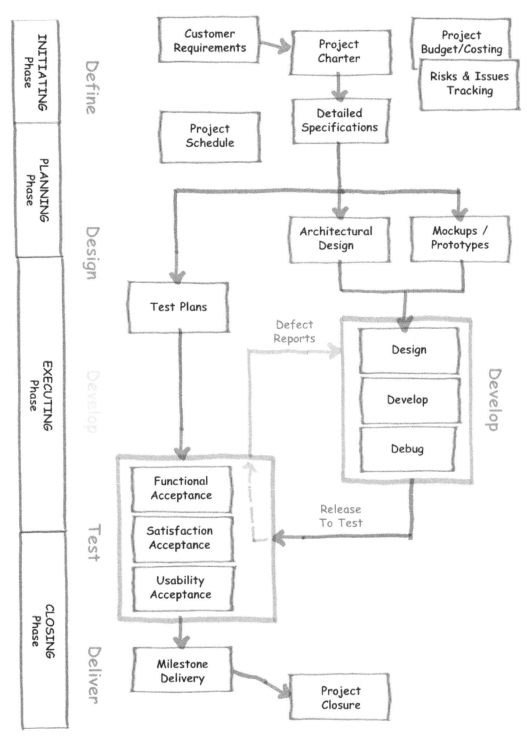

Figure 22: A minimalist set of documents, deliverables, and processes that are necessary to have a chance of delivering the project successfully.

3.2.1. Document Hierarchy

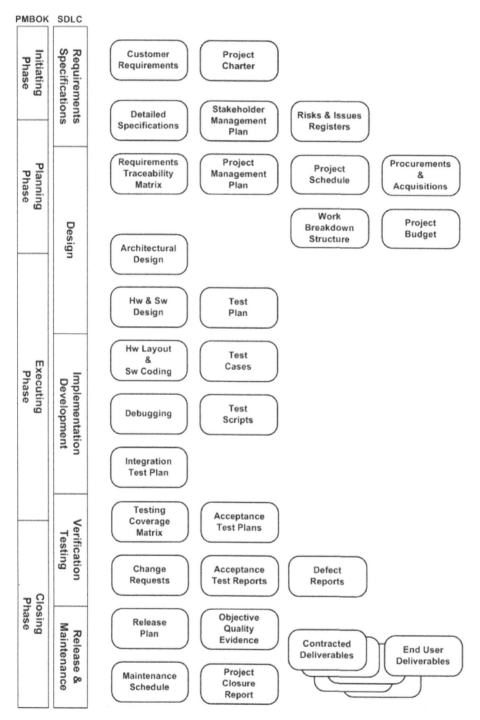

Figure 23: Some of the documents for a PMBOK® based SDLC project.

Project Documentation ... *a necessity, not a curse*

A part of the SDLC project's journey from start to finish is the hierarchy of documentation (or selection of documents) that should be produced along the way, see [Figure 23].

- ❖ **Customer Requirements** ... is a "**wish list**" of the features & functionality that the customer thinks that they "**want**" implemented, and what they perceive as the benefits of undertaking the project; however, this list may not necessarily be a complete representation of what are the customer's real "**needs**", nor what are the "**desired outcomes**". ... *"For an internal project this document may be referred to as the Business Requirements".*

- ❖ **Project Charter** ... is the summation of the analysis & appraisal conducted during the Initiating Phase so as to provide enough details & insight into the proposed project to **enable a responsible assessment** and subsequently a **due-diligent decision** on whether to officially sign-off on the **authorization to commence** the project. ... *"May also be referred to as a Tasking Statement, a Statement Of Work, a Project Initiation Document, or part of the Project Management Plan."*

- ❖ **Detailed Specifications** ... is a "**what will be delivered list**" of the features & functionality that the performing organization considers as **in-scope** of the project; and as such, contains sufficient explanation information to clarify what will be delivered to the customer organization (but doesn't contain the proposed solution). ... *"May also be referred to as a Functional Specification, Detailed Business Requirements Specification, or a Scope Statement."*

- ❖ **Requirements Traceability Matrix** (RTM) ... is an itemized "**check list**" to confirm the "**scope coverage**" of the Detailed Specifications against the Customer Requirements, clearly indicating what is "**in-scope**" and what is "**out-of-scope**".

- ❖ **Project Management Plan** ... documents "**the rules**" for managing the project, and provides "**a guide book of plays**" on how to achieve the project's objectives, and/or a common reference point to the project's subsidiary management documents.

- **Work Breakdown Structure** (WBS) … is a hierarchical decomposition of work packages that "have to be done" so as to deliver what "was agreed to be done".

- **Project Schedule** … is a (graphical) **representation of the relationships between** the sequential & parallel breakdown of the packages of [**Scope**d] work to be done, versus the [**Time**] expected for each to be done, when specific [**People**] & [**Resources**] are assigned to do each one; and, thereby determining the dates for **noteable milestones** on the path to the project's completion.

    "A spreadsheet does not make a credible project schedule because it doesn't effectively represent the relationship between Scope & Time."

- **Project Budget** … establishes the project's **expected expenditure** and thereby is calculated **proof** of the project's **financial viability**, so as to **understand the cash flow** ramifications of conducting the project.

- **Risks & Issues Register** … is a documented **analysis & assessment of the Risks** & Issues currently, possibly, and previously confronting the project; which could prevent the project from achieving the stated objectives and desired outcomes.

- **Stakeholder Management Plan** … is a documented **analysis & assessment of the** project's primary & secondary **stakeholders**, so as to determine the best ways to manage these stakeholder needs, wants, concerns, expectations, and perceptions.

    "Though, don't be surprised, for small to medium sized projects / organizations, when this is done as 'a needs be' ad-hoc activity."

- **Procurements & Acquisitions Plan** … is a documented **analysis & assessment of the** [**Resources**] materials, products, services, and personnel **from outside** of the project that need to be acquired at certain points in the project's progress, so as to be able to proceed with those tasked activities of the work packages.

- **Architectural Design** … is a documented analysis & assessment of the proposed technical solutions.

- **Test Plan** … is a documented **analysis & assessment of how to evaluate** the conformance of the actual deliverables against the agreed [Scope] of deliverables.

- **Acceptance Test Plans** … are documented plans (and instructions) on how to verify & validate that specific features & functionality do conform (i.e. operate) in accordance with the agreed Acceptance Criteria and the Detailed Specifications.

- **Testing Coverage Matrix** (TCM) … is an itemized **"check list"** to **confirm** the **"scope coverage"** of the Acceptance Test Plans against the Detailed Specifications, clearly indicating what has and has not been tested, and as an indicator of PASS:FAIL [Quality] of the features & functionality that have been verified as operating in accordance with the agreed Acceptance Criteria.

    "The Testing Coverage Matrix can also be referred to as a Testing Traceability Matrix (TTM), Requirements Traceability Verification Matrix (RTVM), or Verification Cross Reference Matrix (VCRM)."

- **Acceptance Test Reports** … documents the **outcomes & findings** of those tests that were and were not conducted, the PASS : FAIL status of each of these tests, and the conclusions made about each of these acceptance test results (where the approved Acceptance Test Plans were used as the guide to those tests that were performed).

- **Change Requests** … details what were the **requested alterations to** the project's [**Baselines**]; predominately [Scope] changes that request modifications to the features & functionality, and/or to the stipulated deliverables.

- **Defect Reports** … details what **non-conformances in functionality & features** were detected during the acceptance testing.

- **Objective Quality Evidence** (OQE) … is the accumulated **"Artefacts Of Proof"** that the project's deliverables do satisfy the mutually agreed Acceptance Criteria, and hence these deliverables should be accepted by the Customer's Representative.

    "Though, the trick is to provide the amount of OQE that will satisfy the relevant project stakeholders … nothing more, nothing less … just

the right amount of proof to satisfy their needs. ... So as to get them to move along to the next stage or phase of the project."

- ❖ **Release Plan** ... details when and how specific milestone releases of the deliverables will be provided to the customer organization, and notes any specific information about the milestone release's deliverables that the customer organization / end-user should know.

- ❖ **Maintenance Schedule** ... is analogous to the **regular periodic maintenance** plan for your new car, so that the customer should not experience any prolonged interruptions to the continued satisfactory use of the project's deliverables.

- ❖ **Project Closure Report** ... **denotes THE END** of the project / milestone release, and the sign-off of the Project Closure Report signifies that the project / phase of the project has concluded.

NOTE: The level of detail contained in these documents will vary depending on the scale & complexity of the project, as well as the processes & procedures that are dictated by the primary stakeholders (and recommended by the project's management). Hence, these documents could be highly detailed, broadly framed, or variations thereof. Thus, choose formats, quantities, and quality levels that are appropriate for the project that is being conducted.

"And, don't go blowing a noticeable portion of the project's budget on copious amounts of documentation, unless this is a mandated deliverable."

 DO NOT forgo documentation due to the need to speed up the project, as this will only result in nightmares of misunderstandings, misinterpretations, and rework. As documentation is vital for communications, and especially as a record of interactions and proof of what was & was not agreed to.

"So, the trick is to know which documentation to forgo, which to skimp on, which to merge together, and which to concentrate on."

The Value Of Documentation ... *the quality and the quantity*

The real value of the documentation produced via & during the project corresponds to the purpose for which it was intended, as:

(1) **A medium of communications** to transfer particular pieces of information amongst the project's various stakeholders (from the past, to the present, and through to those who are yet to be) involved with the project and its outputs / outcomes. Primarily, **documentation is a bridge between those who implement the project and the stakeholders who expect [Quality] deliverables for the [Scope] to be covered.**

(2) **To catch things that could be forgotten, under-examined, or misinterpreted**; i.e. the documentation has to be written to a sufficient level of detail to prevent misunderstandings, misconceptions, and forgetfulness. Where the level of detail that is included in each document is very much dependent on the "needs & wants" of the stakeholders who will primarily use that particular piece of documentation.

(3) **To orchestrate the coordination** of [People], [Resources], [Scope] of requirements, [Time], and the financial [Costs] involved with the project.

(4) **To comply** with the agreed [Quality] Acceptance Criteria for the project's deliverables, as per Quality Assurance processes & Quality Control procedures, and **Quality Auditing**. As certain documentation will be specified as a mandatory project deliverable, and thus necessary for project conformance & acceptance.

 The production of documentation should contribute to the project's success and not be a burden to its progress nor hinder its successful delivery. That is, **documentation for documentation sake, unnecessarily consumes project [Time] & [Cost], and limits the effective utilization of [People].**

"How demoralizing it is to write a document, knowing that it won't be read nor used, but is solely there as a bureaucratic checklist item. ... But, we do it to cover our bums when disputes do arise. ... And, without documentation there would be chaos."

3.2.2. Gating & Governance

A **"due diligence" performing organization will conclude each project phase** (and even sub-phases) **with a formal Phase Completion Review**, i.e. **"Gate Review", conducted by the Project Steering Committee** to evaluate, reassess, and decide whether:

☑ The **project is still viable** (strategically and priority wise) **to be continued**, or
does the project **need to be re-envisioned**, or
does the project **need to be paused** for the interim, or
does the project **need to be terminated** henceforth?

☑ The project is **in an acceptable state to advance** onto the next project phase?
For example; to move on from the Initiating Phase to the Planning Phase.

☑ To give **"Go or No Go" permission** to either proceed or terminate the project.

These points can also be used to formalize the "handoff" (i.e. handover) of one phase to the next phase.

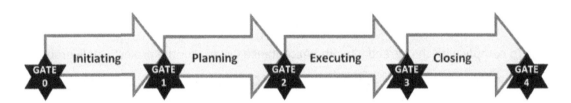

Figure 24: Relationship between Project Phases and project governance "Gates".

Governance ... is the **"Stewardship of Responsibility"** to honestly evaluate the project's **state** when **compared to the performing organization's business objectives & priorities, taking into consideration the current situation & prevailing circumstances.**

Thus, while the **Project Implementation Team** focuses on *"the doing"*, the Project Manager (and the **Project Working Group**) focus on *"making sure that it gets done right"*, and the **Project Steering Committee** focus on the **oversight** of *"deciding whether it is being done for the right reasons"* so as to **provide a beneficial Return On Investment** to the performing organization and to the involved stakeholders.

3.3. Initiating Phase

3.3.1. Purpose

The purpose of the Initiating Phase is to define a potential new project (or a proposed new phase to an existing project) **based on the "needs & wants" of the customer, to evaluate the proposed project's viability, and to then obtain formal signed off approval & authorization to commence the project**.

> Which brings us to an addendum to "Adaptive & Proactive" SDLC Project Management.
>
> **RULE 12^B: At the start there are ideas; but, no real plans on how to move forwards.**

3.3.2. What has to be answered?

For the project to have any realistic chance of success, the Initiating Phase has to provide due-diligent answers to the following questions:

(0) **WHAT** is the **PROBLEM** that **THEY** (the customer) are trying to **RESOLVE**?

(1) What do **THEY** (the customer) **WANT**?

(2) What will **WE** (the implementer) **GIVE** (to the customer)?
What will **WE** (the implementer) **NOT** be **GIVING** (to the customer)?

(3) Approximately **HOW LONG** do **WE** (the implementer) **EXPECT** it to **TAKE**?
What will it **COST US** (the implementer) **TO GIVE** it (to the customer)?

(4) What will **WE** (the implementer) **GET IN RETURN** (from the customer)?

(5) What is **AT STAKE FOR US** (the implementer)?

The **answers to these above questions** must be **recorded in** some formalized document to officially accept or reject the undertaking of the project; i.e. "the **Project Charter**".

3.3.3. QUESTION: what problem to resolve?

This question is such an obvious question, yet it is a question that can be so easily overlooked by the performing organization (and especially forgotten about by the members of the Project Team); because, *"the answer must have already been considered?"* prior to involving the performing organization with this project.

(1) **What is the problem?** ... Not, what are the symptoms, and not, what are the causes.

(2) **What changes to their business** is the customer organization wanting to induce?
See [Section 9.4] **Business Change Management**, [Section 9.6] **Benefits Realization**.

(3) **What is the benefit** to the customer organization by resolving this problem, and **will this resolution add value to their business?** ... **Is it really worth solving**; not, it would be a good thing to do / have, when an acceptable workaround would suffice.

(4) **Is there a time criticality aspect to the finding of a resolution**, when compared to using a temporary / permanent workaround fix?

(5) **What are the <Risks> involved with & without a resolution** being put in place?

(6) **What would need to be changed** to resolve this problem, and
what is the [Scope] of this problem?

(7) **What is envisioned as the potential solution(s)** to this problem?

(8) Hypothetically, **what will be the potential costs incurred with resolving** this problem?

(9) Does the **Return On Investment justify** the [Time], [Cost], [People], [Resources], and <Risks> involved with **putting the potential solution(s) in place?**
For example: Does it make sense to implement a million dollar solution to resolve a problem, when there is currently a workaround that expends $200K per year?

 It is very easy for a "technically minded" Project Team and Project Manager to be blinded to the actual business value that is being impaired by the problem, compared to the cost of the solution to be implemented.

EXAMPLE CASE ... *Based on the Payback Period, is this project worth doing?*

Let's consider that problem with the potential $1M solution compared to a $200K yearly workaround expenditure.

The Real Cost of the worker's "bums on seats" cost of continually doing the workaround.

Real Costs = Staff Wages + Operating Costs + Material & Services Used

Real Costs Per Year = $200K

So, by expending $1M on the project to deliver the resolution then this would save $200K per year; hence, ($1M ÷ $200K / yr.) is 5 years for the financial advantages gained to exceed the amount invested into the resolution; i.e. a "Payback Period".

Payback Period = Cost Of Investment ÷ Annual Gain From Investment

However, the system implemented happens to have ongoing IT administration support, as well as the yearly maintenance & support contract with the system's vendor; hence, this will need to be taken into consideration, which happens to be an extra $30K / yr. So, $30k / yr. × 5 yr. = $150k; hence, an extra year of $200K savings would be required to "break even", such that the cost of implementation plus the accumulated year maintenance & support cost is less than the accumulated yearly saving on the problem's expenditure. Which means that the Payback Period is now in the 6th year.

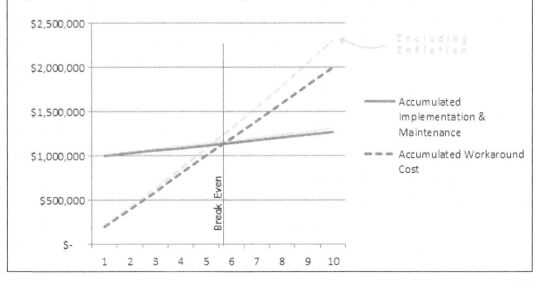

But wait, **a dollar today has a greater buying capacity than that of a dollar next year**; hence, will need to **include** some consideration for the **"time value of money"** (i.e. **inflation**, say of 3%). Though, not for the $1M implementation cost because that was incurred entirely in the first year, but this inflation should be included for everything else in the following years.

Hmm, but those members of staff who happen to be expended on that $200K / yr. workaround worth of effort, their time could be better spent on revenue raising activities. Say, that $200K / yr. workaround has permanently involved two people (i.e. $52 / hr). However, those same staff members could have been charged out for revenue raising tasks at a rate of $100 / hr; thus, a **Lost Opportunity Cost** of $100 / hr × 40 hr/wk × 48 wk/yr × 2 staff = $384K / yr.

Therefore, the yearly **Potential Cost to the organization of lost revenue without the project's solution being implemented** is:

Potential Cost = Real Cost + Lost Opportunity Cost

Potential Cost Per Year = $200K + $384K

Potential Cost Per Year = $584K

Hence, (($1M + n*$30K) ÷ n*$584K / yr.) is 1.7 years, including the inflation percentage.

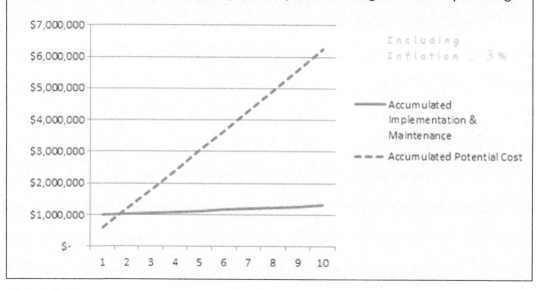

In this example case, **when only the yearly Real Cost of the workaround solution ($200K) is considered** then the 5-6 year payback period **on the implementation's investment** (plus ongoing maintenance & support) may **make this project appear as not** a **worthwhile** endeavour. However, **when the Lost Opportunity Cost is also taken into consideration** then the subsequent 1-2 year payback period **makes this project a lot more appealing**.

Something else to watch out for, is the Project Implementation Team (and possibly the Project Manager) losing focus on the real problem that the customer organization (via the Customer's Representatives) is trying to resolve, as it is very easy to get enthralled with the technical intricacies of the define – design – develop – test. Hence, during the project's implementation, the Project Manager has to **keep the Project Implementation Team's mind on the problem to be resolved and not solely on the technical merits of the solution**.

WARNING: be wary of defining a solution that is in search of an appropriate problem to solve.

Therefore, **it is essential during the incubation stage of the project's life that the customer organization** (and subsequently the performing organization) **obtain a good understanding as to the problem that they are trying to resolve**; and, **what is the intended purpose of the project's deliverables**, so that these will be fit-for-use.

Which brings us to the next rule of "Adaptive & Proactive" SDLC Project Management

RULE 13:   Project management's responsibility is not the solving of the technical problems; rather, it's the resolution of business problems and the creation of business opportunities.

3.3.4. QUESTION: what do they want?

This question is an **overview of the project's [Scope]** and captures some preliminary details about the Customer Requirements.

Thus, the project implementer can obtain this information by conducting a "Top Down Analysis", as illustrated below in [Figure 25].

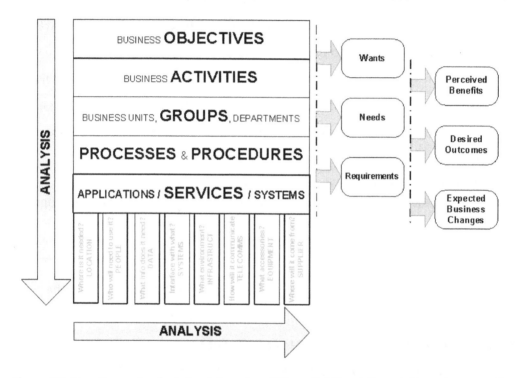

Figure 25: Top Down Analysis to determine "Wants", "Needs", and "Requirements".

Via this **Top Down Analysis** the following can be **determined**:

(1) The customer organization will have specific **Business Objectives** (and priorities) that they are trying to achieve. ... And, thereby **defining** what are their **"Wants"**.

(2) The customer organization will have to undertake specific **Business Activities** to achieve these objectives. ... And, thereby **defining** some of their **"Needs"**.

(3) These business activities will have to be performed by specific **Business Groups** within the customer organization. ... And, thereby **defining** the rest of their **"Needs"**.

(4) These business groups will have certain **Processes & Procedures**, as well as the utilization of **Specific Services** (applications & systems) to execute these activities. ... And, thereby **defining** what are **their** "**Requirements**".

While questioning the customer's representatives, the interviewer(s) will obtain;

(5) an insight of what are the customer's "**Perceived Benefits**" of conducting this project, an understanding of what are the customer's "**Desired Outcomes**", and giving,

(6) both parties realistic "**Expect**ations as to the **Business Changes**" due to the project.

 For the moment, **DO NOT question,** "*when do they need it by*" **[Time]** and "*how much are they prepared to spend*" **[Cost]**; because, this could **inadvertently create inappropriate [Time] & [Cost]** "*expectations*" for the representatives from both organizations.

Which brings us to the next rule of "Adaptive & Proactive" SDLC Project Management.

RULE 14: Misunderstandings and Misinterpretations are project killers.

RULE 14ᴬ: A "similar idea" ain't close enough to the "same thing", for it not to end in flames and blame.

RULE 14ᴮ: An assumption not based on sound evidential facts is the mother of a pending screw-up.

That is, for the project to have a realistic chance of being completed successfully, then the **representatives from both the customer & performing organizations must have a** "**common understanding**" and "**mental alignment**" with respect to; the **interpretations of the requirements**, what are the **desired deliverables**, the **intended milestones**, and the **scale of the project** to be undertaken.   ... "*Hence, the project's stakeholders must be engaged (aka 'stakeholder socialization') to obtain their buy-in on the project, before signing off on the project structure and deliverables.*"

3.3.5. QUESTION: what will we give & not give them?

The answer should be **broad enough to enable an understanding** of the performing organization's response to the customer's requirements, and to provide a "**generalized indication**" of how it is proposed that the project will be implemented.

DO NOT include detailed architectural designs, design concepts, or technical descriptions in the response to the customer.

Because, if **too great a level of detail is included** (and the project is signed off with this information in place) then **potentially the project could be hamstrung by these, when** in retrospect **these were not the best options** for the successful execution of the project.

At this point also consider:

- ☐ **Project Scope Boundaries** ... list (at a high-level) what is within the project's domain ("**In Scope**"), and those (high-level) things that will be "**Out-Of-Scope**".

- ☐ **Tangible Measures** ... list (at a high-level) those things to be used to **gauge whether the project has achieved its objectives**.

- ☐ **Project Acceptance Criteria** ... state (at a high-level) the **determinants for accepting the project's deliverables** and the associated "**Objective Quality Evidence (OQE)**" that is to be presented as proof of conformance.

I would recommend that, **a few senior implementers** be brought together for a quick whiteboard **brainstorm**ing session **to verify that what is to be proposed is technically feasible.**

Before, the customer gets informed, and then expects the impossible, when in hindsight it was obvious that this was technically unachievable.

3.3.6. QUESTION: how long is it expected to take, and what will it cost us?

Now that a reasonable understanding of the customer's requirements and an overview of the project's [Scope] has been obtained, then, based on **past experiences** and **historical data**, make "reasonable **estimations**" of the project's; **[Time] duration** and the **[People] & [Resources]** that would be required to undertake the project, and subsequently the **[Costs]** to implement the project. ... And, the **<Risks> involved** with this project; see [Figure 26].

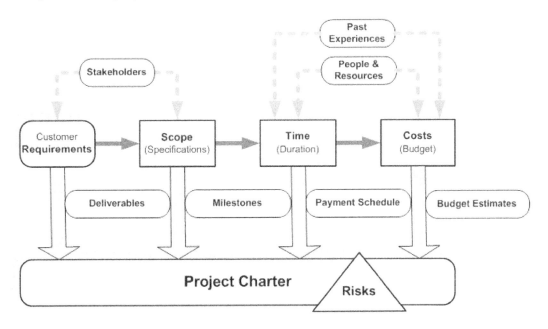

Figure 26: The Initiating Phase as a process to create the Project Charter.

For the [Cost] estimation, **calculate the marked-up "Sell Price"** by **factoring in a Safety Margin** for such things as; Escalated Costs (i.e. exchange rates, inflation), Cost Contingency (for providing warranty support and the repair of latent defects), and a Project Risk buffer for the Level Of Effort. ... And, a **Management Reserve** for those unforeseen risks.

Also, include a **Profit Margin** so as to ensure that this project will most likely be a "profitable endeavour" for the performing organization.

See [Section 5.4.2] and [Section 5.4.3] for more information on the costing of a project.

Similarly for the [Time] estimation, **calculate buffered "delivery milestones"** by factoring in **duration safety margins** for such things as; the lead-time for the procurement of componentry, incorporating "Quality Assurance" processes, undertaking "Quality Control" procedures, producing the necessary documentation and "Objective Quality Evidence", and the performing organization's standardized **percentage for Project Risk**. ... And, include an additional overall percentage of [Time] for ongoing project management, business analysis, and technical authority support (i.e. subject matter expertise).

Thereby, ensuring that this project can be delivered to the customer in a **realistic amount of [Time]**, without burning out the performing organization (nor angering the customer organization due to unexpected long delays to receive their project deliverables).

3.3.7.   QUESTION: what will we get in return?

Establish a "**Payment Schedule**" of the **financial remunerations associated with each particular deliverable and release milestone**; that would result in "**positive cash flow**" for the performing organization. Where, the **accumulated** amount of customer payments (**REVENUE**) up to that point would **exceed** the **accumulated EXPENSE** to the performing organization of conducting the project **to that same point**.

I would recommend that, **each delivery milestone be targeted at providing the customer with features & functionality that can be used to evaluate the project's progress** towards the agreed objectives. So as to, **ensure that there won't be any disagreements over** whether an **associated progress payment** should be rendered (immediately) given the provided "proof of progress".

Which brings us to the next rule of "Adaptive & Proactive" SDLC Project Management.

RULE 15:   What is the point of doing something, if there is nothing positive to be gained (tangibly and/or intangibly).

Project Estimating When the Scope is Not Clearly Understood

During the Initiating Phase, one of the problems that is encountered when determining realistic [Cost] & [Time] estimations is that the complete [Scope] of the project may not be clearly understood, let alone fully defined. Hence, it is not possible to provide a price and duration quotation that "in hindsight" could be described as reasonable & sensible.

A solution to this dilemma is to, firstly, come to a common understanding with the customer representatives as to the "baseline functionality" of the system / application to be delivered. That is, **what is the obvious essential functionality that must be present for the deliverable to satisfy the purpose for which it was intended**? This can usually be determine relatively quickly, by simply holding a brainstorming session (or two) with the customer representatives and those available Subject Matter Experts (including your own).

> *For example, with today's mobile phone it would be; (1) make & receive calls, (2) send & receive text messages, (3) retain a list of contacts, (4) send & receive emails, (5) listen to music, (6) take & view photos, (7) large touchscreen display, and (8) easy one handed use.*

Once this "baseline functionality" is agreed to and signed off then with your experienced implementers, brainstorm those prominent | primary Use Cases, and as a group, quickly sketch architectural ideas on how to implement each of these.

The objective of this in-house brainstorming session is for these implementers to quickly obtain an understanding of the proposed project, to get a feel for the technical complexities & difficulties involved, and to grasp the scale of the undertaking. All the while, you as the Project Manager, know that they will be thinking primarily about the possible technical solutions to the project's [Scope], and secondly, how long they think it would take them to implement what they have worked out [Time]. However, what they won't be thinking about is the [Costs] involved, as that is your (management) problem.

Hence with these experienced implementers, come up with means by which to weight the size, complexity, and possible duration for each of these major chunks of work; and ideally, what order would these be implemented, so as to sketch out a rough overview schedule for the project.

While these SMEs' estimates for the Level of Effort involved will be rough "guestimate" figures; these will be *"a hell of a lot more accurate"* than those numbers dreamed up by Sales & Senior Management.

Having now used the estimated Level Of Effort numbers to calculate the project's budget also include Safety Margins (of Escalated Costs, Cost Contingency, and Risk Buffer), Management Reserve, and Profit Margin, see [Section 5.4.2]. ... And, only then present the Customer's Representatives with the price quotation.

This price quotation can be based on:

(1) **Fixed Price** – the amount to be paid is not going to change from that stated in the quotation, irrespective of whether the actual costs incurred are lesser or greater.

   A modification to this being **Fixed Price with Economic Price Adjustments**.

(2) **Cost Reimbursement** – the amount paid is the costs incurred plus some additional percentage or fixed margin. This could also include **Performance Incentives**.

(3) **Time & Materials** – the amount to be paid is based on the charge-out rate for each person involved plus the sell-price for the materials used.

For a project where the [Scope] is not clearly understood (and/or the technologies involved are a relative unknown), **then a Time & Materials contract is advisable for at least the Initiating & Planning Phases**; then may be later on, switching over to a Fixed Price or Cost Reimbursement for the Executing & Closing Phases. From a performing organization's perspective, **a Fixed Price contract is more appropriate when the project to be undertaken is very similar to previous projects that** the organization **routinely conducts**.

3.3.8. QUESTION: what is at stake for us?

Privately & confidentially to the performing organization, the answer to this question is a **feasibility study, business case**, and/or **cost-benefit analysis** considering such things as:

- [ ] What is the "**Return On Investment** (ROI)"?

    ROI = (Gain From Investment – Cost Of Investment) / Cost Of Investment

- [ ] What "**strategic value**" will this project provide?

- [ ] What are the project's "**intangible benefits**"?

- [ ] What are the **benefits** to other projects or other groups in the organization?

- [ ] What are the **risks involved** with (or with not) undertaking the project?

- [ ] What are the **alternatives available** to not undertaking this project?

So, ... **does this proposed project meet the performing organization's "selection criteria" when compared to the alternatives?**

Does undertaking this project make "rational sense" by providing value to the performing organization's business and its owners / shareholders?

3.3.9. What to do with these answers? ... Gate

These answers would then be accumulated into the **Project Charter, with enough details and insight to enable a responsible assessment & subsequent decision at a Gate Review meeting** (with the representatives from both the customer & performing organizations) **on whether to; proceed with, change, pause, or terminate the proposed project**. If accepted then the Project Charter would be **officially signed off** to authorize the commencement of the project, and the advancement onto the in-depth Planning Phase.

 Before any substantial **work** can **commence** on the project, the **Project Charter must** have **been** officially **approved** and **signed off**.

DO NOT commit to a project without exercising appropriate levels of managerial & technical "due diligence".

That is, **DO NOT ALLOW ONE PART OF THE ORGANIZATION TO MAKE COMMITMENTS THAT OTHER PARTS OF THE ORGANIZATION CANNOT REALISTICALLY DELIVER.**

Because, the lack of appropriate levels of due diligence can result in unrealistic customer expectations which cannot be met without significant impact on the performing organization; i.e. a burden on [Cost] & [Time], possible financial penalties, and definitely reputational damage.

DO NOT base [Cost] & [Time] quotations on a principle of "*whatever values are necessary to win the business*" instead of those quotations being derived by sensible & logical estimates.

Because, if a customer (and your own senior management) put their faith in these ridiculous prices & unrealistic delivery dates then **the project will probably end in a catastrophic failure due to project blowouts and/or poor [Quality]** of the deliverables that resulted from *"just getting there"*.

DO NOT assume that the [Cost] & [Time] of an under-estimated "*win at all costs*" quotation can be recouped with contract deviations, time extensions, and future maintenance agreements.

Because, eventually some party will feel that they are being taken unfair advantage of, and hence this "losing strategy" will **end up in a disgruntled and combative relationship for those involved parties.**

Which brings us to an addendum to "Adaptive & Proactive" SDLC Project Management.

RULE 15^A: Do not place short term gains ahead of the long term viability to survive.

COFFEE BREAK DISCUSSION ... *Battle Lines between Sales and Development.*

One of the real-world SDLC problems that you, as project manager, will encounter is that sometimes you will be handed a project where the contractual agreements between the Performing Organization and the Customer Organization will have already been put in place. ... And frustratingly, the Sales Department | Business Development Group will have already negotiated the Contract Price & Duration, having had minimal (possibly zero) consultation with the Development Group. Subsequently, when you, as the Project Manager, start planning out this project, you quickly realize that you are holding a ticking time-bomb of a disaster, as the **[Cost] & [Time]** that were told to the customer, have **no resemblance to a feasible reality**. … … … … … … … … … … Yet you, as this project's manager, will be held responsible for the project's successful delivery.

Well, it is time to do something that many a project manager has contemplated doing, but very few have been prepared to do. ... And that is, "*dig your heels in, and say, NO*", and **state Why Not with quantifiable evidence to justify your stance**.

It is a given that this act will not make you very popular with members of the Sales Department | Business Development Group, and potentially with your own bosses. But, the consequences of not speaking up before the project has really commenced is that you (and more importantly, your employer) will be left completely exposed; as there is a real risk that the company is going to get into trouble due to these grossly under-estimated [Cost] & [Time] values. … … … While, the sales persons who signed off on the contract for this project, will collect their commission for a pending disaster.

You need to, with the backing of your fellow project managers, to encourage your Development Group senior managers to **push back on the Sales Department** | Business Development Group, and (if possible) **get them to change their focus from** "*booked-in revenue of potential sales*" to instead focusing on "*actual revenues of delivered products & services*".   … "*And yes, this is going to be a struggle, but a struggle that is necessary for the benefit of the organization.*"

EXAMPLE CASE ... *A Catch 22 situation, and which comes first, the chicken or the egg?*
Ah, that "Catch 22" situation of the Initiating Phase, where you are dammed if you do and dammed if you don't. ... How can the performing organization be honestly expected to provide "due diligence" of [Costs] and [Time] estimates to the customer organization without having firstly undertaken in-depth waterfall planning of the entire project, and having not allocated actual [People] and [Resources] to the known [Scope] of the tasks that have to be implemented? ... Given that, there are so many things that are still unknown, both requirements wise and technical know-how.
Most projects cannot be accurately planned, scheduled, and costed to the nth degree prior to the signed off commitment to commence; however, the budget and schedule can be guesstimated to *"within the ballpark"* (of at least 20-25% range) of that which would be produced by a detailed analysis; i.e. a **Rough Order of Magnitude (ROM)**.
Firstly, there must be some overall idea / concept of what the project's OUTPUT would *"kinda look like"*, and what are the business objectives & priorities of the project.
Let's hop into a time machine and go back to the early / mid-1990s, when mobile phones were *"dumb feature phones"* that had small displays, tiny physical dialling push-buttons, and very limited capability for users to install applications. *OH, I got this great idea !!! ... How about, a "SMART, phone" which has this big-as screen, no physical dialling keys and instead uses virtual keys on that touchscreen display, and (here is the clincher) the user can install any applications they want.* Note, in this reality, this kind of product does not exist, yet. ... And, this ain't no mega-corporation, just a small-to-medium sized independent company, and the "business strategy" is to build this first-of-type ground-breaking product with the intension of eventually selling the production rights onto a mega-corporation.
So, how would you do this project ?? ... Not, how was it actually done.

The first thing to note in this example case is that the customer is contained within the performing organization, being the owners and shareholders of that company.

Secondly, there must be some [Time] period by when this first product has to get to market (even if only in a limited production run) so as to start producing a positive cash flow for the company before it runs out of seed | venture capital. … And also, the limited "*window of opportunity*" to beat any other possible competitors, and thus attract those mega-corporation suitors.

(1) The project's overall idea & business strategy has already been established and the concept has been sketched & mocked-up; so that, there is "common understanding" and "mental alignment" between those who will do the project and the customer.

(2) List the major features & functionality [Scope] to be included in this product.

> ☑ Initiate & receive telephone calls.
> ☑ Send & receive simple text messages.
> ☑ Send & receive emails.
> ☑ Retain a list of contact phone numbers (and email addresses).
> ☐ Browse the internet.
> ☑ Listen to digital music files stored on the phone (via headphone or speaker).
> ☐ Listen to radio (via headphone or speaker).
> ☑ Take a photo and view locally on the device (send as an email attachment).
> ☐ Take a video and view locally on the device (send as an email attachment).
> ☐ Act as a GPS navigation device showing up-to-date position information.
> ☐ Play 2D games (and 1st & 3rd person 3D action games). … And, of course,
> ☑ The large touchscreen display. … plus, be easily carried in one hand.

That above list of major features & functionality is rather standard for today's real-world mobile phones; however, go back to the mid-1990s and these requirements essentially describe a super-compact very-portable telecommunications PC.

(3) With the Customer Requirements defined, and the customer's "needs & wants" understood, as well as the customer's "desired outcomes" and "perceived benefits" known. The next thing to do is order those major features & functionality into a sequence of "Release Milestones", so as to eventually arrive at the final deliverable.

Proof-Of-Concept (internal use only)	1.. Develop the "flatbed" engineering platform with all of the underlining hardware components so that the future Prototype Units will be able to function properly. 2.. Develop the software hardware drivers. 3.. Develop the embedded real-time operating system. 4.. Develop the initiating & receiving of telephone calls. 5.. Initiate & receive telephone calls via the touch screen.
Prototypes (internal use only)	1.. Develop the "ugly phone" engineering platform with all of the underlining hardware components so that the future Evaluation Units will be able to function properly. 2.. Port the hardware drivers and the operating system. 3.. Initiate & receive telephone calls via the touch screen. 4.. Retain a list of contact phone numbers (email address). 5.. Develop the create, send, receive, view of simple text messages. 6.. Develop internet access (and browsing the web). 7.. Develop the create, send, receive, view of emails.
Evaluation (external confidential use)	1.. Develop the "pretty phone" platform with all of the underlining hardware components so that the Evaluation / Demonstration Units will function properly. 2.. Port the software hardware drivers & operating system. 3.. Regression test all of the prototype's functionality. 4.. Environmental & telecoms certification testing. 5.. Develop the playing of stored music files (and the radio). 6.. Develop the taking of photos (and videos), store, view locally, and send as media message (email attachment).
Demonstration (public use)	1.. Improve usability & responsiveness, while rectifying defects found via the Evaluation Units operations. 2.. Prepare for mass production.
Mass Production (market availability)	1.. Outsource the production of the Final Product Units. 2.. Distribute & market those produced units for sale.

You will notice that **not all of the "wish-list" features & functionality were included**, because **not all of these are required upfront**, so as **to achieve** the company's **"beachhead objective"** of "establishing a foothold" in this new smart-phone market.

Also, it is a given that a real mobile phone developer will have a whole heap of other things to include in that list of Release Milestones, but this is a much cut down example case.

(4) For each of those previously listed Release Milestones stages of the project then estimate; the types and numbers of [People] that would be required at what operating [Cost], what types and quantities of [Resources] services & materials would be required at what standardized [Cost], and based on past experiences what [Time] duration can be expected for each of those major activities.

Proof-Of-Concept	1.. 3x Sw Engineers, 3x Hw Engineers, 2x Testers, 1x PM:TA. 2.. 2x months duration. 3.. 5x hardware systems (engineering – flatbed units).
Prototypes	1.. 10x Sw Engineers, 3x Hw Engineers, 2x Testers, 2x PM:TA. 2.. 4x months duration. 3.. 12x hardware systems (engineering – bespoke phone).
Evaluation	1.. 20x Sw Engineers, 5x Hw Engineers, 5x Testers, 3x PM:TA. 2.. 3x months duration. 3.. 20x hardware systems (production bespoke phone).
Demonstration	1.. 10x Sw Engineers, 3x Hw Engineers, 5x Testers, 3x PM:TA. 2.. 2x months duration. 3.. 100x hardware systems (small production run).
Mass Production	1.. 2x Sw Engineers, 1x Hw Engineers, 3x Tester, 1x PM:TA. 2.. 2x months duration. 3.. 100,000x hardware systems (production run).

The numbers above are very rough "butt-plucked" figures (for the purpose of this example); however, **Subject Matter Experts in the particular fields of endeavor can often provide reasonably accurate [Time], [Cost], and quantity estimates.**

(5) Put together some [Cost] estimates {including pre-built-in profit & safety margins}:

1x person operating cost = $200/hr X 38 hrs/wk = $7,600 per person per week
   {including the "bums on seat" costs of the work location and equipment}.
1x hardware system – engineering – flatbed units = $25,000 per unit
1x hardware system – engineering – bespoke phone units = $10,000 per unit
1x hardware system – production – bespoke phone units = $3,000 per unit
1x hardware system – production – small run phone unit = $1,000 per unit
1x hardware system – production – run phone unit = $450 per unit
   {including manufacturing, packaging, logistics, distribution, marketing}.
1x Environmental & telecoms certification testing = $100,000 per test run.

Proof-Of-Concept	o 9x People X 8.67 weeks duration	= $592,800
	o 5x Hardware systems (eng flatbed units)	= $125,000
		$717,800
Prototypes	o 17x People X 17.33 weeks duration	= $2,239,467
	o 12x Hardware systems (eng bespoke phone) =	$120,000
		$2,359,467
Evaluation	o 33x People X 13.00 weeks duration	= $3,260,400
	o 20x Hardware systems (pro bespoke phone) =	$60,000
	o 1x Enviro & Telecom Certification Testing =	$100,000
		$3,420,400
Demonstration	o 21x People X 8.67 weeks duration	= $1,383,200
	o 100x hardware systems (pro small run)	= $100,000
		$1,483,200
	TOTAL COST TO GET TO MASS PRODUCTION	**= $7,980,867**
Mass Production (outsourced)	o 7x People X 8.67 weeks duration	= $461,067
	o 100,000x hardware systems (pro run)	= $45,000,000
		$45,461,067
	TOTAL COST FOR MASS PRODUCTION	**= $45,461,067**
	TOTAL COST TO GET TO MARKET	**= $53,441,933**

For this example case project, it is estimated that to get from the "Initial Concept" through to the "Ready To Manufacture" stage is going to take approximately:

- 11 months [Time],
- $7,475,867 [Cost] of [People],
- $505,000   [Cost] of [Resources] materials & services.

$7,980,867 Total Cost To Engineer … with a 25% ROM margin **equals $7M to $9M.**

It is estimated that to get from "Ready To Manufacture" to having 100,000 units outsource produced and then delivered to the retail channels will approximately take:

- 2 months   [Time],
- $461,067   [Cost] of [People],
- $45,000,000 [Cost] of [Resources] materials & services.

$45,461,067 Total Cost To Manufacture … 25% ROM margin **equals $40M to $51M.**

Hence for this project, the **Total Cost To Get To Market** is estimated at **$47M to $60M**.

ID	Task Name	Start	Finish	Duration	Q3 14	Q4 14	Q1 15	Q2 15	Q3 15
1	Proof-Of Concept	01/07/2014	29/08/2014	8.67w	■				
2	Prototypes	01/09/2014	30/12/2014	17.33w	■	■			
3	Evaluation	30/12/2014	30/03/2015	13w			■		
4	Demonstration	31/03/2015	29/05/2015	8.67w				■	
5	READY TO MANUFACTURE	01/06/2015	01/06/2015	0w				♦	
6	Mass Production	01/06/2015	30/07/2015	8.67w					■
7	PRODUCT AVAILABILITY	03/08/2015	03/08/2015	0w					♦

The Initiation Phase should enable the distinction between 1 – 3 – 6 – 12+ months **duration** projects, multi thousand – million **budget** projects, and 10 – 50 – 100+ **people** projects. … **If such distinctions cannot be logically established then the project will most likely FAIL.**

Based on these estimated calculations from the Initiating Phase, the customer (in this case the performing organization's owners / executive management) may conclude that they can realistically cover the $7M to $9M for the "Total Cost To Engineer".

However, the additional $40M to $51M for the "Total Cost To Get To Market" is, to be honest, *"completely out of their league"*. Hence, the organization's owners / executive management may conclude that:

(a) They have 10-12 months from when the project starts to when the project arrives at "Ready To Manufacture", for them to:

- ☐ **Find a Business Partner** to provide the investment and/or the capability necessary to undertake the Mass Production stage.

- ☐ **Obtain Venture Capital** for them alone to undertake Mass Production.

(b) They have to have available and **release more of their Own Capital** by the start of the Mass Production stage.

(c) They have to have **merged** with a similar sized organization, or **be acquired** ("bought out") by a larger organization, who has the financial clout to have the Mass Production stage undertaken.

(d) They should **only take** the project to **as far as** the conclusion of the Prototype or Evaluation stage. ... OR, whenever the [Time] and/or [Cost] budget expires. ... And, **then figure out how to advance the project on from there**.

(e) They should *"forget the whole thing"* as it is an **unfeasible endeavour** for them.

RULE 16: **The Initiating Phase is the last opportunity to terminate the project without incurring significant financial penalties and reputational damage.**

Well, that is the Initiating part of the journey completed.

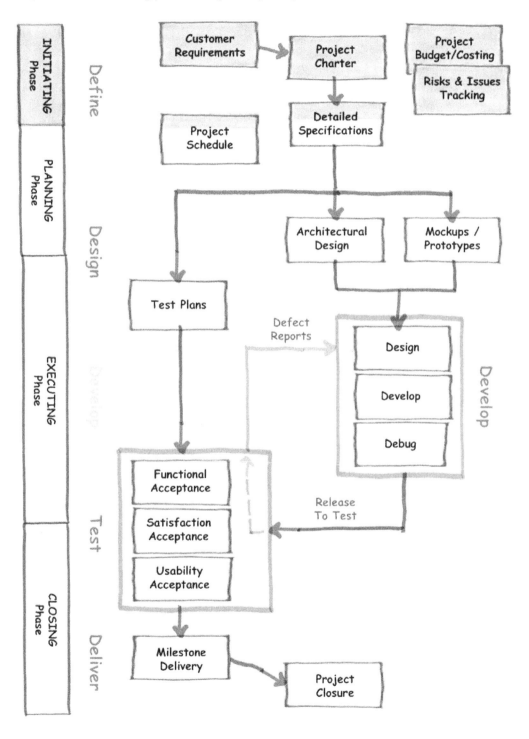

3.4. Planning Phase

3.4.1. Purpose

The purpose of the Planning Phase is to juggle all of the possibilities, prioritizing the probables, recording them for posterity, and thereby:

(1) **Refining** the project's objectives, requirements, and scope boundary.

(2) **Defining** exactly what has to be done, and how it will all be coordinated.

(3) **Communicating** the resultant plans.

3.4.2. What has to be answered?

For the project to have any realistic chance of success, the Planning Phase has to provide due-diligent answers to the following questions:

(1) **WE** (the implementer) **NEED WHAT, WHEN, and HOW MUCH** of it?

(2) How will **WE** (the implementer) **KNOW WE GOT IT RIGHT**?

(3) What do **WE** (the implementer) **THINK CAN GO WRONG**?

The **answers to these above questions** would be **established & recorded in** documents such as; the Detailed Specifications, (detailed) Project Schedule, (detailed) Budget Costings, Risks & Issues Register, (initial) Architectural Designs, and Mockups & Prototypes.

"Also, recorded in a Requirements Traceability Matrix (RTM), Work Breakdown Structure (WBS), Project Management Plan (PMP), etc."

Yes, ... there potentially is a lot of documents that could be produced during the Planning Phase so as to **receive bipartisan sign-off & authorization to advance the project** into the Executing Phase. However, which particular documents is dependent on the performing organization's processes & procedures, the customer's stipulations, and the project's size.

"THE PLANNING PHASE IS A DOCUMENTED THOUGHT EXERCISE TO WALK THROUGH THE IMPLEMENTATION OF THE ENTIRE PROJECT FROM START TO FINISH."

3.4.3. QUESTION: we need >WHAT< , when, and how much?

This question **refines the project's [Scope]** from being general **to being a lot more specific** in its details; with the subsequent answers being documented **in a (version controlled) Detailed Specifications**. *"aka, a Functional Specifications or Scope Statement."*

The Detailed Specifications should be **created via an iterative process**, see [Figure 27], where the **Customer Requirements are (re) evaluated, (re) questioned, (re) defined** with the input from the project's various stakeholders, relevant Subject Matter Experts, and the processes & procedures that the organization has stipulated for undertaking its projects.

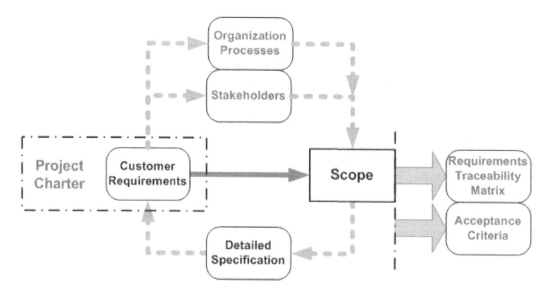

Figure 27: Process to refine the project's [Scope] and the Detailed Specifications.

Additionally, a **Requirements Traceability Matrix (RTM)** and an **Acceptance Criteria** list should also be **produced as a result of creating the project's Detailed Specifications**.

Where, the **Requirements Traceability Matrix (RTM)** would be used as an **itemized "check list"** to confirm the **"Scope Coverage"** of the **Detailed Specifications** against the **Customer Requirements**, clearly indicating what is **"In-Scope"** and **"Out-Of-Scope"**. Where **any feature or functionality that is not included in the approved Detailed Specifications** [Scope Baseline], **or not as described in this approved Detailed Specifications would be deemed as outside the project's [Scope]**; i.e. outside the **"Scope Boundary"**.

During the process of establishing / writing the Detailed Specifications, one may find that it is advantageous to take the project's [Scope], that has been found so far, and decompose this into separate work activities that would have to be done in order to produce the project's required deliverables.

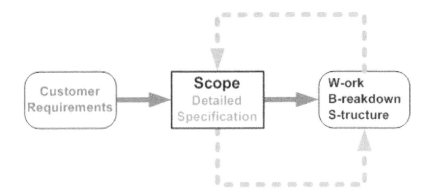

Figure 28: Process to decompose the project's [Scope] into work packages.

By **progressively breaking down** the major **features & functionality** in the Detailed Specifications, see [Figure 28], and stacking these as virtual boxes, a block structure of **related "work packages" would be built up**; i.e. a **Work Breakdown Structure (WBS)**. The feedback creation of the Detailed Specifications and the WBS **may uncover things yet to be realized**, and hence result in the discovery of additional features & functionality.

Which brings us to the next rule of "Adaptive & Proactive" SDLC Project Management.

RULE 17: A forgotten or misplaced something can, later on, be a project wrecker.

The Detailed Specifications is about the [Scope] of the requirements and NOT about the solution. Hence, it **DOES NOT contain the design**; rather, it should contain the information upon which the solutions are built.

With that warning noted, ... the **Detailed Specifications should contain sufficient explanation information to enable all of the involved project's stakeholders to correctly "interpret & understand"** the deliverables. Hence, **mock-ups, architectural designs,** and **prototypes** may be produced to **"aid in clarification"** of the Detailed Specifications.

"But, definitely not covertly implementing the project, in advance."

 I suggest, you think of the **Customer Requirements** as a *"wish list"*, and the approved **Detailed Specifications** (and the Requirements Traceability Matrix) as a *"what will be delivered list"*.

Only when the Detailed Specifications has been officially signed off by the authorized representatives from both the customer & performing organizations, ... and thereby establishing the project's [Scope Baseline], ... **would the Executing Phase of the project be allowed to commence.** ... *"This also applicable to agile projects, where there also has to be a good preliminary understanding of the project's scope".*

However, during the life of the project there will probably be a necessity to change the definition of some features & functionality in the approved Detailed Specifications, as well as to add (and subtract) other features & functionality. To have such **changes made to the [Scope Baseline] would require** that a formalized **Baseline Change Request process be undertaken**, else "Scope Creep" and "Scope Shrinkage" will covertly occur. [Section 2.6.3].

Which brings us to an addendum to "Adaptive & Proactive" SDLC Project Management.

RULE 17^A: Things discovered as missing at the END are a lot more costly to retro-fit than to discover these at the START.

COFFEE BREAK DISCUSSION ... *Agile, where to draw the line on formal scope change?*

The reason that it was decided to do the project using an Agile life cycle was because it was known in advance that changes were going to have to be made to the definition of the features & functionality. Given that, the customer was only going to know what they really wanted once they started to use those progressive deliverables. So, how can a "formalized" Baseline Change Request process be utilized, let alone be necessary.

Well, during the Initiating Phase of the project, ... see [Section 3.3] and especially the Example Case at the end of that Section, ... it was agreed between the representatives from both the customer & performing organizations that the project was authorized to be undertaken based on a defined set of project constraints (i.e. baselines) related to either; an agreed [Time] duration, an agreed [Cost] budget, an agreed set of [Scope] features & functionality, an agreed level of [Quality] acceptance criteria, an agreed assignment of [People], and/or an agreed allotment of [Resources].

What if, in the previous smart-phone example case, the proposed change was to *"just double the screen size from 4 inches to 8 inches"* and thereby effectively turning the project from producing a smart-phone into producing a smart-tablet with telephone capabilities. Such **a change** would have **a massive *"ripple effect"* on the project's [Time], [Cost], [Resources], [People]**, on other **[Scope]**, and **[Quality]**. Potentially, this **change could reverse the validity & viability of the project being undertaken.**

What if, in the previous smart-phone example case, the proposed change was to *"just double the number of allowable concurrently running user applications"*.

Hence, **even for agile projects, if the change effects the signed off project constraints then a "formalized" Baseline Change Request process** ~~should~~ **must be followed.**

Which brings us to the next rule of "Adaptive & Proactive" SDLC Project Management.

RULE 18: Scope Creep will kill a project, and Scope Shrinkage is just as deadly.

3.4.4. QUESTION: we need what, >WHEN<, and how much?

This question **refines when in [Time] different bits of** the project's **[Scope] will get done**; with the subsequent answers being documented in the **project schedule**, *e.g. Gantt Chart*.

There are various techniques to producing a project schedule; though, the essential element is to take the Detailed Specifications (and the Work Breakdown Structure – WBS, if one was produced) and **progressively elaborate the schedule** with those activities to be undertaken, **until confidence is high that no** "primary" **project activities remain undiscovered**, see [Figure 29].

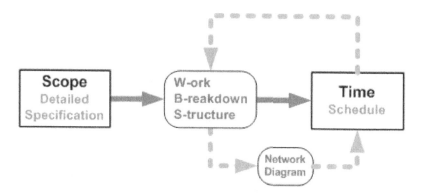

Figure 29: The Planning Process Model to create the "1st Pass" Project Schedule.

Some people / scheduling applications, also create a **Network Diagram** to **represent** the **precedence relationship** between those activities to be undertaken; thereby, determining the **"critical path"** of **sequentially linked tasks that results in the longest duration to complete the project**. Should a task on this "critical path" take longer than was planned, then the project's duration would definitely be increased.

 Personally, I create a 1st Pass "superman" schedule, where I pretend that I have only one person (me) to do all of the work, then I imagine that person working through the entire project, **encountering all of the activities that need to be done**. ... And, in that schedule I map out the execution path.

Once the creation of the 1st Pass Schedule has discovered all of the "primary" activities *(though not considering task durations nor milestone dates)*, then the next steps are to:

1.. Clarify the project's required '**Start Date**' and/or the mandatory '**End Date**'.

2.. **Estimate the [People] required** for each of those activities. ... skill level & quantities.

3.. **Estimate the [Resources] required** for each of those activities. ... grade & quantities.

4.. **Estimate the [Time] duration** for each of those ~~activities~~ tasks.

5.. Build a **logical sequence of sequential & parallel [Scope] tasks**.

6.. **Add** appropriate amounts of **contingency buffering to the Critical Path**, and to those feeder paths where it is expected that some (componentry) delay is likely to occur.

7.. **Progressively elaborate** the 2nd Pass Schedule, as per [Figure 30], until it is in a form that is ready to be presented to the project's stakeholders.

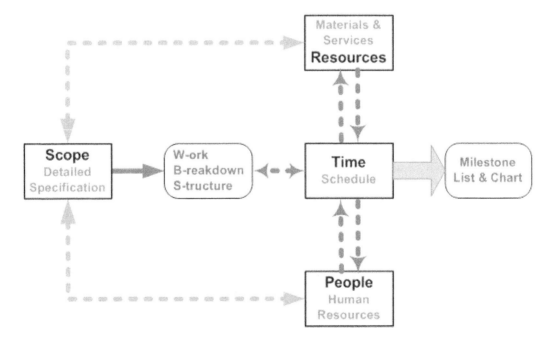

Figure 30: The Planning Process Model to create the "2nd Pass" Project Schedule.

Noting that, **the availability of the necessary [People] and the appropriate [Resources] will have direct effects on the project schedule's [Time] durations**. ... And, the [People] and/or [Resources] that are available on the open-market which is accessible by the performing organization, will influence the [Scope] that can be effectively implemented.

 Make sure that there is specific tasks included in the schedule for project management and technical authority support of the project's undertaking; else these will not be included in the project's estimated costing and subsequently not budgeted for.

I would suggest, 10-15% of the implementers' total effort be equated to this.

 Update the schedule's calendar to reflect those days when project tasks may or may not be worked on, due to; public holidays (*especially national holiday breaks*), [People] annual leave, [People] full time or part-time availability, and when specialist [Resources] will and will not be available.

Now that the 2nd Pass Schedule has been produced, then extract a **list of** those **Milestone Dates** that denote when significant events in the project's life will be completed.

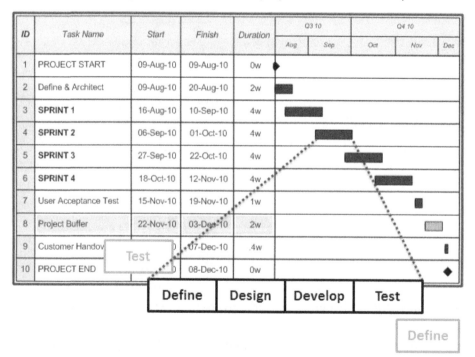

Figure 31: A waterfall schedule with the inclusion of an agile implementation.

THEORY ... *Only two types of projects.*

All projects (or project releases) irrespective of their size and complexity can be rationalized down into one of two generic types (or combination of):

- **Start Date** ... is where the **project / release has a specified starting date from when the work will commence** and at some time in the future it will finish.

    Start Date projects are scheduled forwards from the specified start date towards a yet to be determined end date.

- **End Date** ... is where the **project / release has a specified unmovable "drop-dead" date that the work absolutely has to be completed by, or ELSE**!

    End Date projects are scheduled backwards from the specified end date back towards a date when the work must be commenced by.

NOTE: If a Start Date project / release is taking too long to deliver then eventually, the customer organization's representatives or the performing organization's senior management will demand that the project / release be transformed into an End Date operation.

An Introduction To Scheduling ... Extracted from *"A Down-To-Earth Guide To SDLC Project Management"*.

The following is by no means a complete set of scheduling techniques, but it should be enough to get the project's schedule started (in MS-Project).

1.. Create a task of zero days duration and give it the same name as your project (also make it a larger bold font so that it will stand out).

    This 'Project:' task is here to make apparent what project this schedule is for when the Gantt Chart is printed out in a physical form.

	Task Name	Duration	Start	Finish	Predecessors
1	Project: Stocking-Stuffer	0 days	Sat 25-12-10	Sat 25-12-10	

2.. In the proceeding rows of the schedule, add some tasks of zero days duration for each of the project's major milestones / releases.

Give each task a meaningful name to identify the associated milestone.

These 'Milestone:' tasks are here to clearly indicate the delivery date for these specific releases without the need to trawl through the entire schedule, rolling-up and rolling-down tasks to determine this information.

		Task Name	Duration	Start	Finish	Predecessors
1		Project: Stocking-Stuffer	0 days	Sat 25-12-10	Sat 25-12-10	
2		Milestone: Release Forwards	0 days	Fri 07-01-11	Fri 07-01-11	6
3		Milestone: Release Backwards	0 days	Mon 10-01-11	Mon 10-01-11	13
4		Milestone: Release Bookends	0 days	Wed 05-01-11	Wed 05-01-11	20

Also, indent all of those milestone tasks under the Project Heading task; thereby getting a total duration (in days) for the project, and obtaining what the project's start date and end date will be.

3.. The first milestone for this example project will be a 'Start Date' mini-project containing four tasks to be completed in sequential order (A1 – A2 – A3 – A4):

- Add a title-task to identify clearly this release.

- On the proceeding rows, add each work-task to be completed, and right-indent these tasks to rollup underneath this release's title-task.

- Assign the 'Duration' for each work-task.

- For each work-task set the 'Predecessors' to the row number of the previous row; e.g. row-10's predecessor is row-9, row-9's predecessor is row-8, row-8's predecessor is row-7.

- Now that this release has been scheduled then add the corresponding 'Milestone:' task to be dependent on this release's title-task; i.e. row-2's predecessor is row-6.

	ⓘ	Task Name	Duration	Start	Finish	Predecessors
1		− Project: Stocking-Stuffer	10 days	Mon 27/12/10	Mon 10/01/11	
2		Milestone: Release Forwards	0 days	Fri 7/01/11	Fri 7/01/11	6
5						
6		− Release: Forwards Demo	10 days	Mon 27/12/10	Fri 7/01/11	
7		Task A1	2 days	Mon 27/12/10	Tue 28/12/10	
8		Task A2	3 days	Wed 29/12/10	Fri 31/12/10	7
9		Task A3	2 days	Mon 3/01/11	Tue 4/01/11	8
10		Task A4	3 days	Wed 5/01/11	Fri 7/01/11	9
11		Milestone	0 days	Fri 7/01/11	Fri 7/01/11	10

This example uses a **Finish-to-Start (FS) dependency** where the predecessor task has to (finish) prior to the successor task being able to (start).

This dependency technique is **used for 'Start Date' projects**.

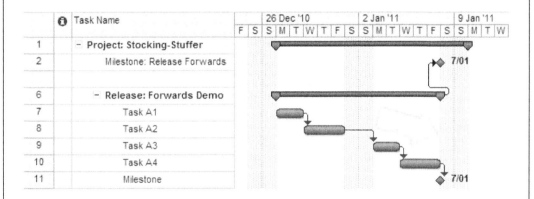

Notice how the diagram above has a nice formation of waterfall activities to be completed sequentially.

This Finish-to-Start (FS) dependency is the most common form of scheduling you will encounter, and is the default technique in Ms-Project. This scheduling technique is often how the "1st Pass" draft of the schedule would be created, so as to layout those tasks which are involved with the project, and then these tasks would be reorganized & paralleled to form the "optimum" flow of work to be undertaken by the implementers.

4.. The second milestone for this example project will be an 'End Date' mini-project containing four tasks that must be completed by a specific 'drop-dead' date (of Monday the 10th of January).

- Add a title-task to identify clearly this release. Though make sure that this title-task is left-indented back to the same level as the title-task and doesn't remain at the default indent level of the previous work-tasks.

- On the proceeding rows, add each work-task to be completed, and right-indent these tasks to rollup underneath this release's title-task.

- Assign the 'Duration' for each work-task.

- For each work-task set the 'Predecessors' to the row number of the next row; e.g. row-16's predecessor is row-17SF, row-15's predecessor is row-16SF, and row-14's predecessor is row-15SF.

- Now that this release has been scheduled then assign the corresponding 'Milestone:' task to be dependent on this release's title-task; i.e. row-3's predecessor is row-13.

	ⓘ	Task Name	Duration	Start	Finish	Predecessors
1		⊟ Project: Stocking-Stuffer	10 days	Mon 27/12/10	Mon 10/01/11	
2		Milestone: Release Forwards	0 days	Fri 7/01/11	Fri 7/01/11	6
3		Milestone: Release Backwards	0 days	Mon 10/01/11	Mon 10/01/11	13
13		⊟ Release: Backwards Demo	10 days	Mon 27/12/10	Mon 10/01/11	
14		Task B1	2 days	Mon 27/12/10	Wed 29/12/10	15SF
15		Task B2	3 days	Wed 29/12/10	Mon 3/01/11	16SF
16		Task B3	2 days	Mon 3/01/11	Wed 5/01/11	17SF
17		Task B4	3 days	Wed 5/01/11	Mon 10/01/11	18SF
18	🗓	Milestone	0 days	Mon 10/01/11	Mon 10/01/11	

> This task has a 'Start No Earlier Than' constraint on Mon 10-01-11.

Notice how the milestone-task on row-18 has in the info column a note about it having a 'Start No Earlier Than' constraint; i.e. a specified End Date.

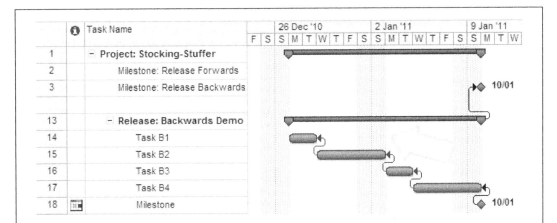

This example uses a **Start-to-Finish (SF) dependency** where **the predecessor has to have (started) before the successor task can (finish)**.

This dependency technique is **used for 'End Date' projects**, where the drop-dead end date would be set as a milestone, and the predecessor tasks would be set as dependent on their successor task, and thereby working backwards to determine when each task and the project needs to commence by in order to have a chance to complete the project by the dictated end date.

5.. The third milestone for this example project will be a mix of the two remaining dependency techniques; i.e. **Start-to-Start (SS)** and **Finish-to-Finish (FF)**.
Think of (SS) and (FF) as the book-ending of tasks.

- Add a title-task to identify clearly the release.

    Make sure that this title-task is left-indented back to the same level as the title-task and doesn't remain at the default indent of the previous work-tasks.

- On the rows proceeding, add each of the work-tasks to be completed, and right-indent these tasks to rollup underneath this release's title-task.

- Assign the 'Duration' for each work-task.

- Set the predecessor, 'Task C3' on row-23 to start at the same time as 'Task C2' on row-22, hence row-23 'Predecessors' is set to row-22SS.

- Set the predecessor, 'Task C5' on row-25 to finish at the same time as 'Task-C4' on row-24, hence row-25 'Predecessors' is set to row-24FF.

- Now that this release has been scheduled then assign the corresponding 'Milestone:' task to be dependent on this release's title-task; i.e. row-4's predecessor is row-20.

	❶	Task Name	Duration	Start	Finish	Predecessors
1		⊟ Project: Stocking-Stuffer	10 days	Mon 27/12/10	Mon 10/01/11	
2		Milestone: Release Forwards	0 days	Fri 7/01/11	Fri 7/01/11	6
3		Milestone: Release Backwards	0 days	Mon 10/01/11	Mon 10/01/11	13
4		Milestone: Release Bookends	0 days	Wed 5/01/11	Wed 5/01/11	20
20		⊟ Release: Bookends Demo	8 days	Mon 27/12/10	Wed 5/01/11	
21		Task C1	3 days	Mon 27/12/10	Wed 29/12/10	
22		Task C2	2 days	Thu 30/12/10	Fri 31/12/10	21
23		Task C3	3 days	Thu 30/12/10	Mon 3/01/11	22SS
24		Task C4	2 days	Tue 4/01/11	Wed 5/01/11	23
25		Task C5	3 days	Mon 3/01/11	Wed 5/01/11	24FF

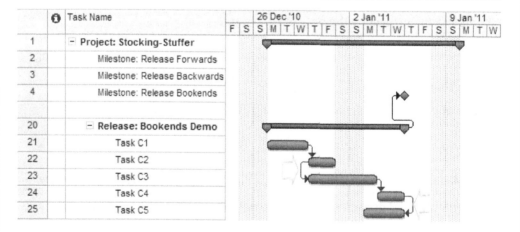

This example uses;

- The **Start-to-Start (SS) dependency** is where the current / predecessor task cannot (Start) until its sibling / successor task is also ready to (Start).

- The **Finish-to-Finish (FF) dependency** is where the current / predecessor task cannot (Finish) until its sibling / successor task is also ready to (Finish).

Below is a combined view of each of the scheduling techniques; Finish-To-Start (FS), then Start-To-Finish (SF), and finally Start-To-Start (SS) and Finish-To-Finish (FF).

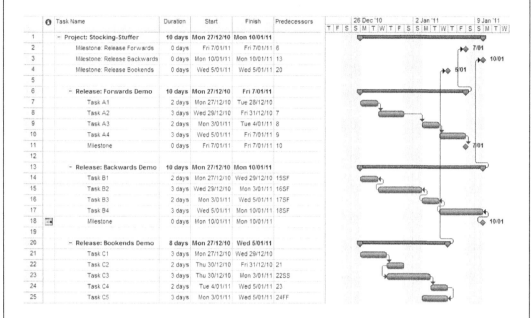

A few other terms should be mentioned at this point:

- ❖ **Lead Time** – is a period of **time before, that the successor can start prior to the predecessor finishing**; i.e. the 2nd task can be underway by a certain amount of time, before the 1st task has finished. A dependency of **[FS] – No. days**.

- ❖ **Lag Time** – is a period of **time after, that the predecessor task has finished prior to the successor task being allowed to commence**; i.e. the 2nd task has to wait a certain amount of time, after the 1st has finished. A dependency of **[FS] + No. days**.

- ❖ **Float / Slack** – is a period of **time that a task could be delayed or exceed its planned time allocation without resulting in the project / release milestone being delivered late**. ... *"Free Float" per task, "Total Float" per milestone.*

- ❖ **Critical Path** – is a **sequence of dependent tasks that if one of these tasks exceeded its planned time allocation then its "knock-on effect" would result in the project / release milestone being delivered late**. Hence, the **Critical Path has no float / slack**.

Adding Buffering To The Schedule ... Extract *"A Down-To-Earth Guide To SDLC Project Management"*.

This previous project schedule was relatively simple but it didn't include any **contingency buffering / reserve for something going wrong.**

> *For example; some safety margin just in case a group of tasks is a bit more technically difficult than was planned for.*

Steps for adding contingency buffering to a schedule:

1.. Make a copy of the existing project schedule.
2.. At the end of the sequence of tasks that require a contingency buffer, add a new task, assign the 'Duration' for this buffer task to the desired safety margin, and set the 'Predecessors' to the row number of the previous row in the sequence.
3.. Optionally, Right-Click this 'Buffering' task's bar and select 'Format Bar' from the menu, and choose an appropriate pattern and colour to differentiate this buffering from those activity tasks.

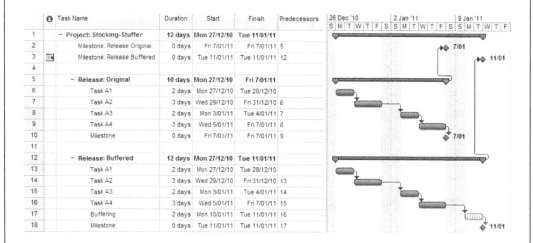

Notice how the Buffered version of the release has a milestone date later than the original version of this release. This is the result of the added buffering.

NOTE: only add buffering to the "critical path", else it could be taken as permission to slacken off on other paths' tasks.

3.4.5. QUESTION: We need what, when, and >HOW MUCH< ?

This question **refines the project's [Cost]** from being general **to being a lot more specific** in its details; with the subsequent answers being documented **in a (version controlled) Budget** spreadsheet. Where the [Costs] are determined from; the schedule's **work [Time] durations**, the **charge rates of those [People]** who will be engaged to work on project tasks, and the **price of materials & services [Resources]** that need to be acquired for the project. Thus, why it was so important to obtain reasonably accurate estimations for the 2nd Pass Schedule, so as to produce a reasonably accurate [Cost Baseline]. See [Section 5.4.2] for costing the project to success and [Section 5.4.3] for certain death.

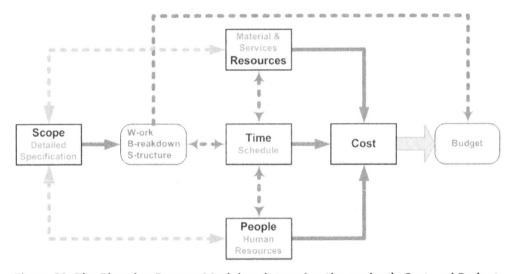

Figure 32: The Planning Process Model to determine the project's Cost and Budget.

During the writing of the Detailed Specifications and subsequently while constructing the project's schedule, a **Work Breakdown Structure (WBS)** was built to help determine all of the project's "primary" tasks. These **WBS Work Packages** can now be **itemized in the Budget** spreadsheet **as aggregates of "Accrued Component Costs"**. … And, the **Accrued Work Packages** could eventually be **assigned "Cost Account Codes"** in the performing organization's **Timesheet and Cost Accounting systems**, so as to track the project's [Time] utilization and [Cost] expenditures against those accrued work packages, see [Figure 32].

The project **Budget** spreadsheet (aka "**Cost Account Authorization**" spreadsheet) should **include** (performing organization standardized) **built-in calculations for Cost Escalations** to determine; **"Sell Price" mark-up**, **safety margins** (*e.g. exchange rates & inflation*), **Cost Contingencies** (*e.g. warranty repairs & latent defects*), and **percentage <Risk> for the type of work**. Thereby, ensuring that this **project will most likely be a profitable endeavour.**

Which brings us to the next rule of "Adaptive & Proactive" SDLC Project Management.

RULE 19: BAD Estimates   = BAD Allocations
BAD Estimates   = BAD Durations
BAD Estimates   = BAD Budget
BAD Allocations = FAILING Project
BAD Durations   = FAILING Project
BAD Budget   = FAILING Project
FAILED Project = BAD Cash Flow
BAD Cash Flow   = Dying Business.

3.4.6. What to do with these answers?

By this stage, the baseline documents should be ready for review, negotiations, final updating, and then **signed off by the authorized representatives from both the customer & performing organizations** (i.e. the Project Steering Committee) **as acceptance & approval of the project's Baselines.** ... And, subsequently as permission to commence the implementation.

- ☐ [Scope Baseline]   ... the approved **Detailed Specifications**,
- ☐ [Time Baseline]   ... the approved **Project Schedule**,
- ☐ [Cost Baseline]   ... the approved **Project Budget^**, ... *not the project's "Overall Budget".*
- ☐ [Quality Baseline]   ... the approved **Acceptance Criteria**.

If changes / deviations are required to any of these **approved baselines, then** some form of **Baseline Change Request (BCR) process** will need to be undertaken; [Section 2.6.3].

EXAMPLE CASE … *A beer coaster ride into planning a way out of this mess.*

Some projects are *"born to failure"*; all because, one would surmise, that the project's Initiating Phase was possibly conceived, contracts negotiated, and strategic plans made *"on the back of a few used beer coasters during an executive luncheon"*.

That is, during the project's Initiating Phase, no **real consideration** was **given to diligently producing a Rough Order of Magnitude (ROM)** that was *"within the ballpark"* **of what would be involved to realistically undertake this project**. Possibly this situation is due to a *"win at all cost"* strategy to *"just get the business"*, then worry about negotiating for Contract Deviations during the Executing Phase of the project.

Subsequently, there are misunderstandings & misinterpretations of the customer's Needs – Wants – Requirements – Expectations, as well as the [Scope] – [Time] – [Cost] – [Quality] – [People] – [Resources] project constraints not aligning with what is realistically achievable given the current situation & prevailing circumstances.

> Consider a project where the customer was given a Fixed Price [Cost Baseline] of $800,000 for the delivery of a product with specific functionality.
>
> However, during the Planning Phase; when the Customer Requirements – Functional Specifications, … i.e. in this case, the customer provided equivalent of a contractual Detailed Specifications [Scope Baseline], … was broken down into a Work Breakdown Structure and all of the involved tasks were [Time] scheduled based on those [People] & [Resources] that would be utilized & available, then the Project Manager discovered that the project would [Cost] something closer to $1,300,000 to deliver, i.e. a 62.5% difference in [Cost]. … And, the [Time] to deliver would be over a year and a half, instead of the 12 months [Time Baseline] that was told to the Customer's Representatives.
>
> *"Fixed Price quoting without some level of diligent preplanning and the stabilization of the requirements is just asking trouble."*

A combat veteran once told me, "*don't get entangled in a fight that you're not prepared to frickin win*". ... And, his favourite, "*you can only piss with the cock you've got*".

While crude expressions, the underlying premise for project management being:

- ☒ **Do not commit to commencing a project that there has been no real consideration of**; the **available [People]** – their skill levels & past experiences, the **available [Resources]** – quantities & grade, the **available [Cost] financials** – liquidity & cash flow, and the **available [Time] duration** to undertake the project.

- ☐ However, **sometimes you have no choice but to go with what's available at that moment and in that particular place**; irrespective of the fact that, these may be inadequate for the job that is currently at hand.

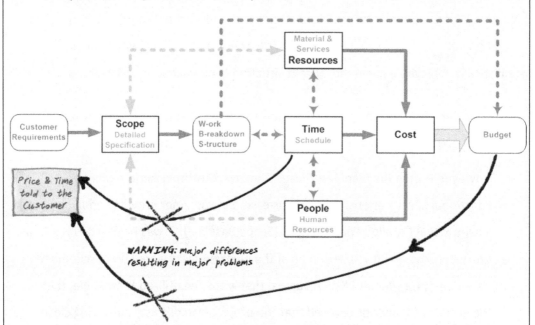

Which brings us to the next rule of "Adaptive & Proactive" SDLC Project Management.

RULE 20: Misaligned project initiation and planning can easily derail the chances of the project ever being a successful endeavour.

3.4.7. Project Variables versus the Baselines

Recall back in [Section 2.7], the push-me pull-me inter-relationship of the project variables and the project constraints (i.e. baselines), and how too many project constraints creates <Risk> on the project being a success. Well, if those [Scope], [Time], [Cost], [Quality] baselines are the project constraints then the project variables would be:

☐ [People] … those persons assigned to work on the project,

☐ [Resources] … those materials & services to be utilized by the project.

However, what if these project variables of [People] and [Resources] were also mandated as being fixed as project constraints? … There would subsequently be great <Risk> of the project ending in failure; due to, a lack of flexibility to continually adjust to the current situation & prevailing circumstances, and thereby not necessarily being able to stably manoeuvre the project towards the desired outcomes. … Unless, alternatively, every aspect of the project was planned out to the nth degree, which in doing so would consume a considerable amount of [Time], [Cost], and [People] effort; let alone that which would be required to micromanage the implementation of such detailed project plans.

"And, this is why those old fashioned waterfall projects were such pains in the butt, because these took so long and consumed so much effort just to get to the end of the planning phase, before being allowed to get to the fun part of implementation. … Which, IMHO, is why the rolling-wave planning of Agile has caught on."

Transiently Flexible Project Constraint Baselines

Yes well, … in reality, what happens is that, while the [Time Baseline] & [Cost Baseline], [Scope Baseline] & [Quality Baseline], and/or [People Baseline] & [Resource Baseline] have been "fixed" as constraints, these **project constraints are in-effect *"transiently flexible"* during the implementation of the Executing Phase. Provided that, when the project is coming to a delivery milestone or is about to finish, then each transitioning project constraint converges to conform with the corresponding approved baseline**.

Consider that, during the progress of project implementation, these baselines will not be entirely complied with, as the deliverables are still being defined – designed – developed – tested. It is only **as the project is coming to a delivery milestone or the finish, then these "transiently flexible" project constraints should be converging on those approved baselines.** If this convergence is not / has not occurred then the project is in real trouble.

Hence, **it is the project's management responsibility to monitor & control these "transiently flexible" project constraints, so as to ensure that these do home in on the agreed baselines**. See [Section 3.5.6] and [Section 3.7.3] on Earned Value Performance Measures (EVPM) and Key Performance Indicators (KPI), and agile Burn Down Charts.

3.4.8. QUESTION: how will we know we got it right?

This question determines whether what has been done so far during the Planning Phase is going to satisfy the project's stakeholders; based on their "Needs", "Wants", "Concerns", "Expectations", and "Perspectives". … And, on their individual "Opinions" of the project's potential for success or pending failure.

"Yep, everyone has an opinion, including those vendors and suppliers who will provide materials and services to the project. So, you had better consider them in addition to your other stakeholders."

Bi-Directional Communications

Therefore, to "Get It Right" will require that all of these stakeholders are managed appropriately, see [Section 2.6.2]. … And, an effective & efficient way of monitoring & controlling these stakeholders is via two-way communications.

> ❖ **Effective Communications** … providing the information in the format that they need, and when they need it.
>
> ❖ **Efficient Communications** … providing only the relevant information they need.

WARNING: A project manager who isn't prepared to spend their time communicating with all of the project's stakeholders is a project manager who is destine to lose control of the project.

This is because, without continuous communication flows and feedback, then the Project Manager and the project stakeholders will not know and understand what is relevant to them, not know what is going on, and not know what is not going on (as they thought it was).

To ensure project success, the Planning Phase must take into consideration the communications aspect, as per [Figure 33].

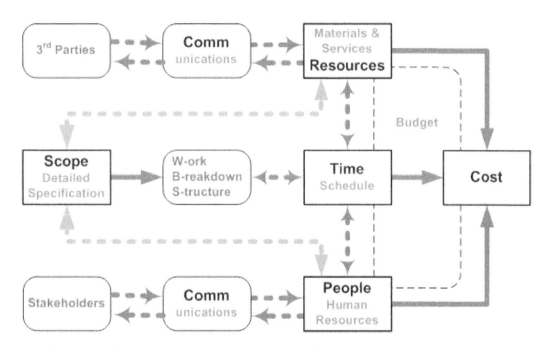

Figure 33: The Planning Process Model including the Communications aspect.

Noting that, the communications are bi-directional between those in the performing organization who are actively involved with the project, and those other stakeholders (internal & external), plus 3rd party vendors & suppliers who are also involved.

With respect to communications, the following has to be considered and planned for:

- **What are the communications methods & mediums** that will be used internally within the Project Team and externally between the performing organization, the customer's organization, and to the project's vendors & suppliers?

    o **Are there standardized communications protocols and formats** to be used? *E.g., document templates, reporting presentation formats.*

- **What are the communications paths and escalation hierarchy**? That is, who talks to who, and who in particular is allowed and not allowed to talk with each other?

- **What are the Points Of Contact (POC)** in the performing organization, the customer's organization, and for each vendor / supplier?

- **What is the frequency of the communication**? That is, how often should the regular communications with each of these parties occur?

- **What tools, applications, and facilities** will be utilized for the communications, ... as well as documentation storage, archiving, and retrieval?

Now that the importance of communications between the project's various stakeholders and 3rd parties has been considered, then the next concern for "Getting It Right" is ensuring that they agree that the project's deliverables have met the approved Detailed Specifications and that these deliverables will be fit for use (as was agreed & defined).

Which brings us to the next rule of "Adaptive & Proactive" SDLC Project Management.

RULE 21: Poor bi-directional communications is a project killer.

RULE 21[A]: A project manager will spend a lot of their time communicating in both written and oral forms.

Quality

From the performing organization's perspective, [Quality] is concerned only with the correctness of the project's measurable & tangible deliverables when compared to the agreed [Scope Baseline], as was defined in the approved Detailed Specifications and in the approved Acceptance Criteria. ... And, **NOT necessarily as the customer had "intended" it to be.** Hence, to produce "quality deliverables" is very much dependent on the definition & description of those deliverables (i.e. agreed [Scope]), as well as ensuring that the representatives from both the customer & performing organizations have the same understanding & interpretation of these deliverables. This also, necessitates that those involved 3rd party vendors & suppliers have the exact same understanding & interpretation on what they have to deliver. See [Figure 34].

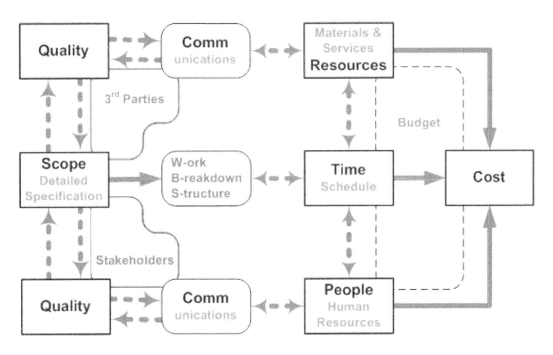

Figure 34: The Planning Process Model including the Quality aspect.

Consequently, poorly defined project [Scope] and/or poor communications with the project's stakeholders and involved 3rd parties will result in poor [Quality] deliverables.

3.4.9. QUESTION: what can go wrong?

With the bulk of the Planning Phase now completed, the next question is concerned with what do we think can go wrong. ... But. ... What could possibly go wrong, given that this project has been planned out so diligently?   See [Section 2.6] on R.I.S.C. Management.

- ❖ **Risk** ... some unforeseen situation happens or the prevailing circumstances change.
- ❖ **Stakeholder** ... Needs, Wants, Concerns, Expectations, or Perspectives *"or Opinions"* change.
- ❖ **Change** ... some agreed baseline has to be amended to better suit the above points.

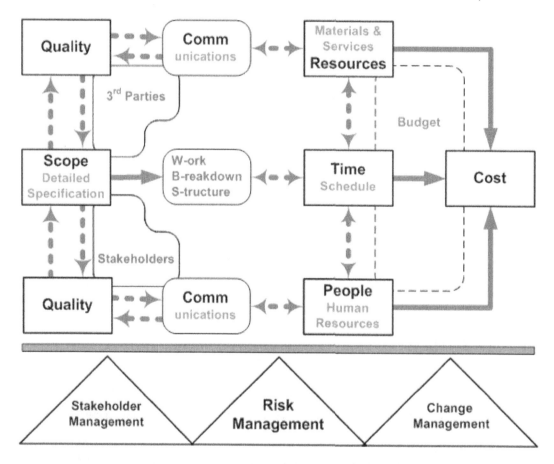

Figure 35: The Planning Process Model including considerations for RISC.

3.4.10. Planning Phase Completion

At this stage, the Planning Phase has essentially been completed, and all that remains now is to hold a **Phase Completion Review** (i.e. a "sanity checkpoint" **Gate Review** meeting) **to decide** whether the project is in **an acceptable state to "ADVANCE" onto the Executing Phase**. ... OR, to **"REVISE", "PAUSE", or "TERMINATE" the project.**

It is crucial that this Phase Completion Review meeting involve representatives from the senior management of the performing organization (i.e. the Project Steering Committee), representatives of the customer organization, the Project Manager, and selected Subject Matter Experts from the Project Working Group (and maybe even from the Project Implementation Team). ... *"And, this also applies to agile based projects."*

If the resultant decision from this Phase Completion Review meeting is permission to proceed (with the current release | milestone) then **ensure that all of the authorization signatures are in place for the** approved Detailed Specifications **[Scope Baseline]**, the approved schedule **[Time Baseline]**, the approved budget **[Cost Baseline]**, and/or the approved Acceptance Criteria **[Quality Baseline]**. ... And, if also mandated as constrained, then the approved **[People Baseline]**, and materials & services **[Resource Baseline]**.

With the Planning Phase signed off, it is time to **hold a "Kick-Off Meeting"** with the Project Working Group, Project Implementation Team members, and selected members of the Project Steering Committee, so as **to let everyone involved know where they stand, and where everyone fits into "The Plan" for moving the project forwards from here**.

> Which brings us to the next rule of "Adaptive & Proactive" SDLC Project Management.
>
> **RULE 22:** **The Planning Phase is the most opportune moment to correct any oversights with the Initiating Phase and lay the foundations for future project success.**

EXAMPLE CASE ... *Planning the [Time & Cost Baselines] for a simple project.*

Consider a software development project with seven [People] allocated; *e.g. a project manager come business analyst, a system designer / platform architect, 4x developers, and a tester.* Where each of these [People] is accounted for as having the same "bums on seats" [Cost] of $100 per hour with a standard workday [Time] duration of 8 hours.

The project [Scope] of tasks and [Time] durations are as follows.

Phase	Task	Duration	Resource
Initiating	Initiation	1 hour	Project Manager
	Conceptualization	5 hours	Project Manager (Analyst)
	Work Approval	2 hours	Project Manager
	Requirements Gathering	1 days	Project Manager
Planning	Planning	3 days	Project Manager (Analyst)
	Design	2 days	System Designer (Architect)
Executing	Development	5 days	Developer 1-4
	Debugging	2 days	Developer 1-3
	Integration	3 days	Developer 1-2
	Verification Testing	2 days	Tester & Developer
Closing	Acceptance Testing	1 day	Tester
	Release / Rollout	1 day	Developer
	Closure & Payment	4 hours	Project Manager

A waterfall project schedule, in [Figure 36], is created for the above sequence of [Scope] tasks, then the corresponding [Time] is scheduled per task, the appropriate [People] are allocated to each of these tasks, and [Costs] assigned to each of these people.

The resultant schedule would form the project's [Time Baseline] and [Cost Baseline] of; one month's duration and a **"Level-Of-Effort (LOE)"** value of $35,600.

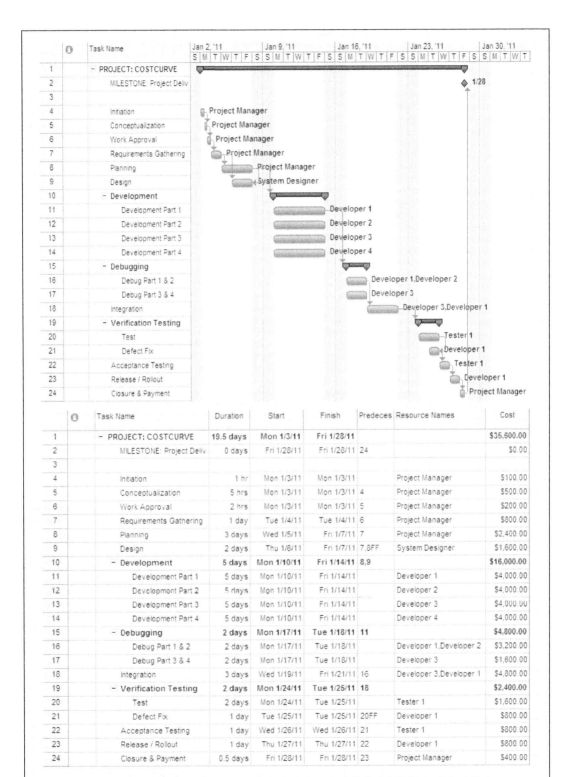

Figure 36: Example – development project with [Time] & [Cost] Baselines.

If these planned [Costs] per [Scope] task were totalled up on a per day basis, then the resultant "Daily Cost vs. Time" graph would look like [Figure 37].

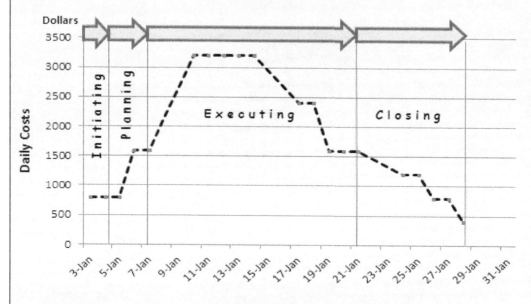

Figure 37: Example – Daily Costs vs. Time.

While adding up those daily Planned [Costs]; if these [Costs] were accumulated day after day over the expected duration of the project then the resultant plotted [Figure 38].

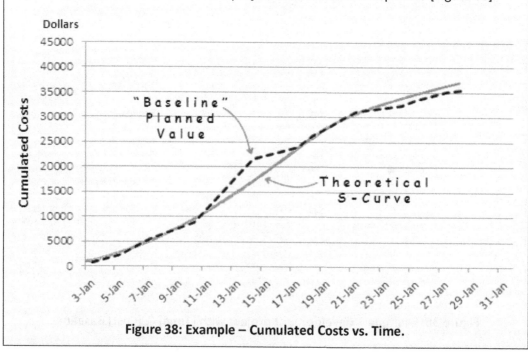

Figure 38: Example – Cumulated Costs vs. Time.

THEORY ... *Why is the plot of the Cumulated Costs vs. Time shaped like an S-Curve?*

This 'S' shape is because, at the beginning of the project there are only a few people involved (i.e. charging time to the project's Cost Account Code); whereas, by the Executing Phase the entire Project Implementation Team is on board. However, as the project begins to conclude then The Team will begin to be dispersed to work on other projects, so that eventually there will be only a few people remaining on the project.

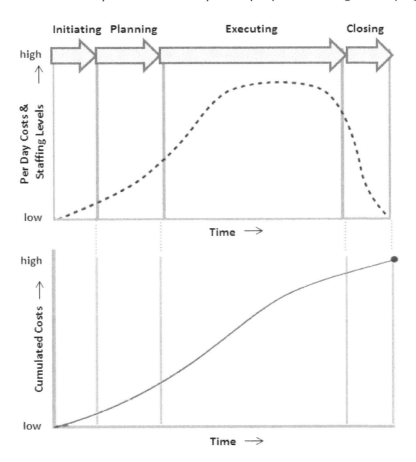

Figure 39: Staffing levels compared to the Cumulated Cost vs. Time ... S-Curve.[1]

Though, for a project with a fixed / obligated work effort *(e.g. a stipulated 40 hours per week)* then the S-Curve would in fact be a rising straight line.

---

[1] Based on PMI 2013, *"A Guide to the Project Management Body of Knowledge (PMBOK® Guide), 5th Edition"*.

Well, that is the Planning part of the journey completed.

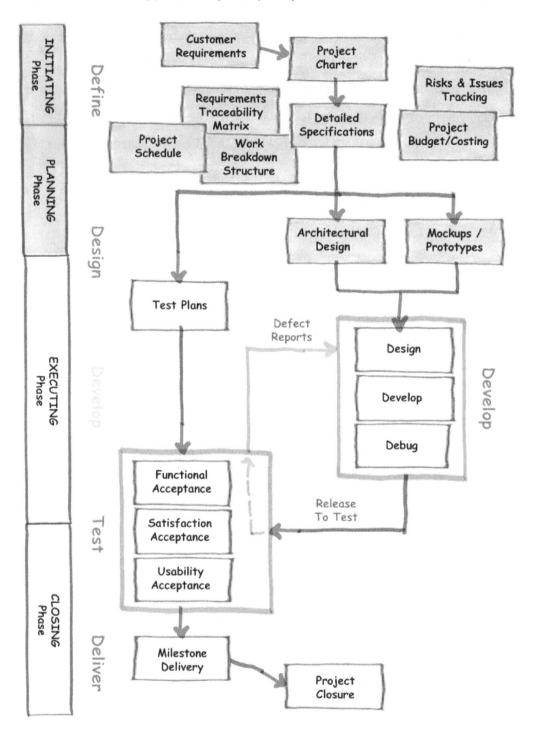

3.5. Executing Phase

3.5.1. Purpose

The purpose of the project's Executing Phase is to; perform those specified tasks to produce the agreed [Scope] of deliverables, as per the agreed [Quality] Acceptance Criteria, within the agreed [Time] frame, and to keep within the agreed [Cost] Budget. Thereby, resulting in a successful project, as per those determinates listed in [Section 1.2].

Though, from the project's management perspective, there is little (to no) hands-on involvement with "implementing" the Executing Phase, other than what is required to "Monitor & Control" the project's progress.

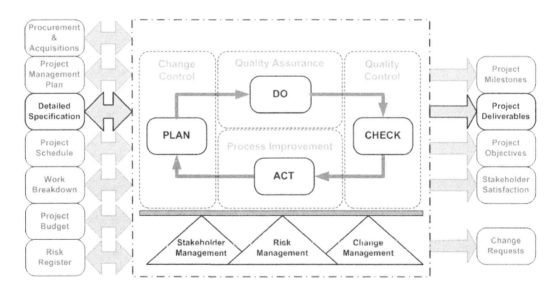

Figure 40: The Executing Phase as a 'PLAN – DO – CHECK – ACT' process model with associated inputs and outputs.

Noting that, the Executing Phase is based around a 'PLAN – DO – CHECK – ACT' process model, see [Section 2.4].

This 'PLAN – DO – CHECK – ACT' process is due to, **the Executing Phase being concerned with "doing" the implementation work**; i.e. the Project Implementation Team's activities. However, **there is a symbiotic relationship** between the other project phases, with the **Monitoring & Control Phase "Overseeing" the implementation** so as to ensure that the **Executing Phase's "Doing" conforms** with the **Planning Phase's 'To Be Done' [Baselines]**.

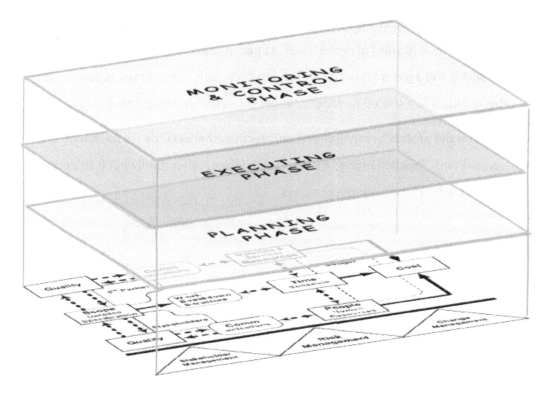

Figure 41: The symbiotic relationship of the Executing Phase with the Planning Phase and the Monitoring & Control Phase.

Which brings us to the next rule of "Adaptive & Proactive" SDLC Project Management.

RULE 23: A project manager's role is not to do; but rather, to make sure it gets done.

RULE 23[A]: within the agreed [Time] duration,
within the agreed [Cost] budget,
within the agreed [Scope] boundary,
with the agreed [Quality] criteria.

3.5.2. What has to be answered?

For the project to have any realistic chance of success, the Executing Phase has to provide due-diligent answers to the following questions:

(1) How exactly **WILL** the Project Implementation **TEAM DO IT**?   ... **PLAN**

(2) Is the Project Implementation **TEAM DOING IT**?   ... **DO**

(3) Is the Project Implementation **TEAM CHECKING IT**?   ... **CHECK**

(4) Is the Project Implementation **TEAM DONE YET**?   ... **ACT**

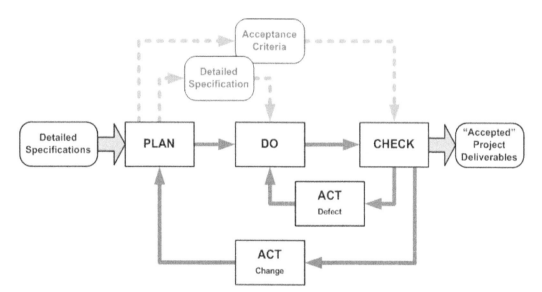

Figure 42: 'PLAN – DO – CHECK – ACT' iterative nature of implementation.

Which brings us to an addendum to "Adaptive & Proactive" SDLC Project Management.

RULE 23[B]: A project manager's role is to balance what was planned to do, versus that which can realistically be done. ... When taking into consideration the current situation & prevailing circumstances.

3.5.3. QUESTION: how exactly will The Team do it?

For the implementers this question refers to such things as; architectural designs, hardware designs, software designs, database schemas, network designs, Use Cases, test plans ... etc. That is, those areas that this book will not be delving into; because, **from the project's management perspective, this question was conceptually / hypothetically / virtually answered during the "walk through" of the Planning Phase**.

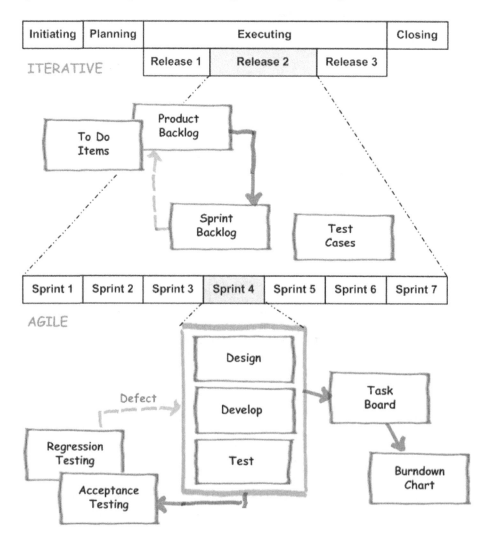

Figure 43: A hybrid mashing of Iterative & Agile project life cycles.

Product Backlog, Sprint Backlog, and Sprint Planning Meeting

With an agile based project, see [Figure 43], this question would entail:

1.. **Determine the approximate number of sprints** that would be involved **for the upcoming milestone release**, and what is **the optimum durations for each of those individual sprints**.

    o Though, NOT imposing a predefined fixed number of sprints and durations, as this could result in the perceived necessity to cram features & functionality into each sprint (especially for the final few sprints). Rather, *"through the wisdom of hindsight"* of being based on the Project Implementation Team's past experiences with this project (and with other projects), the "**velocity**" of how many 'To Do Items' were being successfully "Done" for those earlier sprints, and what is the upcoming roadmap of project milestone events / organizational events / state & national holidays, and 3rd Party considerations.

2.. **Determine the overall objective** ("goal") **for each proposed sprints so each adds value to the whole release**; *e.g. lay the foundations, put up the scaffolding, construct the structural walls, construct the associated "basic" rooms per section, ... etc.*

3.. At the start of each sprint, **hold a 'Sprint Planning Meeting'** with representatives from the Project Implementation Team and the 'Product Owner'. The Project Implementation Team would then **choose the 'To Do Items'** to go into the **"Sprint Backlog"** list for this sprint.

    o The **Product Owner** (i.e. the representative of the customer and the primary stakeholders) would **present a prioritized set of high-level requirements** that they "would like" to be delivered, based on these requirements perceived value to them the customer (i.e. Return On Investment). This prioritized list forms the '**Product Backlog**', which is independent of the 'Sprint Backlog'.

    o The **Project Implementation Team's representatives** (i.e. their Subject Matter Experts, Team Leader, guys & gals, 'Scrum Master', dudes who know) would select a reasonable number of 'To Do Items' (features & functionality) that they are prepared to commit to delivering by the end date of this sprint, but only for this current sprint (and not for any other future sprints).

4.. With **the current Sprint Backlog** having now been established, this **forms the "Scope of Deliverables" for this current sprint** (and only this particular sprint).

Noting that, **it is NOT the project's management, but rather the members of the Project Implementation Team that decide which 'To Do Items'** (features & functionality) go into **the Sprint Backlog, and** what is **the amount of effort that will be required to complete each selected 'To Do Item'**. Hence, it is **the Project Implementation Team members that commit to delivering the sprint**. Whereas, the project's management purpose (in The Team's view) is to remove the barriers and clear the path for their successful delivery.

3.5.4. QUESTION: is The Team doing it?

For the implementers this question refers to such things as; software coding, hardware schematics design, hardware layout, prototyping, pre-production samples, debugging, module & component testing ... etc.

With an agile based project, **"Tasks" are NOT assigned specifically to anyone**; rather, the **individual Project Implementation Team members autonomously pick, from the Sprint Backlog** of 'To Do Items', that item which they intend to ~~implement~~ deliver.

Task Board

The progress of their work on individual 'To Do Items' is often **represented** (either physically on a wall or virtually on a computer screen) via the use of a '**Task Board**', see [Figure 8] reproduced in this section. Where the Task Board is lined-off into sections for; 'To Do' (i.e. the Sprint Backlog), 'Work In Progress' (WIP), and 'Done'. ... And, the 'To Do Items' card passes from left-to-right across the board as it is worked on, until it has been verified as having been successfully 'Done'. ... Though, these last two columns could be further subdivided into 'Develop' and 'Test'; where, the 'Develop' portion is for the

Quality Assurance of "doing it right", and the '**Test**' portion is for the **Quality Control of "verifying it's been done right"**.

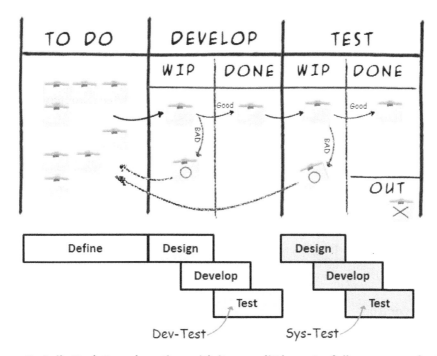

Figure 8: Agile Task Board motion with its own little waterfall sequences for both Development and Testing.

If an implementer is having a problem with a 'To Do Item' then they may place that item back into the 'To Do' section, for another implementer to choose to do. Alternatively, that implementer and another team member may decide to work cooperatively to resolve the problem (*e.g. pair programming*) so as to move the item along, that item may then be passed onto another team member(s) to carry it further forward, possibly passing the item onto another team member(s) to get it finally across the 'Done' line.

"Though, does it really matter how the implementers decide to work together to get that sprint's 'To Do Items' from one side of the Task Board to the other side. Provided that, how they choose to do it is effective & efficient to their needs, while ensuring the production of Quality deliverables that do conform to the agreed Acceptance Criteria (or extract thereof), and that they don't interfere with the performance of other project team members and other people."

Each 'To Do Item' that has **not been completed** (i.e. is **not ready to be deployed** or is **not ready to be demonstrated**) **by the end date of the current sprint**, then this incomplete item would be **removed from the current Sprint Backlog** (and "stubbed out" so as to be functionally isolated from execution in the completed deliverable), and this incomplete item would be placed into the 'Product Backlog' with all of those other outstanding features & functionality (that will possibly be chosen for implementation in a later sprint).

THEORY ... *The difference between Quality Assurance and Quality Control.*

- **Quality Assurance** .. is concerned with the **processes & procedures used to produce the deliverables**, so that the resultant deliverables **WILL CONFORM** to the agreed Acceptance Criteria. That is, "**keeping failures from reaching the deliverables**".

- **Quality Control** ... is concerned with **evaluating & verifying that the deliverables DO CONFORM** to the agreed Acceptance Criteria. That is, "**keeping failures from reaching the customer**".

"Though, it is not unusual, to find projects where there is an active attempt near the end of the release to inspect quality into the deliverables; rather than, to have actively designed and built quality in during the implementation of those deliverables."

Agreed, ... they firstly worried about "getting it done", after which they worry about "getting it right". Unfortunately, this approach to [Quality] often results in a lot of rework while trying to get the deliverables to meet the agreed Acceptance Criteria.

Which brings us to the next rule of "Adaptive & Proactive" SDLC Project Management.

RULE 24: Quality Control "inspection" after the facts is no substitute for Quality Assurance "prevention" during the acts.

3.5.5. QUESTION: is The Team checking it?

For the implementers this question refers to such testing as; debug, unit, module, integration, system, functional – satisfaction – usability acceptance testing.

With an agile based project, using a Task Board as per [Figure 8], **suggest the division into the sections of 'Develop' and 'Test'**, as well as the further subdivisions into 'WIP' and 'DONE'; so as to, **demarcate those activities of Quality Assurance and Quality Control**.

Hence, this demarcated agile task board would have regions for:

☐ **'Develop' implementation of the individual 'To Do Item'** would have its own phases of Design – Develop – Tested by the implementer. Where, the testing would entail **debug to system testing** of the implemented functionality **in an implementers only 'Dev-Test' environment**; thereby, performing an implementer's product verification.

☐ **'Test' implementation of the individual 'To Do Item'** would have its own phases of Design – Develop – Tested by the tester. Where, the design and development would involve test plans, test cases, test harnesses, test data, and test scripts that would be used to verify the correct operation of the implemented functionality (from both an individual unit's perspective as well as being part of multiple-coverage test cases). ... And, the tester could be another implementer who was not involved with the implementation of that particular 'To Do Item' that is being verification tested.

Subsequently, this testing would entail **system testing, various acceptance testing, and automated regression testing** of the completed 'To Do Items' **in a testers only 'Sys-Test' environment**.

- o If an individual 'To Do Item' is found to not function as defined, then its corresponding "card" (with the required changes or defect noted down on this card) would be placed back into the 'To Do' section of the Sprint Backlog.

- o If it is decided in hindsight that it would be best not to implement a 'To Do Item' during this current sprint, then that item would be moved to the 'OUT' of sprint area; essentially, "sin-binning" it from the Sprint Backlog. ... Though, the **"personalized agile methodology"** being used for this particular project may allow another similar effort-sized 'To Do Item' from the Product Backlog to be moved into the Sprint Backlog as a substitute item; thereby, retaining the level of effort that is required and the work being done for this current sprint.

"A point about automated regression testing ... this testing does NOT verify that the features & functionality is as described in the approved Detailed Specifications; rather, the purpose of automated regression testing is to verify that the application / system still functions as it did previously when the test scripts were written. ... There is no point in running automated regression tests while the targeted portions of the application / system is still being extensively developed. This is because, the tester will spend a significant amount of their time re-modifying test scripts that were broken by the nightly source-code build. ... Hence, automated regression testing needs to be targeted at relatively stable portions of the application / system."

Which brings us to an addendum to "Adaptive & Proactive" SDLC Project Management.

RULE 24[A]: Quality Assurance is only as good as the "proficient" people implementing what has to be done.

RULE 24[B]: Quality Control is only as good as the "independent" people confirming what should have been done.

RULE 24[C]: Quality Assurance & Quality Control are continuously improving processes, and NOT tacked-on after thoughts of something else to do.

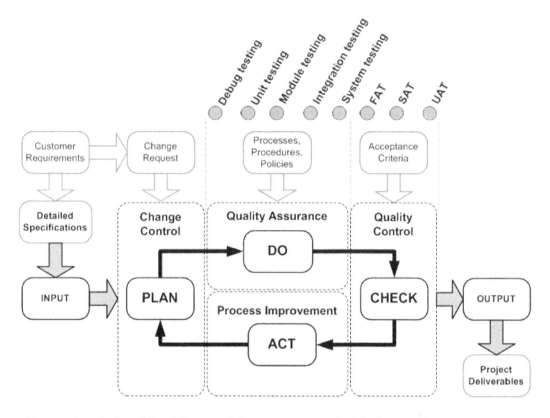

Figure 44: Relationship of 'PLAN – DO – CHECK – ACT' with the Quality Assurance and Quality Control testing suites.

- ❖ **FAT – "Functional" Acceptance Testing** … reproducible verification that every in-scope feature & functionality described in the approved Detailed Specifications has been successfully implemented in accordance with the agreed Acceptance Criteria.

- ❖ **SAT – "Satisfaction" Acceptance Testing** … retesting of those features & functionality found to be non-conforming during the FAT, and verifying that it is "satisfactory" to provide the deliverables to the customer, and/or that it has been deployed "satisfactorily" to the customer's site.

- ❖ **UAT – "Usability" Acceptance Testing** … Testing that the deliverables "in-situ" operation (at the intended deployment location) is functioning as was commissioned, and hence the deliverables are ready for (on-going) end-user usage.

Also known as "**FAT – Factory** Acceptance Test", "**SAT – Site** Acceptance Test", and "**UAT – User** Acceptance Test".

Quality is NOT about producing project deliverables that precisely meet the customer's "needs & wants". Rather, **quality is delivering exactly what was agreed to between the performing organization and the Customer's Representative(s), as recorded in the signed off & approved Detailed Specifications.** Hence, [Quality] deliverables are bounded to the **conformance with the approved Detailed Specifications and the agreed Acceptance Criteria.** ... And, **NOT to the Customer Requirements.**

3.5.6. QUESTION: is The Team done yet?

For the implementers this question refers to such things as; **coverage** of the Requirements Traceability Matrix, Acceptance Test results, Earned Value Performance Measures converging on ONE, see [Section 3.6.5], and the Post Implementation Review being signed off. Though, these areas will be covered in more depth in the Monitoring & Control Phase.

Burn Down Chart

With an agile based project, **The Team's progress towards the finish** of the current sprint (i.e. emptying the Sprint Backlog) is **represented via a "Burn Down Chart"**, see [Figure 45].

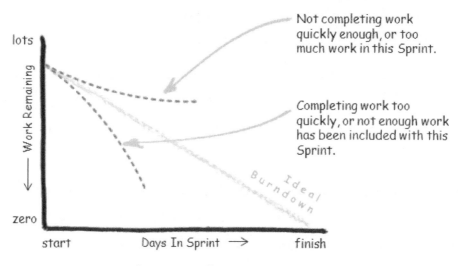

Figure 45: Agile Burn Down Chart.

Where the agile **Burn Down Chart is concerned with the daily updated count of the "still open To Do Items" [Scope] for the current sprint versus the work [Time] remaining**.

Noting that, other methodologies such as **Earned Value Performance Measures (EVPM)** have similar progress metrics and representations. For example, the **Earned Value S-Curve** which is **concerned with the accumulated [Cost] versus the work [Time] remaining**.

THEORY … *Similarity between the Agile Burn Down Chart and Earned Value S-Curve.*

While **the Burn Down Chart is NOT an Earned Value S-Curve**, it does have similarities if the Burn Down Chart's vertical axis was inverted; see [Figure 46].

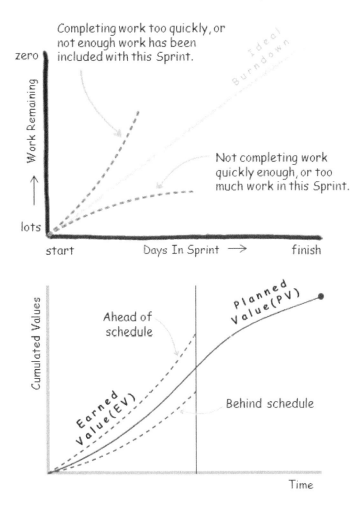

Figure 46: Comparison of an inverted Burn Down Chart to Earned Value S-Curve.

137

> The difference being, the **Burn Down Chart is concerned with [Scope] versus [Time] of the work that has been done** towards delivering usable product portions to the customer that could right now earn revenue for the performing organization. Whereas, the **S-Curve is concerned with the [Cost] versus [Time] of the effort** that the performing organization has expended in the pursuit of delivering that revenue producing product.

Scrum Stand-up Meeting

An agile method of establishing "the feel" of **The Team's progress towards the finish** of the current sprint is **via the daily "Scrum" "Stand-up Meeting"** (preferably at the same time of the day, and ideally, directly in front of The Team's communal Task Board and Burn Down Chart). This meeting is literally a stand up on their feet meeting, where all of the meeting participants stand-up for the entire **time-limited 15 minutes** of the meeting.

During this stand-up meeting in a circular fashion, each team member summarizes:

1.. **What did I achieve** during the last 12 working hours?   *... "next person's turn."*

2.. **What is "blocking" me** from doing my work in the next 12 working hours, and have I identified any <Risks> to the sprint's timely completion?   *... "next."*

3.. **What do I plan to do** in the next 12 working hours (a work day and a bit)?   *... "next."*

Noting that, the **purpose of the stand-up meeting is "centralized information exchange"** (i.e. information revealing not explaining) between all of the members of the Project Implementation Team (not status reporting to management), and as **"a forum by which to ask for help"**. ... And, the **stand-up meeting is NOT a "let's discuss the issues to resolution" arena**; as any such discussion about an issue and its resolution should be done as an independent get together with only those people who really need to be involved.

 Personally, I like to have a whiteboard present, so as to note down those issues to be discussed / resolved outside of the stand-up meeting. ... And, I like to rotate who (in The Team) is the daily stand-up meeting master.

COFFEE BREAK DISCUSSION ... *Agile, it's my scrum and I'll run it how I want to.*

With Agile methodologies, such as with Scrum, there are characteristic "shoulds":

- ☐ There "should" be three core roles of; (1) the **Product Owner** who represents the interests of the primary stakeholders, (2) the **Scrum Master** who ensures that the Scrum processes are followed and acts to keep the scrum process going but doesn't manage people, and (3) the **Development Team** who implements the 'To Do Items' to be delivered. ... I personally, prefer a different arrangement of; (1) the Project Manager who represents the interests of the primary stakeholders, removes the barriers and clears the path for The Team to successfully deliver, while focusing on delivering the "big picture" of the current release and future releases, (2) The Team **Leader** who acts as the technical mentor, the senior voice of The Team, and focuses on the overall delivery of the current release, and (3) the **Development Team** who implements the 'To Do Items' to be delivered. In a military analogy, the Project Manager is the lieutenant officer, The Team Leader is the platoon sergeant, and the Development Team via their own internally established pecking order will know who are their corporals and privates based on their experience and skill levels. Noting that, with any group of people, a natural ordering of leaders & followers will occur; thus, choose a team structure that works for the project and for the Project Team.

- ☐ In the daily scrum meeting each team member "should" say what they did in the last day, what they intend to do for today, and what their blocking issues are. However, I contend that the revealing of a blocking issue could very well change someone's (or someone else's) perspective on what they should be doing today. Hence, do a **round-robin of**; **what** each person **did** (i.e. *Recognize – Reassess*), then **what** each person's **blocking issues** are (i.e. *Reassess – Revise*), then **what** each person intends **to do today** until tomorrow's meeting (i.e. *Revise – Reapply*). See [Section 2.5.1].

- ☐ **Rotate among all of The Team members** (juniors and seniors) **who will be today's Daily Meeting Master**, i.e. who takes the whiteboard notes for the meeting, and who controls if & when someone has deviated outside the bounds of the meeting.

3.5.7. What to do with these answers?

These answers would serve to provide the information necessary for the project's management to retain "**diligent oversight**" (i.e. Monitoring & Control) of the project's progress towards the agreed objectives for the conclusion of the current release | sprint.

Some Implementation Things To Watch Out For

There are a few things the project's management needs to be on the lookout for:

- ☒ **Not looking at the big picture** ... where implementer(s) only consider the feature & functionality as it needs to be for the current release | sprint. Hence, the implemented solution is only appropriate for now (*e.g. hardcoded*), and has to be mostly / completely trashed when being incorporated into later releases | sprints.

- ☒ **Painting outside the lines** ... where implementer(s) put in extra features & functionality in addition to that which was [Scoped] for the current release | sprint, just because they would *"have to do that bit in the future, so why not do it now"*. Hence, most possibly, taking longer [Time] & [Costing] more to implement & test, as well as adding extra complexity that is not necessary to achieve the current objectives.

- ☒ **While I'm here, I'll fix that up** ... where implementer(s) while working on a particular feature & functionality decide to *"refactor"* *"improve"* the surrounding software code / hardware design. Effectively, null & voiding previous debug & testing of that refactored portion; thereby, taking longer [Time] & [Costing] more to implement & test, and adding complexity that is not necessary to achieve the current objectives. As they say, *"If it ain't broken then don't touch it"*, *"unless it is about to break"*.

- ☒ **Building a limousine and not a daily driver** ... where implementer(s) over-engineer / gold-plate the implemented solution, so that it's quality grade exceeds that which is necessary / required to achieve the current objectives. Hence, most definitely, taking longer [Time] & [Costing] more to implement & test.

3.5.8. Executing Phase Completion

Post Implementation Review

At this stage, the Executing Phase has essentially been completed (i.e. the agile sprint is coming to an end), and hence some form of "**Post Implementation Review (PIR)**" needs to be **held to confer that what was supposed to have been implemented has in fact been done**, and to determine what remains **outstanding** of that which was planned to be done.

This Post Implementation Review also serves as **a sanity check of the exit & entry criteria that must be satisfied in order for the project to be allowed to proceed to the commencement of the Closing Phase**; i.e. a checklist review of the "readiness indicators". Subsequently, the outcome of this review should be an informed & responsible decision to proceed, rework, or halt this project | milestone release.

Thus, it is vital that this Post Implementation Review meeting involve; representatives from the senior management of the performing organization's Project Steering Committee, representation for the customer organization, the Project Manager, selected Subject Matter Experts from the Project Working Group, and representatives from the Project Implementation Team.

Sprint Review Meeting

With an agile based project, this Post Implementation Review (PIR) would be the **Sprint Review Meeting**, where the Product Owner would represent the interests of the primary stakeholders (i.e. the Project Steering Committee, and the customer organization). At this meeting, a demonstration of the work done may also be presented; however, such a demonstration should have stable & graceful exits from 'To Do Items' that were not completed (and for features & functionality yet to be done).

At the conclusion of the sprint, the Task Board and the Sprint Backlog would be cleared and an empty one used for the next sprint. … And, what was not done would go back into the Product Backlog, so as to be candidates for implementation in the next sprint.

A PERSONAL EXPERIENCE ... *Be only the coach, not the captain-coach.*

Most possibly you started your SDLC career as an implementer; where you probably did the equivalent of a "working apprenticeship" straight after completing your higher education studies, then over the following years you worked your way up through the technical implementer's ranks, culminating in being made team leader, and now you are being (or have been) promoted into project management. ... And, this is when your perspective on work will have to change, else <Risk> not succeeding.

No longer will you be *"getting your hands dirty doing things"*; rather, you will delegate others to *"get dirty on your behalf"*, then watch how they do their assigned work, and issue corrective instructions when required. This is, **a big change from the "hard technical" skills of an implementer to the "soft inter-personal" skills of a manager**.

If you have the illusion that you can successfully do both the project management and the technical implementer roles at the same time (for a prolonged period) then don't be surprised to discover that, **as the project's manager, you spend about 80-90% of your time planning, coordinating, and communicating in various forms**. Hence, how can you do much constructive implementation work in the remaining 10-20% of the day?

I have witnessed project managers who believed that they could fulfil both roles simultaneously, but alas this only resulted in them working extremely long hours, producing work substandard to their pre-promotional norms, being continually in a rush to meet promised commitments, and in some cases - wrecking their home life.

The truth is that; **being a project manager is not like being a team leader**.

No longer are you a "doer", but rather you are a thinker, planner, analyst, strategist, coordinator, logistics, and even a people-person councillor. Analogously, you are no longer the captain of The Team, but rather the coach who calls out suggestions from the sidelines, thinks through the game's progress, and mulls over the game's strategies.

Project management is more about "interpersonal skills", than it is technical expertise.

EXAMPLE CASE … *Failure of trying to be both the technical lead and project manager.*

Once upon a time, there was a senior implementer – team leader who tried to simultaneously perform both the technical leadership role of doing tasks, and the project management functionality of orchestrating the completion of scheduled tasks.

Initially the project progressed along "nicely" with task after task marked as 100% done. The Project Implementation Team was transformed to operate in a hierarchical structure with The Team leader at the apex of the pyramid, as the "project leader".

However, as the project progressed into the core of the implementation work, tasks started to slip the scheduled completion dates. Yet, this project leader felt the need to report that those slipped-tasks had been done, even though these slipped-tasks were still being worked on as the next set of scheduled tasks were to begin. Eventually this covert "backfilling" of slipped-tasks built-up to the point where tasks that had not truly commenced were reported as partially completed and almost finished. … Then CRASH; the facade of mistruths came tumbling down, and this "nicely progressing" project and this "highly productive" Project Implementation Team were found-out to be in real trouble (with only a relatively short time before the major delivery milestone).

This "*cooking of the books*" (misrepresentation of the truth) by the project leader was not initially a deliberate act of deception towards the project's senior management. Rather, this all started out as just trying to keep-up with acting as both the project's senior technical implementer and being the part-time project manager responsible for progress. But as things slipped out of control, then an implementer's "fire-fighting" instinct kicked in, until the overworked team could no longer keep the issues contained.

In the worst case; the Project Implementation Team can self-destruct due to internal accusations of blame, and the project leader can become secretive & defensive when a full-time Project Manager is assigned to determine what has gone wrong with this project. … Then, the real dilemma is how to get this project back on track, yet not <Risk> losing this once excellent technical leader and the Project Implementation Team.

Well, that is the Executing part of the journey completed.

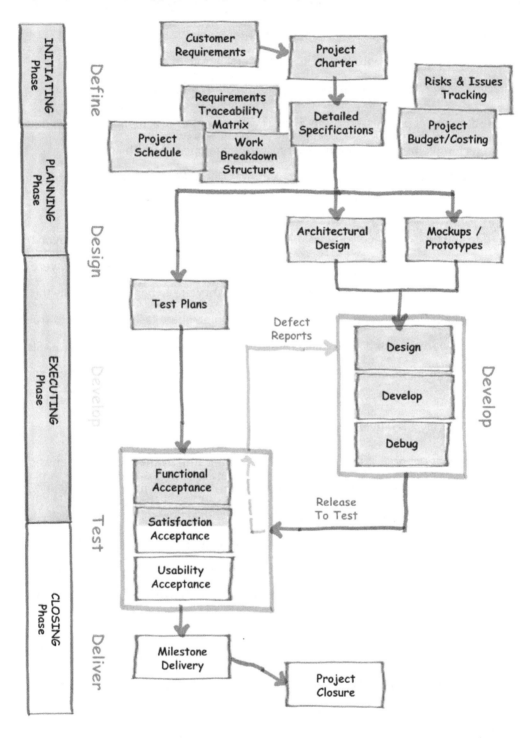

3.6. Monitoring & Control Phase

3.6.1. Purpose

The purpose of the **Monitoring & Control Phase** is to **"Oversee"** the implementation of the project; so as to, **ensure that the Executing Phase's "Doing" conforms** with the **Planning Phase's 'To Be Done' [Baselines]**.

That is, **Project Monitoring VERIFIES** that the Executing Phase is **progressing in accordance with the agreed project baselines** (that were established during the Planning Phase). Whereas, **Project Control** ensures the **CONFORMANCE with those agreed project baselines**.

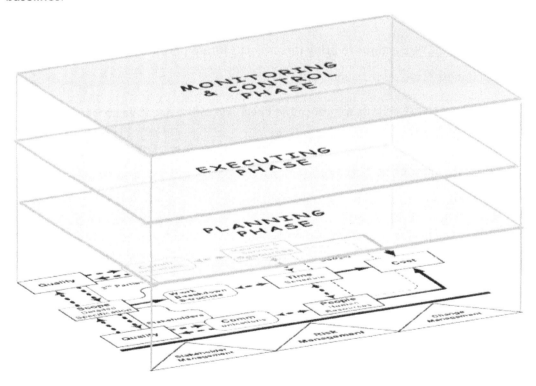

Figure 41: The symbiotic relationship of the Executing Phase with the Planning Phase and the Monitoring & Control Phase.

Monitoring & control is achieved by:

1.. **CONTINUAL tracking & reviewing the project's progress & performance by quantifiable COMPARISON** of the **actual results against those planned baselines**. Traditionally this concentrates on the [Scope], [Time], and [Cost] measurements.

2.. **IDENTIFICATION of any deviations from the agreed project baselines**, then **EVALUATION of the causes for such deviations** or the potential causes for future deviations (i.e. "**variance analysis**"), and if necessary, initiation of actions in **RESPONSE to such deviations**.

3.. **COMMUNICATE the project's status** and its expected future situation to the relevant project's stakeholders (especially to those who "*definitely need to know*").

Based on the above listed steps, the following decisions can be made in **RESPONSE to each deviation** away from an agreed project baseline, see [Figure 47]:

- Take **CORRECTIVE actions to bring** the project's future performance **back into line** with the expected performance.

- Take **PREVENTIVE actions to reduce** the likelihood of the **negative consequences** associated with that particular deviation.

- Undertake **REPAIRS to fix or replace** those components of the project that are **not conforming** to the expected performance.

- **UPDATE the project's documentation & plans** to accurately represent the current situation and the true state of the project.

- Seek authorized approval via a baseline **CHANGE REQUEST to modify those already approved baselines to something mutually agreed as more appropriate** (when taking into consideration the project's current situation & prevailing circumstances).

Though, one thing for sure, **monitoring & control involves a significant amount of Risk Management, Stakeholder Management, and Change Management** (R.I.S.C.).

3.6.2. A Continual Iterative Process

The project's variables & constraint baselines are going to require regular and periodic monitoring & control; hence the presence of two feedback loops, in [Figure 47] below.

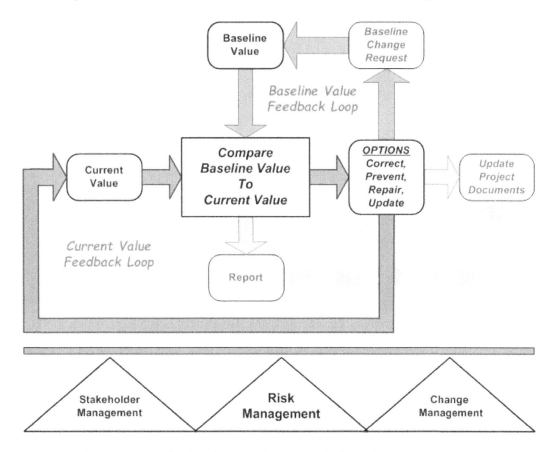

Figure 47: Monitoring & Control as a simple iterative process model.

(1) The "**current value feedback loop**" would be updated on a weekly basis (via the Weekly Progress Report), and deals with the current value of that project variable (or "*transiently flexible*" project constraint) when compared to the planned baselines.

(2) The "**baseline value feedback loop**" would be updated less frequently (possibly at the beginning & end of each agile sprint, release cycle, or via the Monthly Project Update Report), and deals with the (planned) baseline value for that same project variable.

Usually, the weekly / fortnightly project monitoring & control is via the comparison of the current values for [Scope], [Time], and [Cost] against the corresponding agreed baselines.

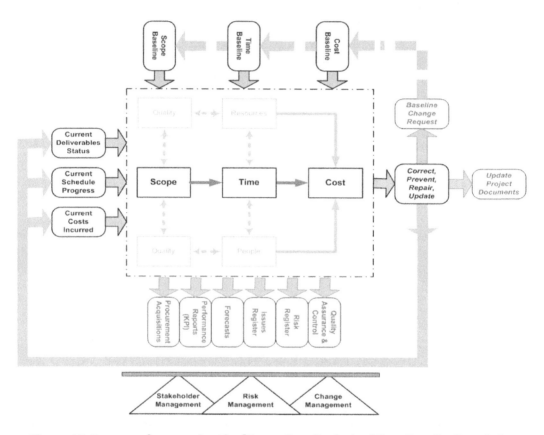

Figure 48: Process of comparing the [Scope Baseline], the [Time Baseline], and the [Cost Baseline] with the current actual performance values.

Which brings us to the next rule of "Adaptive & Proactive" SDLC Project Management.

RULE 25: Keep one eye on the road, keep the other eye on the gauges, and both hands on the wheel.

3.6.3. [Scope] Monitoring & Control

The purposes of [Scope] Monitoring & Control are to:

(1) **Confirm that only those agreed features & functionality [Scope] are being included in the appropriate deliverables** as was specified in the [Scope Baseline]; i.e. keeping an eye on what is & is not going into each and every release | sprint.

(2) **Ensure that only those approved changes are being incorporated into the relevant deliverables**, and to ensure that unapproved changes are not being snuck into releases; i.e. watching out for "scope creep" and "scope shrinkage".

(3) **Manage the scope change process**, so as to review & approve requested changes, coordinate the incorporation of those approved changes into the [Scope Baseline] for the relevant deliverables, and verifying that each approved change is present in the corresponding release(s).

(4) **Determine & report on** how any **variations between the implemented [Scope] and the [Scope Baseline]** will affect the successful outcome of the project.

Noting that, **the scale of the difference between the [Scope Baseline] and the implemented [Scope], plus the cause & effects of such variations, will influence how best to respond to such deviations**. Hence, if there is only a small deviation then only a small response should be required; whereas, a large deviation may require drastic measures to be taken to get the project "*back on course*".

[Scope] Monitoring & Control does NOT verify that the features & functionality delivered actually meets the purpose for which the customer "intended". This type of "purpose-ability" verification is the concern of [Quality] Monitoring & Control (and Change Requests).

That is, **[Scope] Monitoring & Control only verifies that the agreed features & functionality are being included in the appropriate deliverables / releases**.

As depicted in [Figure 49] below, any changes to the [Scope], either officially approved or unapproved, will have a direct effect on the [Time]; as well as, a ripple effect on the [People], [Resources], [Quality]. ... And, subsequently a cascading effect on the [Cost].

"Alas, there ain't no such thing as an innocuous scope change."

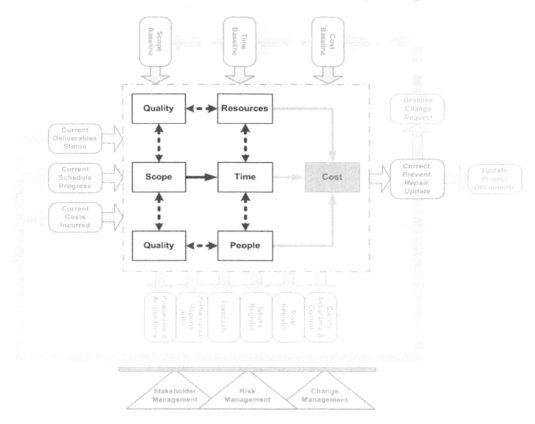

Figure 49: The direct effect of [Scope] on [Time] and the ripple effect on [People], [Resources], [Quality] ... and, a cascading effect on the [Cost].

A common way of tracking this relationship between [Scope], [People] and [Resources] with the project's progress over [Time] is via the project schedule; i.e. the Gantt Chart.

 Be wary of this demarcation between [Scope] and [Quality], as there has been many a disagreement between the customer & performing organizations **due to a deliverable conforming to the specifications yet it did not correspond to the purpose for which the customer had "intended".**

3.6.4. [Time] Monitoring & Control

The purposes of [Time] Monitoring & Control are to:

(1) **Keep an eye on the project's [Time] progress** by comparing over the same time period (*e.g. weekly*) those completed & work-in-progress tasks against their counterparts in the [Time Baseline] schedule.

(2) **Manage changes to the [Time Baseline]** due to changes in the [Scope Baseline], as well as due to the changing situation & circumstances confronting the project.

(3) **Perform variance analysis**, and thereby determine the **CAUSE** of any timing & duration variations between the actual and baseline scheduled tasks. … And, how will any such variations **AFFECT** the project's successful outcome.   … *"cause & affect."*

(4) When the variance is such as to require the **performance of corrective & preventive actions** then:
    - **DECIDE what actions** are necessary to bring the project's [Time] duration and delivery dates back into line with the agreed [Time Baseline] milestones, via:
        - **"Crashing"** the schedule by compressing the time allocation for tasks (only) on the critical path; i.e. performing "C.P.R." of throwing additional Cash, People, and Resources at the problem.
        - **"Fast-tracking"** by performing some tasks in parallel and glossing-over those tasks that are not evident in the deliverables.
        - **"Scope minimization"** (aka "de-scoping") where certain features & functionality will not be included in the time-sensitive release.
    - **REPORT** to the relevant stakeholders about what are the chosen actions and what are the reasons for taking such actions.
    - **EXECUTE** those chosen (and authorized) actions.
    - **CONFIRM** that those chosen (and authorized) actions have been implemented.

While **changes to the [Time] will NOT have** (and should not have) **a direct effect on the project's [Scope], these [Time] changes will directly affect the project's [Cost]**. ... And, have a ripple effect on the [People] & [Resources] which will subsequently have an effect on the [Quality] aspect of the [Scope] that is implemented; see [Figure 50] below.

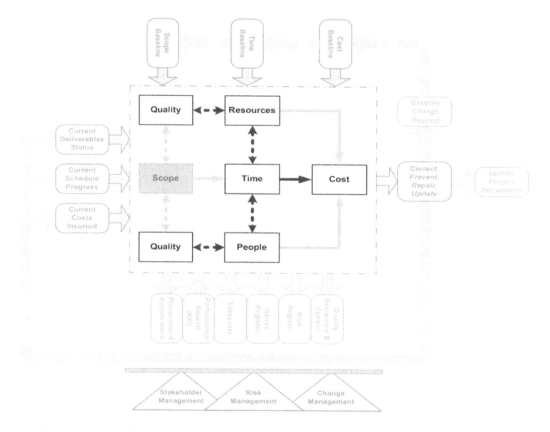

Figure 50: The direct effect of [Time] on [Cost] and the ripple effect on [People], [Resources], and [Quality].

Changes to [Time] should never directly affect the [Scope].

Rather, if the project's duration has to be reduced (or contained) then the representatives of the performing & customer organizations will **need to negotiate scope changes, else covert scope shrinkage may occur,** ... as some members of the Project Implementation Team may try to stay within the stipulated [Time] frame by shortcutting, via what to them *"seemed to be a good idea at the time"*.

3.6.5. [Cost] Monitoring & Control

The purposes of [Cost] Monitoring & Control are to:

(1) **Keep an eye on the project's [Costs]** by comparing over the same time-period (*e.g. weekly*) the actual costs against the authorized [Cost Baseline].

(2) **Manage changes to the [Cost Baseline] budget** due to changes in the [Scope], [Time], [People], and/or [Resources], as well as, due to the changing financial circumstances confronting the project.

(3) **Perform variance analysis**, and thereby determine the **CAUSE** of any cost variations between the actual spend and the budget. … And, how will any such variations **AFFECT** the project's successful outcome.

(4) When the variance is such as to require the **performance of corrective & preventive actions** then:

- o **DECIDE** on what actions are necessary to bring the project's [Costs] back into line with the approved [Cost Baseline].
- o **REPORT** to the relevant stakeholders about what are the chosen actions and what are the reasons for taking such actions.
- o **EXECUTE** those chosen (and authorized) actions.
- o **CONFIRM** that the chosen (and authorized) actions have been implemented.

Noting that, [Scope] changes (and/or problems with implementing such [Scope]) will result in more [Time] being worked by [People] and/or more [Resources] being consumed for the actual progress made when compared against those planned baselines. Subsequently, these increases will result in a greater financial outlay [Cost] than was accounted for, and possibly more than was prepared to be coped with during that particular [Time] period. Such **excessive financial outlays could mean that the organization experiences a "negative cash flow crisis"** that could **consequently affect the business's solvency**.

Hence, the **bi-directional relationship of [Time]** with both the **[People] & [Resources]** does have a direct flow-on effect on the project's **[Costs]**.

"IMHO, monitoring & control is all about managing the direct & indirect effect that changing project variables have on each other."

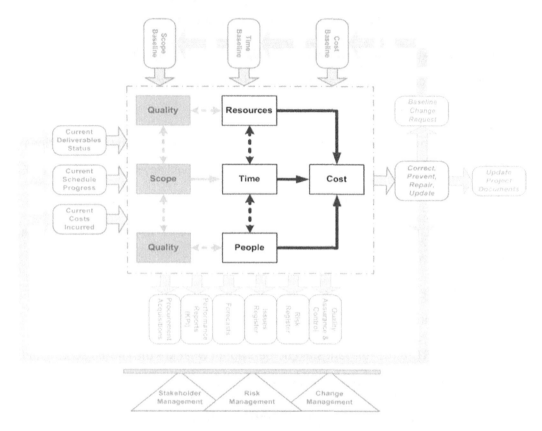

Figure 51: The direct effect of [Time] – [People] – [Resources] on [Cost].

$$COST_{TO\ DATE} = (PEOPLE_{RATES} \times TIME_{USED})$$
$$+$$
$$(RESOURCE_{UNITS} \times AMOUNT_{USED})$$
$$+$$
$$(FIXED\ COSTS_{TO\ DATE})$$

Earned Value – Measures   … as depicted in [Figure 52].

- ❖ **Planned Value (PV)** … is the baselined budgeted value in terms of monetary units (and/or labour hours) that were planned to be incurred up to the current status date (when the measurements were taken) for the performance of a planned number of scheduled tasks | WBS work-packages. That is, **(PV) is what was planned to have been spend to date, to get the project to where it was planned to be by now**. (PV) is also known as, **Budgeted Cost of Work Scheduled (BCWS)**.

    For example; in 6-days of this 10-day project we planned to have spent $7000 and be 60% of the way through the project. In this case, PV = $7000.

- ❖ **Earned Value (EV)** … is the budgeted "Expected Value" in terms of monetary units (and/or labour hours) that were anticipated / planned to have been incurred up to the status date for the work actually performed on scheduled tasks | WBS work-packages (including completed and work-in-progress). That is, **(EV) is what was planned to have been spent to date, to get the project to where it really is now.** (EV) is also known as, **Budgeted Cost of Work Performed (BCWP)**.

    For example; so far, in 6-days of a 10-day project, we have only got 43% of the way through the project, and for this 43% we planned to have only spent a corresponding $5500. In this case, EV = $5500.

- ❖ **Actual Cost (AC)** … is the actual "real" cost in terms of monetary units (and/or labour hours) incurred up to the current status date (when the measurements were taken) for the work that was actually performed on scheduled tasks | WBS work-packages (including both completed and work-in-progress). That is, **(AC) is what has actually been spent to date, to get the project to where it really is now.** (AC) is also known as, **Actual Cost of Work Performed (ACWP)**.

    For example; so far in 6-days of a 10-day project we've spent $8742 to get 43% of the way through the project. In this case, AC = $8742.

Is the accumulated **Actual Cost (AC)** of all of the work that has been undertaken, less or greater than the **Planned Value (PV)** of all of the work that was planned to be done ?? Is the project's performance above or below the [Time & Cost Baselines], and how does the **Estimate At Completion (EAC)** compare with the **Budget At Completion (BAC)** ??

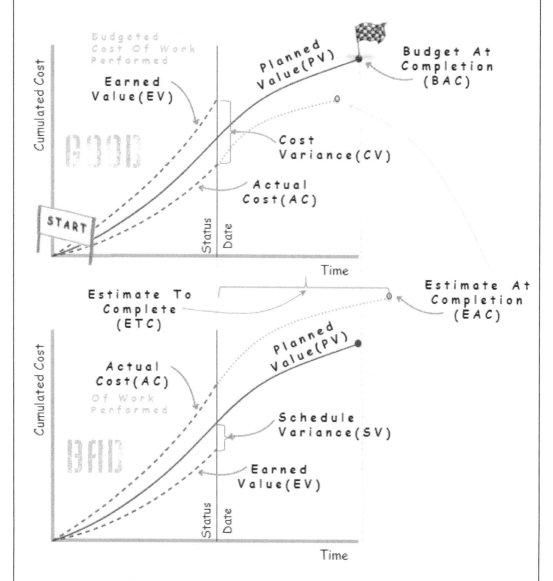

Figure 52: Actual Cost (AC), Earned Value (EV), and Planned Value (PV) versus time mapped as Earned Value Management S-Curves.[2]

---

[2] Based on PMI 2013, "*A Guide to the Project Management Body of Knowledge (PMBOK Guide), 5th Edition*".

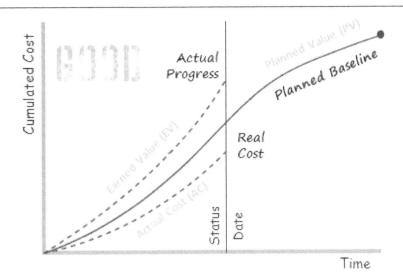

Actual Cost (AC) less than the Planned Value (PV) is potentially GOOD, because the project is "**under budget**".

Earned Value (EV) greater than the Planned Value (PV) is probably GOOD, because the project is "**ahead of schedule**".

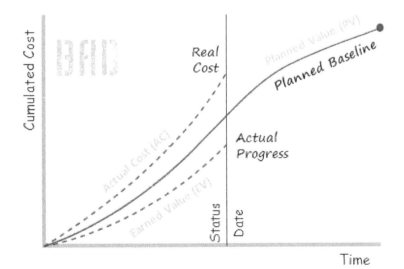

Actual Cost (AC) greater than the Planned Value (PV) is potentially BAD, because the project is "**over budget**".

Earned Value (EV) less than the Planned Value (PV) is probably BAD, because the project is "**behind schedule**".

While **Earned Value Measures** are excellent for tracking the performance of a project's progress, this technique is **"practically useless"** if there are **"noticeable discrepancies"** **between the work effort recorded** (in the timesheet system) **and the actual work effort that has really been done.** ... OR, if the percentage complete (as stated for scheduled tasks) is not representative of the actual progress made. Because, the project won't be progressing as per the amount of work effort indicates that it should have.

Which brings us to the next rule of "Adaptive & Proactive" SDLC Project Management.

RULE 26:   **BAD Data IN**   **= BAD Information**
                   BAD Information = BAD Analysis
                   BAD Analysis   **= BAD Decisions**
                   BAD Decisions   **= Dying Business.**

So, just how well is the project tracking towards those agreed [Time & Cost Baselines] ?? And consequently, are the project's deliverables going to be provided by the agreed [Time] milestone dates, while remaining within the allocated [Cost] budget ??

EXAMPLE CASE ... *Did we achieve the [Time & Cost Baselines] for a simple project.*

Consider the development project Example Case at the back of [Section 3.4.10].

Once the project was underway, the Execution Phase ran into *"a few technical snags"* and subsequently the project's implementation dragged on by *"a coupla days"*; i.e. task 11 through task 14 all increased from 5 days to 7 days in duration (which is 40% longer for these tasks than was originally planned for in the [Time Baseline] Schedule).

As a result, the Work-In-Progress Schedule [Figure 53] now extends out beyond the milestone date in the original Baseline Schedule [Figure 36]. ... And, notice how in [Figure 53] when compared to [Figure 36], that the project's total [Costs] has increased (by 18%) from $35,600 to $42,000 and the total [Time] duration has also increased (by 10%) from 19.5 days to 21.5 days.

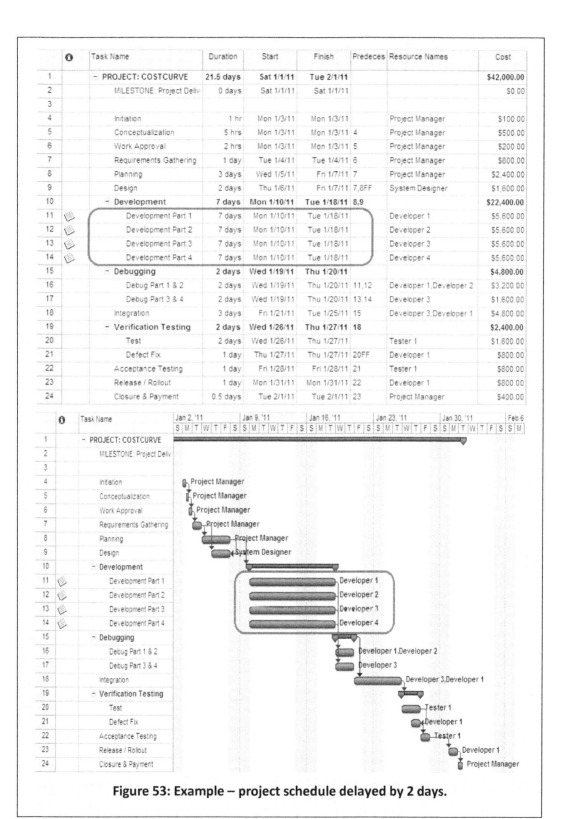

Figure 53: Example – project schedule delayed by 2 days.

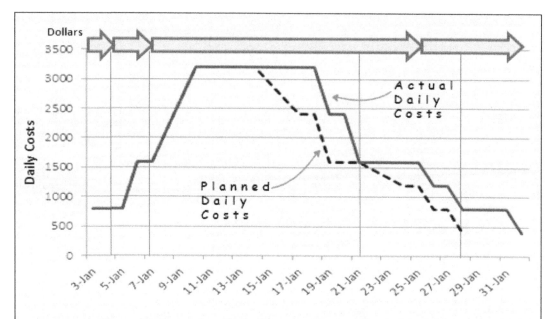

Figure 54: Example – Daily Costs vs. Time with a 2 day delay.

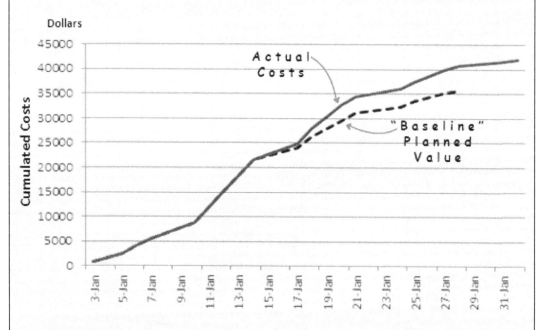

Figure 55: Example – the actual Cumulated-Cost vs. Time with a 2 day delay.

While in this Example Case, those 2 days delay only resulted in a $6,400 increase to the [Total Cost] and an extension to the [Total Time]; what if, this was for a lot larger project where the delays were for multiple tasks that extended the project by several months.

Earned Value – Performance Management

- **Cost Variance (CV = EV - AC)** … calculates the difference between the Earned Value and the Actual Cost of the work performed up to the current status date. That is, **(CV) is how much over or under the "Expected Value" of the [Cost Baseline] Budget to get the project to the point where it really is now.**

    For example; in 6-days of this 10-day project we have actually spent $8742 for those tasks that were actually performed (i.e. to get 43% of the way through the project), while we planned to have only spent $5500 on those particular tasks.
    CV = EV - AC = $5500 - $8742 = - $3242   where, **NEGATIVE CV IS BAD**.

- **Schedule Variance (SV = EV - PV)** … calculates the difference between the Earned Value and what was the Planned Value for the same point in the project's duration. That is, **(SV) is how far ahead or behind the "Expected Value" of the Scheduled [Time Baseline] of where we really are now, against where we planned to be by now.**

    For example; in 6-days of this 10-day project, we planned to have completed $7000 worth of tasks, but we have only completed $5500 worth of tasks.
    SV = EV - PV = $5500 - $7000 = - $1500   where, **NEGATIVE SV IS BAD**.

- **Cost Performance Index (CPI = EV / AC)** … determines up to the status date whether the project is currently exceeding its allocated budget by calculating the ratio of the "Expected Value" (Earned Value) against the "ACtual value" (Actual Cost). That is, **(CPI) denotes how good or bad is the project's performance when compared to the [Cost Baseline] budget** (in terms of monetary units / labour hours), **to get the project to where it really is now?**

    For example; in 6-days of this 10-day project we've spent $8742, when we expected to have spent $5500 for those tasks that were actually performed.
    CPI = EV / AC = $5500 / $8742 = 0.63   where, **CPI LESS THAN ONE IS BAD.**

CPI of less than one (< 1.0) indicates a **Cost Overrun** for the work performed so far. **CPI of greater than one** (> 1.0) indicates a **Cost Underrun** for the work performed. CPI GREATER THAN ONE IS GOOD. … Or, it could mean that we are not working at full capacity, and if the "budgetary burn-rate" is so much greater than ONE then this can be because we are currently not doing much work on the project.

- **Schedule Performance Index (SPI = EV / PV)** … determines how well the project is currently progressing through the scheduled tasks | WBS work-packages by calculating the ratio of the Earned Value against the Planned Value, for the same point in the project's duration. That is, **(SPI) denotes how good or bad is the project's performance when compared to the [Time Baseline] schedule** (in terms of monetary units / labour hours), **where the project really is now, against where it was planned to be by now?**

    For example; in 6-days of this 10-day project we planned to have spent $7000, but based on those scheduled tasks that were actually performed we have achieved "earned" only $5500 worth of work.

    SPI = EV / PV = $5500 / $7000 = 0.79   where, **SPI LESS THAN ONE IS BAD.**

SPI of LESS than ONE (< 1.0) denotes **less progress made** than was planned for.
SPI of GREATER than ONE (> 1.0) … **more progress made** than was planned for.
SPI GREATER THAN ONE IS GOOD. … Or, it could mean that we are exceeding our working capacity, and if the budgetary burn-rate is frighteningly scary then this can be because we are currently working so far in advance of the planned schedule.

"However, these CPI and SPI performance measures of GOOD and BAD are only as useful as the accuracy of the quantified data used to determine these. For example; incorrect percentage complete on scheduled tasks and overstated / understated hours worked during the period will result in a misrepresentation of the project's current true condition. Though, there will be a significant correction in CPI and SPI values as the truth of actual progress becomes available."

Earned Value – Key Performance Indicators (KPI)

What if on a regular basis (*e.g. weekly*) the values of the Cost Performance Index (**CPI**) and the Schedule Performance Index (**SPI**) were **plotted** independently as the Y-Axis **against** the project's run **time** on the X-Axis, so that **the "trend" of the movement towards or away from the respective CPI and SPI "ideal values" of ONE** was traced.

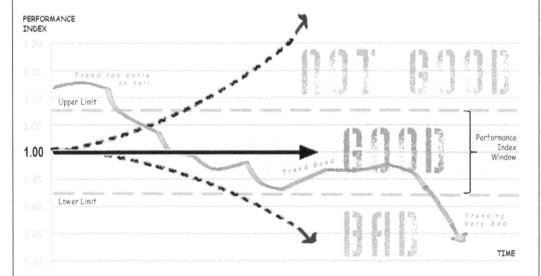

Figure 56: Performance Index Window with upper & lower ranges.

Independently adjustable value range windows should be used, for both CPI and SPI. For example; a relative loose window of 0.90 to 1.10, or a tighter window of 0.95 to 1.05. As **long as the current performance index values are within the prescribed "Performance Index Window" then all is well**, but outside of this window then all is not good. Though, also **consider the continued "trend" of that performance index**.

If the current value of the **SPI exceeds the upper limit** of the prescribed "Schedule Performance Index Window" **then possibly the schedule has too much buffering "fat" included**. ... OR, it could mean that the **[People] & [Resources]** are being **worked excessively** to get more done than what was originally planned for. Conversely, if the **SPI** is **below the lower limit** of the Schedule Performance Index Window **then the percentage of work completed is falling behind** that which was planned for.

If the current value of the **CPI exceeds the upper limit** of the prescribed "Cost Performance Index Window" **then possibly the budget has been over estimated "inflated"**. ... OR, it could mean that the **[People] & [Resources]** are being **under-utilized**, and/or the expected amount of [Time] is not being booked against the project's job code. Conversely, if the **CPI is below the lower limit** of the Cost Performance Index Window **then the cost of the work completed is exceeding** that which was planned for, and/or there is non-related work being booked against this project's job code.

If at present, **both the SPI and the CPI were within the "good" zones** of their respective (independent) Performance Index Windows **then** it would be reasonable to assume that there are **currently no performance problems** with this project's progress.

With the performance index values being calculated regularly, then these values could be affiliated with traffic light **Key Performance Indicators (KPIs)** based on a comparison with the tight & loose Performance Index Windows.

INDICATOR	PROJECT ALFA	PROJECT BRAVO	PROJECT CHARLIE	PROJECT DELTA
PROJECT MANAGER	PAUL M.	JOHN L.	GEORGE H.	RINGO S.
CPI	1.03	0.97	0.91	0.22
SPI	0.99	0.98	0.94	0.85
CPI TREND	⇒	⇒	⇘	⇒
SPI TREND	⇘	⇗	⇘	⇗
MILESTONE	GREEN	GREEN	YELLOW	RED

Figure 57: Key Performance Indicators (KPI) Project Dashboard.

Subsequently, all of the currently active projects' KPI values could be summarized in a single Dashboard, [Figure 57], which is then inspected by the performing organization's senior management to identify which projects they need to pay more attention to.

Earned Value – Percentage Complete Measures

Unfortunately, those previously demonstrated Earned Value Performance Measures judge a project's progress against packages of [Time] & [Cost] hours of effort spent, and not specifically against the percentage of tasks completed.

- **Planned Value (PV = BAC x Planned % Complete)** ... calculates Planned Value (PV) as the Budget At Completion (BAC) multiplied by that Planned Percentage of work Completed (including work in progress) up to the current status date.

    For example; a project of 5 days [Time] duration is composed of 4x independent tasks [Scope] to be undertaken by 4x [People], where the allocated budget [Cost] is $1,000 per day per [Person] and their associated [Resources] utilized.

    Hence, Budget At Completion (BAC) = 5 (days) x 4 (people) x $1,000 = $20,000 which equals $5,000 per task.

    At the conclusion of the third day, it was planned (according to the schedule) that the project would be 60% complete, then PV = $20,000 x 60% = $12,000

- **Earned Value (EV = BAC x Actual % Complete)** ... calculates Earned Value (EV) as the Budget At Completion (BAC) multiplied by that Actual Percentage of Work Completed (including work in progress) up to the current status date.

    At the conclusion of the third day, things are not exactly going to the plan of being 60% complete on each individual task, instead the 1st task is 20% done, the 2nd task is 60% done, the 3rd task is 40% done, and the 4th task is 80% done.

    (Task 1) EV1 = $5,000 x 20% = $1,000   ... (Task 2) EV2 = $5,000 x 60% = $3,000

    (Task 3) EV3 = $5,000 x 40% = $2,000   ... (Task 4) EV4 = $5,000 x 80% = $4,000

    EV = EV1 + EV2 + EV3 + EV4 ... EV = $1,000 + $3,000 + $2,000 + $4,000 = $10,000

In this example, the project is in a bit of trouble because at the current status date, the Earned Value (EV = $10,000) is less than the Planned Value (PV = $12,000).

Earned Value – Forecasting

- **Budget At Completion (BAC)** ... What is the expected total value (in terms of monetary units and/or labour hours) when all of the project's planned tasks (aka "work-packages") are completed? That is, **what is the total Planned Value when the project is finished?** ... Go back and examine [Figure 52].

- **Estimate To Complete (ETC)** ... From where the project has got up to (by the current status date), what do we calculate is required (in terms of monetary units and/or labour hours) to finish off the project's remaining planned tasks (work-packages) that are still to be completed? That is, **what has to spend from here on in, to get this project to the finish?**

- **Estimate At Completion (EAC = AC + BAC – EV)** ... Based on the project's actual progress up to the current status date, what is estimated to be the total value (in terms of monetary units and/or labour hours) when all of the project's planned tasks (work-packages) are finally completed? That is, **what do we now calculate as going to be the total cost when this project is finished?** ... See [Figure 52].

  Or, if continuing at the current rate of performance then ... **EAC = BAC / CPI**

- **To Complete Performance Index (TCPI = (BAC - EV) / (BAC - AC))** ... Compares the value of the remaining work to the value of the remaining budget. That is, **what cost performance is required from now on, to deliver the remaining work within the planned budget?**

> Which brings us to an addendum to "Adaptive & Proactive" SDLC Project Management.
>
> **RULE 26[A]: The understanding of the project's current situation is only as good as the accuracy, reliability, and relevance of the gauges used to represent what is happening now.**

Management Reserve ... *the project budget vs. the overall budget*

Back in [Section 3.4.6], there was the point about the [Cost Baseline] being the approved Project Budget, ... well, this is not entirely correct.

While the [Cost Baseline] is the Budget At Completion | "Project Budget" as viewed by the Project Manager and the Project Team; this Project Budget is actually less than the "Overall Budget" that would be allocated to the project. This is because, risk adverse **senior management at the performing organization should have held back additional funds** (i.e. **"Management Reserve") just-in-case the project encounters "unknown unknowns" that happen to fall within the [Scope] boundary of the project.**

That is, this **Management Reserve is not included in the [Cost Baseline]**; but rather, this reserve is a "just-in-case hidden" **additional allocation** of funds **on top of the "Project Budget"** to **form** the project's **"Overall Budget"**. See [Figure 58].

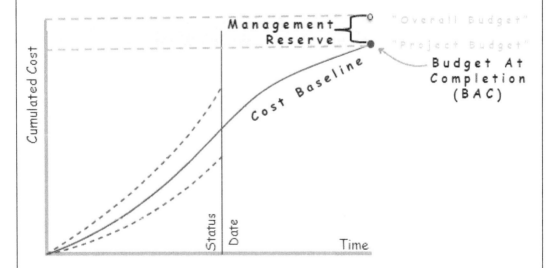

Figure 58: Relationship of the Management Reserve to the Project's budgets.

NOTE: It is not the Project Manager nor the implementers who decide when to utilized this reserve; rather, it is the senior management at the performing organization who decide when and how much of the management reserve is to be released.

Burn Rates ... *cash flow IN vs. OUT*

Each month | quarter year | half year | full financial year, senior management at the performing organization will inject money into the project (i.e. the Project Budget) to be used to pay for the upcoming assignment of [People] and allocation of [Resources]. Subsequently, during the following [Time] period, actual [Costs] will be incurred for those [People] & [Resources] that were used to transform the [Scope] into [Quality] deliverables. These Actual Costs will be a drain on the project's budget; and, if the rate at which money is going out (i.e. **Burn Rate = 1 / CPI**) exceeds the rate at which money is put in, then the performing organization is going to experience a "**negative cash flow**", where the organization will have to respond by drawing on its financial reserves.

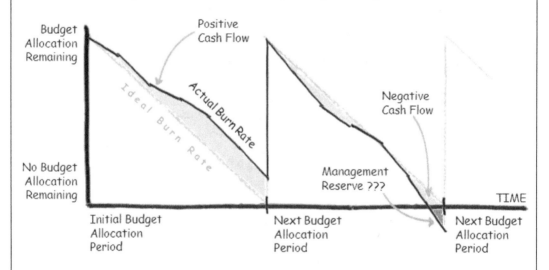

Figure 59: Positive & Negative Cash Flow compared to Budget Burn Rates.

So, ... provided that the actual burn rate does not exceed the ideal burn rate, see the first half of [Figure 59], then all is well; but, if the actual burn rate does exceed the ideal burn rate, see the second half of [Figure 59], then the project and subsequently the performing organization would be experiencing a cash flow problem. ... And, if this cash flow issue was to coincide with the need to payout other financial claims, then the performing organization could find itself with **"liquidity" and "solvency" problems**.

In addition to the monitoring & control of [Scope], [Time], and [Cost], there is also the monitoring & control of those other project variables of [Quality], [People], [Resources], and R.I.S.C. Management.

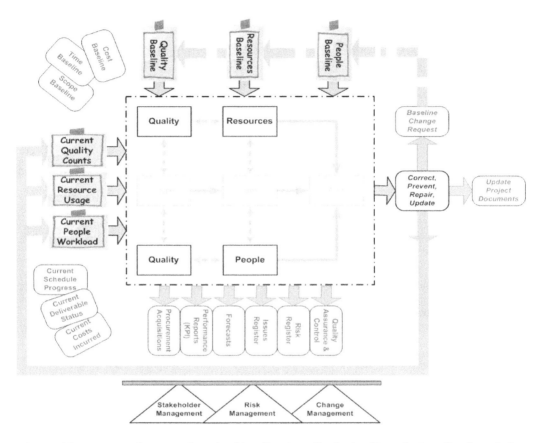

Figure 60: Process of comparing the [Quality Baseline], the [People Baseline], and the [Resources Baseline] with the current actual performance values.

That is, **monitoring & control needs to ensure that, over [Time], there is an effective & efficient utilization of those assigned [People] and allocated [Resources] that are used to transform the project's [Scope] into "satisfactory" [Quality] deliverables.**

Though, this transformation is more akin to a 'PLAN – DO – CHECK – ACT' process of; examining the at-hand planning of what to do next, doing that next thing, checking that it was done properly, re-acting accordingly if what was done does not correspond to what was planned to be done, and repeating this cycle again & again.

169

3.6.6. [Quality] Monitoring & Control

The purposes of [Quality] Monitoring & Control are to:

(1) Ensure that the **produced deliverables are "going to conform"** with the agreed [Scope] as defined in the approved Detailed Specifications; i.e. **Quality Assurance**.

(2) Ensure that the **produced deliverables "do conform"** with the agreed [Scope] as defined in the approved Detailed Specifications and the agreed Acceptance Criteria; i.e. **Quality Control**.

(3) Ensure that the **processes & procedures being utilized during the implementation are continuously improved upon**, so as to reduce waste, eliminate / minimize those activities that don't add value, and to apply applicable industrial "best practices"; i.e. **Process Improvement** (aka "Continuous Improvement").

(4) Ensure that **reviewed & approved Change Requests & Process Improvements are to be implemented** in a timely fashion; i.e. **Change Control**.

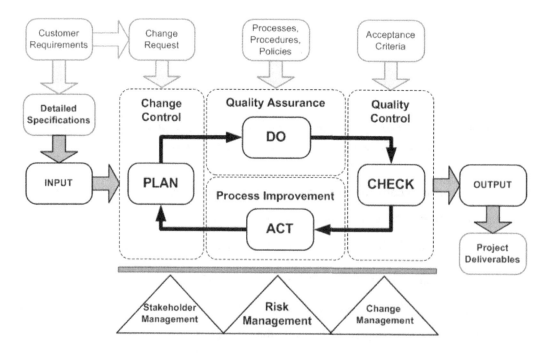

Figure 61: Monitoring & control as a 'PLAN – DO – CHECK – ACT' process.

Quality is NOT the pursuit of "**perfection**"; as to do so, "*is to chase moving goal posts*". That is, **meeting the customer's "needs & wants" exactly as they "intended" and not as was "agreed to"** (as per the signed off Detailed Specifications), **is a sure way to blowout the project's [Cost] budget** and/or the **[Time] duration**. As well as, **resulting in the over-utilization of assigned [People] and the over-consumption of allocated [Resources].**

However, delivering what the customer perceives as "quality deliverables" is essential for the project to be deemed a success; hence, **the dilemma of how to balance [Scope] & [Quality] versus [Time] & [Cost]**, see [Section 2.8.2].

"Quality does not mean ~~exceeding the Customer Requirements~~ *... exceeding the approved Detailed Specifications; i.e. a "gold plated" higher grade output, if not specified, is not the desired level of quality output."*

- ❖ **Grade** – is the level or category of the end-product due to **the number of features that it contains, or its characteristics when compared to its competitors and siblings**. For example; the "top of the line" versus the "entry level" model.
- ❖ **Quality** – is how well the "end-product" **meets its specifications and is fit for use**.

The grade of the deliverable will be affected by increasing or decreasing the project's [Scope]. However, these **increases / decreases in [Scope] will not necessarily affect** the **[Quality]** of the deliverable; what will most probably affect [Quality] is those [People] assigned and the [Resources] utilized.

As noted in the theory block on the next page, see [Figure 62], increasing [Costs] does not necessarily increase the [Quality]. Though, increasing the [Quality] will probably increase the [Cost] via; the flow-on effect of any necessitated [Scope] changes & defect repairs, the involvement of additional [People] & [Resources], and the additional [Time] required to implement such improvements in [Quality].

COST DOES NOT REFLECT THE QUALITY

Notice that, none of those Earned Value [Cost] measures, in [Section 3.6.5], have dealt with the [Quality] of the project's deliverables. ... And, as such, here is a counterintuitive realization, that ... **"THE COST DOES NOT REFLECT THE QUALITY"**.

That is, there is no direct path from [Cost] to [Quality]; as high costs does not necessarily ensure high quality deliverables, and conversely, low costs does not mean low quality.

Rather, **[Quality] is achieved via the intermediaries of [Scope], [People], [Resources]**.

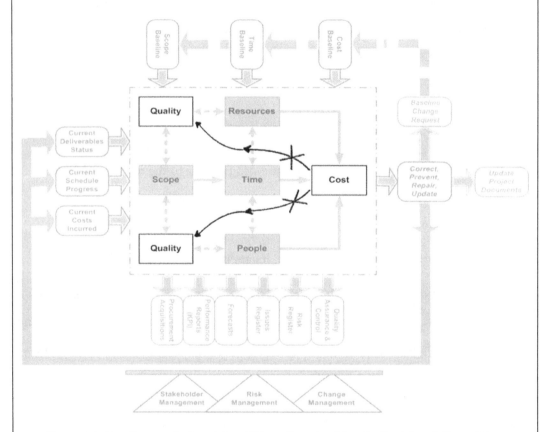

Figure 62: No direct relationship of [Costs] to the [Quality] of the deliverables.

Though, [Cost] cutting is often done by reducing [People] and/or [Resources], which if the [Scope] is to remain constant then this results in an increase in the actual [Time] required. ... And, if the [Time] is also to remain constant then [Quality] takes the hit.

Cost Of Quality

There is a flow-on relationship between [Quality] and [Cost]; this relationship is known as the **"Cost of Quality", which is the total cost of all efforts pertaining to achieving quality over the entire life of the project**. That is, the "Cost of Quality" equals the "Cost Of Doing It Right" plus the "Cost Of Getting It Right".

$$\text{COST OF QUALITY} = \underbrace{(\text{COST OF CONFORMANCE})}_{\text{COST OF DOING IT RIGHT}} + \underbrace{(\text{COST OF NON-CONFORMANCE})}_{\text{COST OF GETTING IT RIGHT}}$$

- ❖ **Cost of Conformance** – the **costs incurred** during the project's life **to avoid failures**.
    - o **Costs of Prevention** (Quality Assurance) – ensures that the project deliverables **are going to conform** to the approved Detailed Specifications and to the agreed Acceptance Criteria; i.e. **keeping failures from reaching the deliverables**.

        For example; training of the Project Implementation Team members so that they have the appropriate skillsets and use appropriate methodologies to produce quality deliverables, providing them with appropriate equipment to produce quality deliverables, unit testing each implemented feature, and the time to document what is necessary to pass knowledge onto those who will be involved with the project or supporting the deliverables in the future.

    - o **Costs of Appraisal** (Quality Control) – verifies that the project deliverables **do conform** to the approved Detailed Specifications and to the agreed Acceptance Criteria; i.e. **keeping failures from reaching the customer**.

        For example; acceptance testing each deliverable against the test specifications, and inspection of the final deliverables.

> ❖ **Cost of Non-Conformance** – the **costs incurred** during & after the life of the project **due to failures that did occur.**
>
>   o **Costs of Internal Failures** – relates to failures, defects, issues, and problems **found by the performing organization**; i.e. internally *"behind closed doors"*. For example; rework, scrapping, refactoring, and "do overs".
>
>   o **Costs of External Failures** – relates to failures, defects, issues, and problems **found by the customer**; i.e. externally *"in public view and scrutiny"*. For example; warranty repairs, product recalls, refunds, hotfixes, service patches, … litigation, class action suits, compensation payments, the negative impact on reputation, and reduced customer goodwill.

Which brings us to the next rule of "Adaptive & Proactive" SDLC Project Management.

RULE 27: **Quality, where doing it right the first time is more cost effective than eventually getting it right.**

Defect Reports versus Change Requests

During the acceptance testing, it may be discovered that the implemented functionality, while conforming to the definition in the approved Detailed Specifications, this functionality does not correspond to the purpose for which the customer had "intended". Subsequently, the Customer's Representatives could submit Defect Reports related to their (re) interpretation and (re) understanding of how such functionality "should operate".

Consequently, **it is the responsibility of the project's management (and the Project Implementation Team) to analyse & compare each reported defect against the approved Detailed Specifications** (and NOT comparisons against the Customer Requirements); thereby, curtailing Scope Changes (and Scope Creep) that are disguised as defect repairs.

Noting that, **an easily quantifiable way of measuring the perceived [Quality]** of the project | release **is via the count of the number of defects that have been reported**.

However, **without an alignment of both the performing & customer organizations' interpretations & understanding of; the features & functionality to be implemented**, what are the deliverables to be produced, and what will be included in each particular release & milestone, **then a significant number of (perceived) Defect Reports are going to be generated** by the Customer's Representatives. Subsequently, the project and the project deliverables could be perceived as having "poor quality", because appropriate due consideration has not been given to the relevance & severity of those defects that have been submitted.

Therefore, so as to not skew the stakeholders' perception of the project's [Quality], then what is required is some consistent process "work flow" for vetting & reassigning Defect Reports and Change Requests, so that these are categorized appropriately. As well as, a process for planning & authorizing the implementation of those approved defect repairs and scope changes.

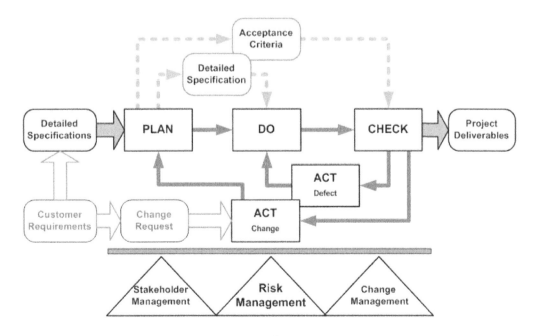

Figure 63: Monitoring & control as a 'PLAN – DO – CHECK – ACT' process.

3.6.7. Change Management

[Section 2.6.3]
[Section 8.3.3]

As part of monitoring & control, the project's management will also have to concern themselves with Change Management, as illustrated below in [Figure 64].

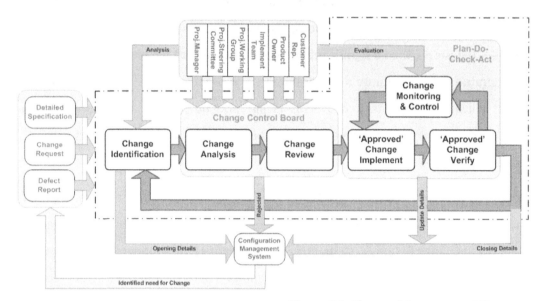

Figure 64: Change Management Process.

1.. **IDENTIFY the change desired for the project.** Whenever a **project stakeholder** believes that some change is necessary for the project to meet the agreed objectives, then they need to **submit to the performing organization** either **a Change Request or a Defect Report** (in a written text-based format and not solely as a verbal statement). The information pertaining to this submission should be **retained in** some form of **Change Management System | Defect Tracking System**.

2.. **ANALYSIS of the received Change Request | Defect Report** to determine; if any **additional information is required for clarification**, whether this Change Request | Defect Report is a **duplicate or very similar to an existing one**, and whether the **Change Request should be a Defect Report |** the **Defect Report should be a Change Request**. Where the **approved Detailed Specifications is used as the arbitrator** of the **validity** of the Change Request | Defect Report, and **NOT the Customer Requirements**.

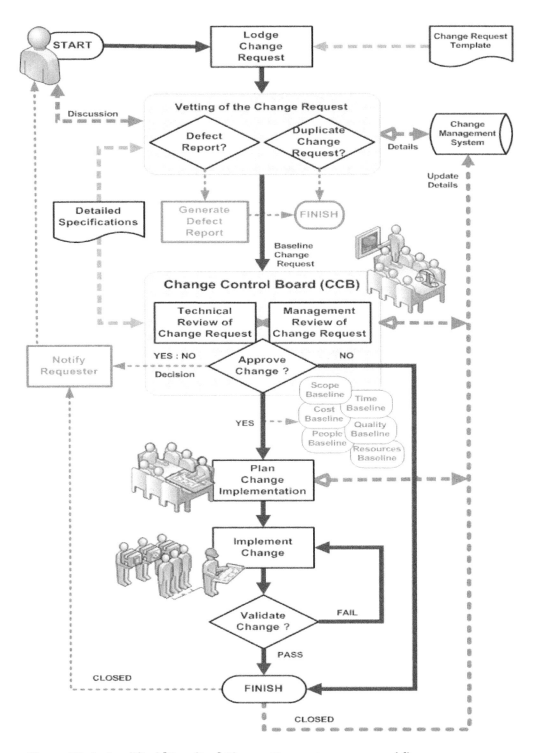

Figure 65: A simplified [Baseline] Change Request process workflow.

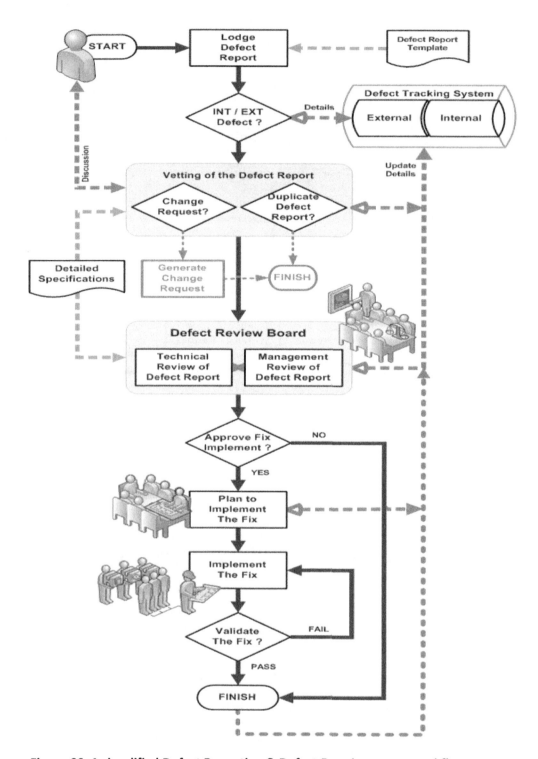

Figure 66: A simplified Defect Reporting & Defect Repair process workflow.

3.. The vetted Change Request | Defect Report would be **REVIEW**ed respectively by a **Change Control Board (CCB) | Defect Review Board** to evaluate whether the Change Request | Defect Report should be **Accepted or Rejected**, and whether instead a reciprocal Defect Report | Change Request should be generated in its place.

The Change Control Board | Defect Review Board would **EVALUATE** all of those accepted outstanding Changes | Repairs that are to be implemented into the current pending release or into some future releases, considering such things as:

- o   Ensuring that, if the Change | Repair was included then what **benefit or opportunity** would result when compared to the (possible negative) impact on the project's [Scope], [Time], [Cost], [Quality], [People], and [Resources]; i.e. what is the "**Return On Investment**".
- o   Ensuring that the Change | Repair will be **feasible** to implement.
- o   Ensuring that the Change | Repair is **implemented at an opportune moment**, so that the Change | Repair **does not become an "ad-hoc tack-on"**.
- o   Ensuring that the Change | Repair will be **implemented properly**.

4.. A decision would be made as to, if and when it would be appropriate to **AUTHORIZE** the implementation of those approved Changes & Repairs.

If the Change Request was accepted then it would be **translated into adjustments** to:

- o   **[Scope Baseline]**   in the project's (re)approved **Detailed Specifications**,
- o   **[Time Baseline]**   in the project's (re)approved **Project Schedule**,
- o   **[Cost Baseline]**   in the project's (re)approved **Project Budget**,
- o   **[Quality Baseline]**   in the project's (re)approved **Acceptance Criteria** / test plans, plus any other affected project documentation, and
- o   **[People Baseline]** & **[Resource Baseline]** in the project's (re)approved **Resourcing Calendar & Allocation Plans** for procurements.

Which brings us to the next rule of "Adaptive & Proactive" SDLC Project Management.

RULE 28: **Change has to be expected, else the project will have to be pre-planned down to the minutest details, then the implementation micromanaged.**

5.. **IMPLEMENT** those approved & authorized Changes | Repairs that are to be incorporated into the project in accordance with the agreed Acceptance Criteria.

6.. **MONITOR** each of those authorized Changes | Repairs, and then via feedback, **CONTROL** the responses to and interactions with each of these Changes | Repairs.

Which brings us to an addendum to "Adaptive & Proactive" SDLC Project Management.

RULE 28^A: **Any Change can inadvertently introduce undesirable side-effects that can Risk the project being a success.**

Change Management has to be undertaken over the entire life of the project.

"As there can be serious consequences of making a change just because it seemed like a good idea at the time."

The lack of a formalized Change Management Process can be a catalyst for problems during the later phases of the project's life cycle.

"So who agreed to that, cause this ain't what I recall."

DO NOT implement the change without some form of (written) agreement as to the recompense that is to be afforded to the affected stakeholders.

"So who exactly is paying for this to be done, cause it ain't me."

3.6.8. Risk Management

[Section 2.6.1]
[Section 8.3.1]

As part of monitoring & control, the project's management will have to concern themselves with Risk Management, as illustrated below in [Figure 67]:

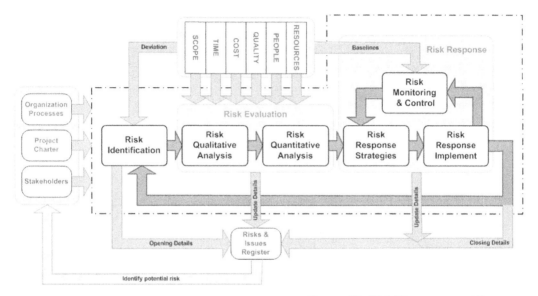

Figure 67: Risk Management Process.

1. **IDENTIFY** the <Risks> associated with the project, via such means as; interviewing, brainstorming, Delphi technique, checklists, project initiating & planning discovery, SWOT analysis, cause & effect diagrams, flow charts, timeline diagrams, and re-evaluations of those previously established project assumptions & hypotheses.

2. **QUALITATIVE Analysis** for each identified <Risk>, **subjectively describe** the severity of
   how much it is going to hurt the project's objectives, and
   how likely is it of happening. Though, qualitative analysis is prone to bias, as it is potentially based on opinions & hearsay, rather than the real facts.

3. **QUANTITATIVE Analysis** for each identified <Risk>, **numerically define** the **Impact** (**Severity** on a 1 to 5 scale of least impact through to most impact) and the **Probability** (**Likelihood** on a 1 to 5 scale of very unlikely through to highly likely) of happening.

Risk Priority = Impact X Probability

Some performing organizations will have predefined numerical descriptors for grading the Impact and the Probability; thereby, clearly & consistently differentiating between each level of Impact & Probability when evaluated against the agreed baselines.

Which brings us to the next rule of "Adaptive & Proactive" SDLC Project Management.

RULE 29: One person's "Calculated Risk" is another person's "Life Gamble".

When analysing each identified <Risk>, a **Risk Matrix** (of a 5x5 grid) maybe used to represent   the **Impact**   / Severity   (along the X-Axis), as in [**Figure 68**] **versus**   the **Probability** / Likelihood   (along the Y-Axis).

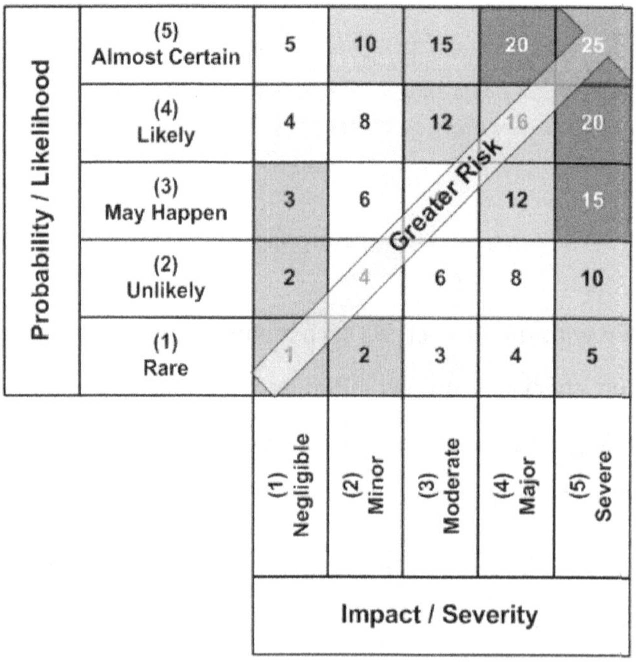

Probability / Likelihood	Impact / Severity				
(5) Almost Certain	5	10	15	20	25
(4) Likely	4	8	12	16	20
(3) May Happen	3	6	9	12	15
(2) Unlikely	2	4	6	8	10
(1) Rare	1	2	3	4	5
	(1) Negligible	(2) Minor	(3) Moderate	(4) Major	(5) Severe

Red = Extreme

Orange = High

Yellow = Medium

Green = Low

Where the width of the coloured bands of Red – Orange – Yellow – Green is influenced by the risk tolerance | risk aversion of the performing organization.

Figure 68: Example of a Risk Matrix for Risk Priority.

With such a **Risk Matrix**, then any identified <Risk> that is calculated as being positioned **in the top right-hand corner is really BAD** and should be **dealt with urgently**; whereas, a <Risk> in the bottom-left corner is most probably not that urgent to be dealt with. However, the position of a <Risk> in the Risk Matrix is not static and throughout the life of the project, a <Risk> could move to a different position on the grid (i.e. become more or less urgent to be dealt with). Alternately, a <Risk> could disappear from the grid (i.e. be resolved or eliminated), and a new <Risk> could appear.

Another technique would be to diagrammatically group by "common topic" categories those related <Risks> in a **Risk Breakdown Structure**, and see if there is a "clustering" of high priority <Risks> that have common cause(s). ... And, by resolving the common cause(s), a concentration of <Risks> could possibly be eliminated at the same time.

The **information about each identified <Risk> and the progress with handling** each <Risk> should be **recorded as a uniquely identifiable entry in a Risks & Issues Register** that is dedicated specifically to this particular project; *e.g. a spreadsheet*. ... And, this register should be **continually updated** (i.e. per incident and weekly) with the latest information and the decisions that have been made relating to each particular <Risk>.

"Risk analysis is the perfect time to seek the advice of (technical) Subject Matter Experts, and not depend solely on one's own opinion."

4.. **PLAN** the Risk Response Strategies for each of the identified <Risks>, to either:

- ☐ **Accept** the <Risk> ... concede that the <Risk> could occur, and as such take a **"passive approach" of doing nothing** versus a **"proactive approach" of doing something in advance** to compensate for the <Risk's> possible impact; *e.g. by adding buffering & contingency reserves that is to be used if and when needed.*

- ☐ **Avoid** the <Risk> ... taking actions to prevent the <Risk> from evolving into an issue, by removing or reducing the cause of that specific <Risk>; *e.g. increasing the project duration [Time], increasing the budget [Cost], reducing the [Scope], increasing the [People] available, increasing the [Resource] allocation, and/or engaging Subject Matter Experts.*

- ☐ **Mitigate** the <Risk> ... take steps to lessen the impact of the <Risk>, take steps to reduce the likelihood of the <Risk> occurring, and/or having contingency plans in place; *e.g. in-depth Quality Control processes, improved Quality Assurance procedures, use of prototypes and proof-of-concept models.*

- ☐ **Transfer** the <Risk> ... to a third party who is more capable of dealing with that type of <Risk>; *e.g. outsourcing / contracting the work to specialists, obtaining insurance.*

- ☐ **Share** the <Risk>   ... where the burden for and of the <Risk> response is distributed amongst the project's various stakeholders.

"Ideally aim to eliminate / prevent the risk, or at least reduce the exposure to the risk."

Where **each Risk Response can be thought of as its own 'PLAN – DO – CHECK – ACT' mini-project** with its own considerations for [Scope] – [Time] – [Cost] – [Quality] – [People] – [Resources]; see [Section 2.7].

Though for each <Risk>, the **time criticality of choosing Risk Response Strategies and** the **subsequent implementation** of one or more **of those chosen strategies** will be greatly **influenced by the "actual urgency"** (based on the calculated Risk Priority), as well as the **"perceived urgency"** (based on the involved stakeholders' perspective). Additionally, the involved stakeholders' **Risk Tolerances will greatly influence their concerns & biases** towards how and when to deal with certain types of <Risks>.

5.. **IMPLEMENT** the approved & authorized chosen Risk Response Strategies for each identified <Risk>.

6.. **MONITOR** the progress of implementing the Risk Response per identified <Risk>, and then via feedback, **CONTROL** the adjustments to those chosen Risk Responses.

Risk Management needs to start during the Initiating & Planning Phases, and NOT wait until the Executing Phase. ... And, Risk Management has to be an **ongoing iterative activity** that is performed over the **entire life of the project**.

"Not just when shit happens, because it can happen at any time, even to the most stable and mundane of projects."

3.6.9. Stakeholder Management [Section 2.6.2] [Section 8.3.2]

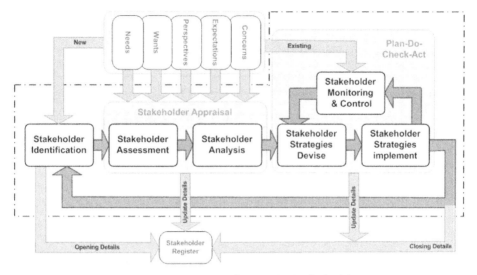

Figure 69: Stakeholder Management Process.

And yes, it is not coincidental that the diagram above in [Figure 69] for Stakeholder Management does look very similar to the Risk Management diagram depicted previously for [Figure 67]. Given that, Risk Management and Stakeholder Management (as well as Change Management) all go through the cyclic stages of identification, assessment / analysis, planning / strategizing, implementation, and monitoring & control. … And, all as part of a larger R.I.S.C. Management process; see [Section 2.6].

Stakeholder Management involves the following steps:

1.. **IDENTIFY** the project's stakeholders and determine whether they are:

- ☐ **Involved with** the project either as an INPUT provider, as a process doer, or as an OUTPUT receiver?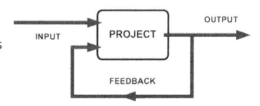

- ☐ **Affected by** the project either directly (immediately) or indirectly (delayed), during the project's implementation, by the outcomes & OUTPUT of the project. … OR, by the project having not already been implemented?

185

☐ **Effect on** the project either directly (immediately) or indirectly (delayed), as an INPUT provider, as a process doer, or as an OUTPUT receiver?

"Watch out for the indirect feedback path, as this often result in surprised fire-fighting situations."

One way of determining the project's stakeholders is to take the Project Life Cycle model being used for this particular project and layout on this model's diagram, those persons involved with the project, as in [Figure 70] and [Figure 71].

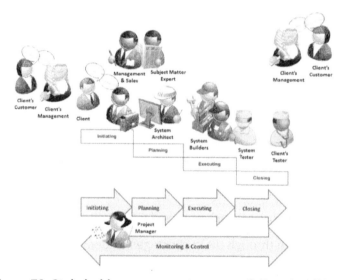

Figure 70: Stakeholders common to a waterfall project life cycle.

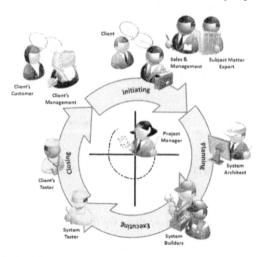

Figure 71: Stakeholders common to an iterative / agile project life cycle.

Which brings us to the next rule of "Adaptive & Proactive" SDLC Project Management.

RULE 30: The key to project success is knowing those Stakeholders who must be addressed, and how soon.

2.. **ASSESS** these identified project's stakeholders and determine what is there:

☐ **Expectations, perspectives, concerns, needs, and wants,** … i.e. "prominence".

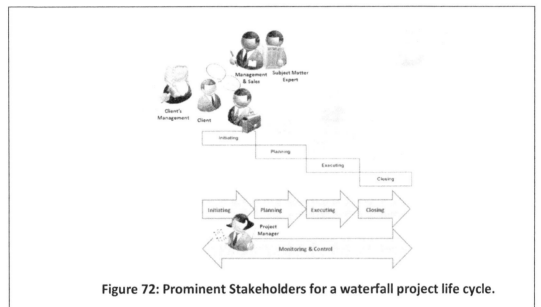

Figure 72: Prominent Stakeholders for a waterfall project life cycle.

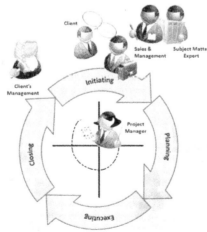

Figure 73: Prominent Stakeholders for an iterative / agile project life cycle.

As depicted in [Figure 72] and [Figure 73] some of these identified stakeholders have "prominence" over the project, and primarily this is during the project's Initiating Phase. Where "prominence" is akin to the summation of that stakeholder's; **power-over, interest-in, position-on, uncertainty-about, influence-over**, and their being **impacted-by** the project's outcomes.

3.. **QUALITATIVE Analyse** these identified stakeholders so as to determine each one's:

- ☐ **Power** over ... their **capability** to get things done **to** either **help or hinder** the project's progress?

- ☐ **Interest** in ... are they **concerned** for the project, or (at present) are they the **least bit interested**?

- ☐ **Position** on ... do they **support** (champion) the project, or are they **oppose** (going as far as striving for its demise), or somewhere in between?

- ☐ **Uncertainty** about ... how likely / **predictable** are they **to react** as expected (either positively or negatively)?

- ☐ **Influence** over ... their **ability to effect change** and their capability to get others to do what they desire (for or against) with respect to the project and its outcomes.

- ☐ **Impact** by ... how will they be **affected by** the project and its outcomes. Though, NOT their impact on the project (see "influence").

4.. **QUANTITATIVE Analyse** each identified stakeholder by **assigning a numerical value** to those qualitative aspects listed above (on a 1 to 5 scale of least is 1 and most is 5).

Tabulate this information to obtain an overall understanding for each of these identified project stakeholders.

Map this information onto the various **Stakeholder Matrices**; **Power versus Interest** [Figure 74], **Power versus Position** [Figure 75], **Power versus Uncertainty** [Figure 76], **Influence versus Impact** [Figure 77], plus other possible combinations thereof.

Noting that, this textbook has used arrangements of Stakeholder Matrices that have the higher risk to the project in the upper right-hand corner and those with the lower risk in the bottom left corner of the grid, just like with the Risk Matrix in [**Figure 68**].

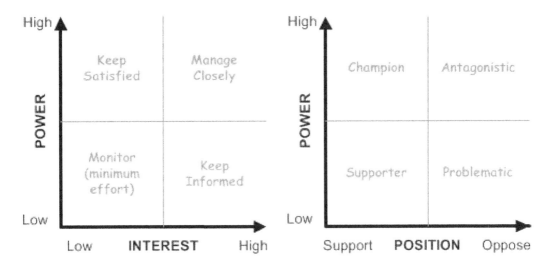

Figure 74: Power versus Interest.

Figure 75: Power versus Position.

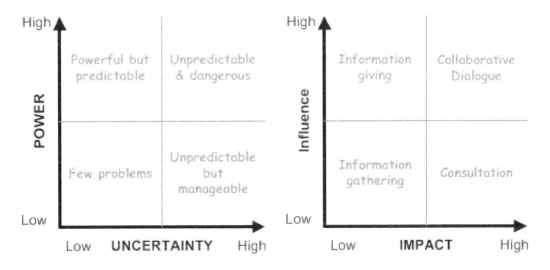

Figure 76: Power versus Uncertainty.

Figure 77: Influence versus Impact.

Though, the stakeholder's position on these matrices is based on the perceptions of the person(s) undertaking the analysis, and this may not represent the project's true relationships with each particular stakeholder.

Which brings us to an addendum to "Adaptive & Proactive" SDLC Project Management.

RULE 30[A]: The stakeholder's view of reality is not necessarily based on factuality; rather, on their belief in what is true.

Notice that, in those Stakeholder Matrices there are grey comments in each quadrant of each grid; these comments either define what sort of actions should be undertaken to handle stakeholders who appear in that particular quadrant, or are comments about the characteristics of stakeholders in that quadrant. Later on, these quadrant comments will be used to decide on the relevant strategies to be used to manage each stakeholder. ... *"Not the usual, ad-hoc management of stakeholders then?"*

5.. **STRATEGIZE** how best to engage & manage these stakeholders so that they will collaboratively participate in the project (ideally for positive outcomes).

Look at where each stakeholder is positioned in each Stakeholder Matrix, and this will provide a collective starting-point on how to build each stakeholder strategy.

- ☐ **Power Vs. Interest** ... keep satisfied, monitor (minimal effort), keep informed, or manage closely (most effort).

- ☐ **Power Vs. Position** ... champion, supporter, problematic, antagonist.

- ☐ **Power Vs. Uncertainty** ... powerful but predictable, few problems, unpredictable but manageable, unpredictable & dangerous.

- ☐ **Influence Vs. Impact** ... information giving, information gathering, consultation, or collaborative dialog.

6.. **IMPLEMENT** those chosen stakeholder strategies giving serious consideration to:

- ☐ **Continuous bi-directional communications** with the stakeholders.

  The lack of **frequent "well grounded" two-way communications** can be a catalyst for misunderstandings, misinterpretation, and mistrust amongst the project stakeholders, as well as a source of much angst and unhappiness.

- ☐ **Continuous collaborative effort to reconcile** the stakeholders' expectations, perspectives, concerns, "needs & wants" with the project's objectives.

    The strategies employed will have to "**build the relationships**" with the project stakeholders **by involving their participation** in the project, and (where appropriate & practical) **by involving them in the decision-making process** (or **at least taking into consideration their "best interests"**).

    Noting that, the more powerful the stakeholders that can be engaged to support the project (especially during the early phases of the project's life) then the more likely it is that the project will be successful. ... And, by winning over the support of a few critical stakeholders, then this can serve as a catalyst for winning over many of the other stakeholders.

- ☐ **Continuous implementation of strategies** that are based on the principle of; **maximizing the positive influences** on the project, while **minimizing those negative impacts** on the project's stakeholders.

    Thereby resulting in resolutions that are **mutually beneficial** to those involved, while **fairly distributing** who receives the **benefits and** who bears the **burdens**.

The execution of the Stakeholder Management Strategies has to be prioritized based on the potential "riskiness" that is associated with each particular stakeholder. ... Also, on occasions, the resolution of the stakeholder's expectations, perspectives, concerns, "needs & wants" will result in [Baseline] Change Requests that must be managed.

7.. **MONITOR** each of these implemented stakeholder strategies, and then via feedback, **CONTROL** the responses to and interactions with each of these identified stakeholders.

 Stakeholders should never be taken for granted; rather, stakeholders **should be managed throughout the entire life of the project**.

"And not just when they kick up a fuss or throw a tantrum."

Quantitative Project Analysis & Review on the Stakeholders Behalf

Periodically during the life of the project, quantitative project analysis & review should be carried out on the primary stakeholders behalf; so that, their representatives can **systematically & logically judge** the project's progress as either **an unequivocal PASS : FAIL**, based on an **analytical comparison of the "planned" and "expected" values against the "actual" values**. Hence, by **putting "cold hard" numerical values to attributes & aspects that describe the project**, then this can mitigate & alleviate these stakeholders' expectations, perspectives, concerns. This analysis & review would entail:

❖ **Project Audit** ... is an infrequent **formalized objective appraisal** of the project's previous activities, processes & procedures. This audit would be conducted by an **independent third party to systematically examine** (by visual inspection of the project artefacts & associated Objective Quality Evidence, and via interviews with the project participants) **whether pre-establish sets of Audit Criteria** (up to the time of the audit) **have or have not been complied with** by the project's participants. So as to, **"historically" determine whether the specified project processes & procedures** (such as recognised industrial quality standards & guidelines) **have been followed**, and whether managerial due-diligence has been exercised.

The outcome of this project audit would be an **Audit Report** containing the auditing party's observations on the project's "apparent conformance" to those specified project processes & procedures, identification of revealed non-conformances & irregularities, notes about these discovered non-conformances, and maybe suggestions on Process Improvements. ... *"but legally, not recommendations."*

❖ **Project Inspection** ... is a regular **semi-formal assessment** of the project's current performance & progress when compared against tangible measures (*e.g. Earned Value Performance Measures, and Key Performance Indicators*). This inspection would be conducted by representatives of the Project Steering Committee, either **Routinely** at specific intervals (*e.g. monthly or quarterly*), triggered by the ~ ~ ~ ~

Triggered by the conclusion of a project phase (*e.g. a "gate review"*), or

Arbitrarily at the discretion of the project's primary stakeholders.

This inspection would look into such things as:

☐ How is the project progressing when compared to the [Scope Baseline], the [Time Baseline], the [Cost Baseline], and the [Quality Baseline]?

☐ How effectively & efficiently has the project utilised its assignment of [People] and allocation [Resources]?

☐ What is the status of project's <Risks & Issues>, and has the Risks & Issues Register been kept up-to-date?

☐ How rigidly have the Quality Assurance processes and the Quality Control procedures been complied with, and has the Objective Quality Evidence been produced that would satisfy the scrutiny of a Project Audit?

☐ What is the uncovered state of the project, and does this correspond to that which has previously been described in the Project Status Reports?

☐ Have the prescribed project management processes & procedures been followed, or is this project being managed ad-hoc *"on a wing and a prayer"*?

☐ How capable is the current project management at supervising and exerting control over the conduct of the project?

The outcome of this project inspection would be instructions (aka "orders") for corrective actions to be undertaken where deemed necessary.

"Project Inspection is checking that the project's planning & undertaking are in acceptable condition to continue on as is, or do more oversight controls need to be applied."

NOTE: Quantitative analysis & review is useful for establishing Quality Control policies & procedures, where the resultant numbers can be correlated & compared in some form of process feedback loop.

3.6.10. [People] Monitoring & Control

While those [People] who implement the project are also stakeholders of the project, these particular [People] will require a more hands-on management approach. Because, the sooner that unexpected utilization of the project's assigned [People] is detected, then the sooner that the project's management will be able to undertake those actions that are necessary to limit / counter the potential damage that could result due to such deviations away from the planned [People Baseline].

Consider that, these [People], ... and for that matter, [Resources], ... will over [Time] transform the project's [Scope] into physical deliverables. However, these **[People] "don't come for free"**, as they will have to be paid for as project [Costs]; plus, there is the associated supporting infrastructure for these [People] that will also add to these [Costs].

Hence, **the utilization of the project's [People] needs to be in line with the planned [People Baseline], else the project will be incurring [Costs] that don't align with the expected progress of [Scope] coverage when compared to the [Time] being taken.**

Task Board & Burn Down Chart

As stated previously, for an agile project, it is the members of the Project Implementation Team who decide what 'To Do Items' [Scope] that they intend to deliver per sprint; and, these [People's] activities per 'To Do Item' can be represented via the Task Board.

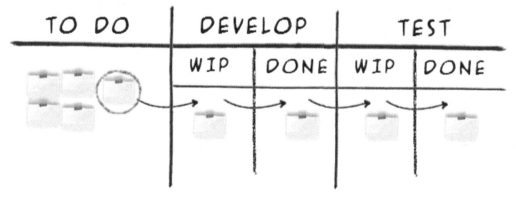

Figure 78: Motion across an agile Task Board.

With an agile based project, instead of updating the schedule with Percentage of Work Completed and calculating Earned Value Performance Measures, the monitoring of the project's progress is via the **Burn Down Chart**.

When an individual 'To Do Item' passes from left-to-right across the Task Board to eventually be verified as DONE in testing, in [Figure 78], then this will reduce the amount of "Work Remaining" and hence this will affect the "Actual Burn Down" line in [Figure 79].

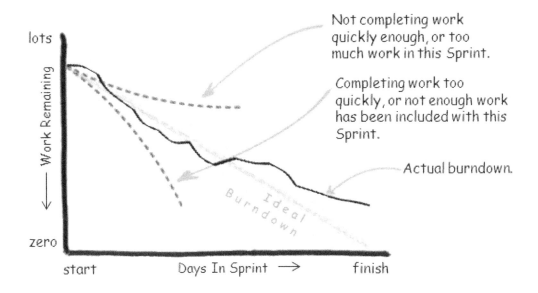

Figure 79: Agile Burn Down Chart with up-to-date data.

Provided that, the "Actual Burn Down" line travels downwards around the straight "Ideal Burn Down" line then the project will be progressing along nicely. However, when the "Actual Burn Down" line starts to move away from the "Ideal Burn Down" line, especially movement up towards the right as illustrated in the second half of [Figure 79], then the Project Implementation Team is having problems with not being able to complete the Sprint Backlog of 'To Do Items' within the allotted [Time] box of this sprint, or supposedly completed 'To Do Items' are being reopened due to defects that are being found.

Alternatively, if there is a noticeable dive-down of the "Actual Burn Down" line, then 'To Do Items' are being closed a lot quicker than was expected, and thus the question has to be raised as to the [Quality] of these deliverables being closed out in such a rush?

And, if the "Actual Burn Down" line suddenly changes direction so as to converge with the "Ideal Burn Down" line, then questions need to be asked as to "how come".

As evidenced by a burn down line that travels relatively horizontally for much of the sprint's duration, then suddenly dives down towards zero at the end, see [Figure 80]. This kind of motion indicates that "agile best practices" are probably not being followed, and that 'To Do Item' statuses aren't being updated, or 'To Do Items' were closed-out in a rush.

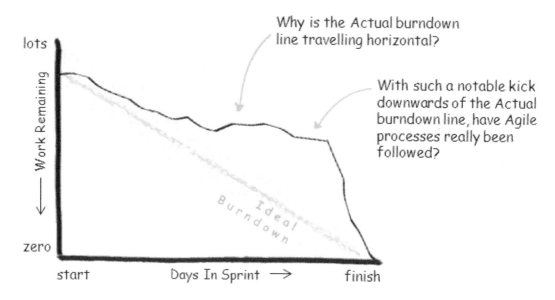

Figure 80: Agile Burn Down Chart with a sudden change of direction.

"And, if this sudden change of direction is happening then the project's management might as well go back to old-fashion inquiring of the percentage completes for iterative scheduled tasks; because, it is apparent that the implementers ain't tracking their own progress via a methodology that is in anyway representative of reality."

Hence, It is essential that each individual Project Implementation Team member updates their 'To Do Item' status on a daily basis, or (preferably) immediately after they change the operational status of a particular 'To Do Item' that they have been working on; ELSE, the Burn Down Chart will not accurately depict the true state of the project's progress. Thus, being ineffectual as a progress tracking tool.

3.6.11. [Resources] Monitoring & Control

As with the [People] assigned to the project, there will be similar expectations on the quantity and types of material & services [Resources] that are utilized by the project. If there is unexpected utilization of the project's allocated [Resources] then the project's management will have to undertake those actions that are necessary to limit / counter the potential damage that could result due to such deviations away from the planned [Resource Baseline]. ... And, these **[Resources]** *"don't come for free"*, **as these have to be paid for as project [Costs]**; plus, there is the associated supporting personnel for these [Resources] that will also add to these [Costs].

Hence, **the utilization of the project's [Resources] needs to be in line with the planned [Resources Baseline], else the project will be incurring [Costs] that don't align with the expected progress of [Scope] coverage when compared to the [Time] taken**.

$$COST_{\text{TO DATE}} = (PEOPLE_{\text{RATES}} \times TIME_{\text{USED}})$$
$$+$$
$$(RESOURCE_{\text{UNITS}} \times AMOUNT_{\text{USED}})$$
$$+$$
$$(FIXED\ COSTS_{\text{TO DATE}})$$

Yet surprisingly, the project's management can sometimes forget about this relationship of the daily [Costs] equating to the amount of [Resources] used, plus the rate of [People] utilized, plus other operating Fixed Costs.

COFFEE BREAK DISCUSSION ... *People are more than just numbered Resources.*

When you read that title, were your thoughts something along the lines of, *"Err der, of course people ain't resources"*. ... Yet, you would be surprised at how many accountant type managers, and senior & executive managers, consider the Project Team members to be nothing more than interchangeable numbered parts that can be so easily procured (and then let go) without any real consideration being given to the humanities aspect of, and what is their individual impact on the "success-ability" of the project. Because, SDLC Projects that are operated with a "People-As-Machines" mindset will fail due to:

- ☐ **A successful project requires the right number of people with appropriate skill sets, and the right intermixing of personalities**; i.e. a blend of both technical minded and socially minded individuals. Consider that, people can be trained to improve their productivity, whereas machines can be upgraded. However, once the project commences the machine's performance will stay at relatively the same level, while the human's performance will vary due to the involved person's ongoing experiences, attitudes, believes, understandings, expectations, and morale. Hence, the characteristics of the [People] performance aspect will change dynamically throughout the life of the project, whereas the characteristics of the inanimate [Resource] performance aspect will remain relatively constant.

- ☐ **A successful project requires positive team morale (and high internal self-drive) to succeed, in spite of the current situation & prevailing circumstances.** That is, "project team morale" is that elusive ingredient that can give the project continued drive (and more importantly, a "faith in success") when things turn bleak.

- ☐ The longer that someone has been working on a project (or at the organization), **when they decide to leave then a significant amount of essential knowledge is leaving with them**. Can the project (and the performing organization) afford to lose such an asset, given that it could potentially take weeks / months to procure a replacement, and then the **several weeks / months for the replacement to come up to speed to be anywhere near as productive** as the person who has departed?

3.7. Closing Phase

3.7.1. Purpose

The purpose of the Closing Phase is to perform those activities that are necessary to officially conclude the project (or to culminate the current milestone release), and thereby;

(1) **Wrapping up** the project's **Executing Phase** (in entirety or just for the current release).

(2) Ensuring that those **[People] & [Resources]** that are no longer needed by the project **are made available** for other activities.

(3) Ensuring that, in a logical sequence, **all of the project stakeholders are communicated with to inform them that the project / release is ending**, stating why it is ending, and from their perspectives when will it conclude.

(4) Ensuring that the **specified & agreed deliverables are handed over** to the appropriate project stakeholders, and **list any outstanding deliverables**. This also includes making available the stipulated & associated **Objective Quality Evidence (OQE)**.

(5) Ensuring that these **deliverables are officially accepted** by the relevant project stakeholders, and **list any deviations from the contracted deliverables**.

(6) Ensuring that the **administrative functions are closed out**; i.e. to finalize the financial accounts, legal obligations, project reports & records, project documentation, intellectual property, and archiving the project's artefacts.

> Which brings us to the next rule of "Adaptive & Proactive" SDLC Project Management.
>
> **RULE 31: Closing out a project is NOT the winding down of that project; rather, it's a mini-project specifically to End.**
>
> *"Yep, not just some ad-hoc muddled through road to abandonment."*

> A project is closed out for one of the following reasons:
>
> ❖ **The project successfully achieved its objectives** … that is, the project's deliverables were handed over to the customer with the agreed [Scope], by the agreed [Time], within the agreed [Cost] budget, and has the agreed level of [Quality] to "satisfy" the customer.
>
> ❖ **The project failed to achieve its objectives**, or at the current rate of progress then the project would take too long or be too costly to eventually achieve its objectives.
>
> ❖ **The project never succeeded nor failed because it is no longer required** … that is, the objectives are no longer feasible given the current situation & prevailing circumstances. *For example; a decline in the economy, a change in the targeted market segment, or a change in the business's direction.*

3.7.2. What has to be answered?

For the project to have any realistic chance of success, the Closing Phase has to provide due-diligent answers to the following questions:

(1) Did **WE** (the implementer) **GET IT RIGHT**?

(2) Does **EVERYONE** (implementers & stakeholders) **AGREE** it is **ALL DONE**?

(3) What did **WE** (the implementer) **LEARN** from this?

(4) Can **EVERYONE** (implementers & stakeholders) say **GOOD BYE** as friends?

> Which brings us to an addendum to "Adaptive & Proactive" SDLC Project Management.
>
> **RULE 31[A]: Project closure requires just as much managerial attention as did the project's initiation and planning.**

3.7.3. QUESTION: did we get it right?

This question refers to:

- [ ] **Does the project's deliverables conform to the agreed [Scope Baseline]**, as was defined in the approved Detailed Specifications (and those approved Change Requests) that were signed off by the representatives of both the customer & performing organizations?
    - Verification that the agreed [Scope] has been covered, and that the project deliverables & associated artefacts have been supplied to the relevant recipients.
- [ ] **Do these project deliverables & associated artefacts conform to the agreed [Quality Baseline]**, as was prescribed in the signed off Acceptance Criteria?
- [ ] **Have these project deliverables & associated artefacts been provided within the agreed [Time Baseline]?** That is, before or after the milestone due-dates, as was laid out in the signed off project schedule / milestone list.
- [ ] **Have these project deliverables & associated artefacts been produced within the agreed [Cost Baseline]?** That is, for more or less than the amount specified in the authorized budget.

Additionally,

- [ ] **Have the assigned [People] been effectively & efficiently utilized?**
- [ ] **Have the allocated [Resources] been effectively & efficiently utilized?**

Satisfying the customer's "needs & wants", or more specifically, the confirmation that the deliverables are *"fit-for-use"* *"as was meant"* are NOT the objectives of the Closing Phase. In that, [Quality] is about producing project deliverables that comply with the agreed [Scope] of what was defined & recorded in the approved Detailed Specifications. ... And, NOT what was *"intended"* per se in the Customer Requirements, nor as was *"mentioned"* solely via word-of-mouth.

[Scope] & [Quality]   ... aka *"Customer Satisfaction"*

It is **the Closing Phase, when** the project's primary stakeholders find out exactly how effectively the project's management and the Project Implementation Team have understood & interpreted what the customer was wanting. Because, **any differences in understanding & interpretation will manifest themselves** now, as disagreements & disputes between the representatives of the performing and customer organizations.

Hence, this is when the **coverage** of what was agreed to, has to be evidentially proven:

☐ The **agreed [Scope Baseline] has been delivered**; as identified in the **Requirements Traceability Matrix (RTM)** which lists **what functionality has & has not been included**.

☐ The **agreed [Quality Baseline] has been achieved**; as identified in the **Testing Coverage Matrix (TCM)** which lists **what functionality has been verified as operating** in accordance with the agreed Acceptance Criteria (and the approved Detailed Specifications), what functionality has been found to be not conforming, and what functionality has not yet been tested.

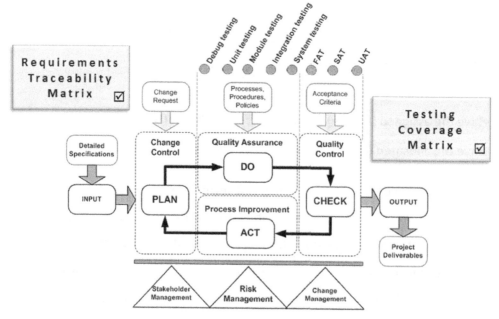

Figure 81: Relationship of the Requirements Traceability Matrix (RTM) to the Testing Coverage Matrix (TCM) to the 'PLAN-DO-CHECK-ACT' model.

☐ The **Acceptance Test Suites** (of the FAT – "Functionality" Acceptance Test, SAT – "Satisfaction" Acceptance Test, UAT – "Usability" Acceptance Test) have **confirmed** that the **agreed [Scope Baseline]** detailed within the approved Detailed Specifications and the **agreed [Quality Baseline]** Acceptance Criteria **have been complied with**.

If a deliverable does confirm with the agreed specifications, but it does not correspond with what the customer *"had meant"* (or not as they now realize *"it should have been"*), then a Change Request must be raised and not the generation of a Defect Report. However, if that functionality is not as was defined in the agreed specifications, then a Defect Report should be raised. See [Section 2.6.3] on Change Management.

[Time] & [Cost] ... aka *"Performance Measures"*

While the conformance to the agreed [Scope & Quality Baselines] will be a deciding factor on the *"customer's satisfaction"* with the project's outcomes; what will greatly interest the performing organization, while still satisfying the customer, is staying within the **"performance measures" of the agreed [Time & Cost Baselines]**.

That is, **was this a *"profitable venture"* for the performing organization**?

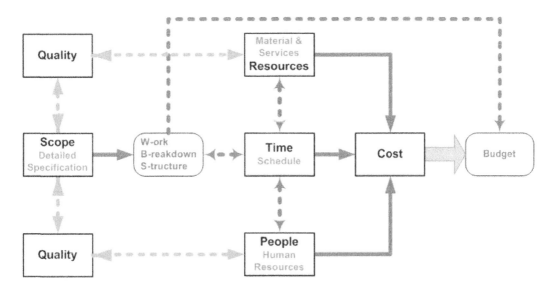

Figure 32: The Planning Process Model to determine the project's Cost and Budget.

3.7.4. QUESTION: does everyone agree it is all done?

To answer this question of, "does EVERYONE AGREE it is ALL DONE" then the following sub-questions need to be replied to with a positive, "YES".

☐ Have all of the **implementation activities** associated with this particular project | milestone release been concluded | **closed out**?

☐ Have all of this particular project | milestone release agreed **baselines constraint been settled "satisfactorily"**?

☐ If non-conforming project deliverables were produced then have **Concessions and Waivers been generated, reviewed, and approved** for these non-conformances?

☐ Have the **project deliverables been transferred** to the appropriate recipients, and has the **acceptance** of each of these deliverables been **signed off** by an authorized representative of the recipient organization?

  o Are there any warranties, licenses and certificates that any party is obligated to provide, as these will have to be turned over to the appropriate recipients? Similarly, are there any software license keys, activation codes, and related manuals & paperwork that also have to be turned over?

☐ Have the agreed **project artefacts & documentation been handed over** to the authorized & appropriate representative of the recipient organization?

  o Are there any sourced materials (i.e. documentation, equipment, information, and data) that were provided to any Third Parties by either the performing or customer organizations, then these will have to be returned and/or responsibly disposed of?

☐ Has the agreed **Objective Quality Evidence (OQE) been produced, cumulated, and signed off** by the relevant quality authorities? ... And, archived accordingly?

- ☐ Has the project's **Intellectual Property** been **accumulated, filtered, archived, and indexed for future reference**? ... And, verified as successfully retrievable?
    - o Is there any documentation, schematics, drawings, and release notes that any Third Party is contractually obligated to provide, as these will have to be turned over (with an acceptable level of quality & presentation to be usable by the intended audience)?

- ☐ Have the project's externally **procured / acquired resources** (*e.g. third party and customer furnished equipment & information*) been **returned** to their authorized & approved recipients? ... And, has this return been recorded as such, so that there is a **clearly traceable "Line Of Custody"** from beginning to end, else who will be **"held accountable"** if these items were to go missing or were to be received in a damaged unusable state?
    - o Has any associated **usage payments** been concluded, and has a **"thank you"** been communicated?

- ☐ Is there **anything else** that is **specific to this project**, to the **industry / market-segment** that the performing organization and/or the customer organization are engaged?

- ☐ Is there **anything else prescribed by national, state, or statutory authorities** that must be taken into consideration?

- ☐ Have those **persons who will no longer be required** to work on this project | milestone release been **timely & reasonably informed in advance that their services will not be required**, and have they been told **what they will be going onto next**?
    - o Has senior management and other project managers been informed that these persons will soon be available for other projects and for other activities?

- ☐ Have those **persons departing** from this project | milestone release **done sufficient handover to their successor** or to whoever remains with the project?

"Not just a 5-15 minute ad-hoc mental dump it in your lap run."

- [ ] Have the relevant project | milestone release's **Timesheet Job Codes** and **Cost Account Codes** been **disabled**? ... And, have the **associated records & reports been updated**?

- [ ] Have **all residual claims** against this project | milestone release been **resolved by when the Final Invoice is submitted** to the customer organization for payment?

    - Have **all Third Party timesheets & financial claims** (for services rendered / goods supplied) that are outstanding been **submitted for approval** by an appropriate representative of the performing organization (or in some cases by the customer organization)?

    - Are there **any outstanding financial obligations** between a Third Party, the performing organization, and/or the customer organization that will have to **be paid-out**? Similarly, are there **any unresolved financial claims** (including liquidated damages) that will have **to be settled**?

    - Are there **any outstanding performance bonuses** for a Third Party and the performing organization that will have to be judged & awarded to the relevant recipient(s)?

- [ ] Have all associated **financial accounts** been finalized, and closed out where necessary?

- [ ] Have all of the **contractual obligations** been fulfilled?

    - Are there any **existing contract agreements** between Third Parties, the performing organization, and/or the customer organization that will have to **be concluded or updated** accordingly?

- [ ] Are there any **outstanding issues** that need to be resolved, and have these been **noted in the Risks & Issues Register**? ... And, have these issues been notified to the appropriate project stakeholders for resolution?

- [ ] Has this project | milestone release been confirmed as having "**satisfactorily**" **complied with the Exit Criteria** (as detailed in the Project Charter)?

"Yep, project closure is one big 'HAS Questioning' checklist of things that have TO BE DONE and CROSSED OFF before THE END."

- ☐ Has a **Post Completion Review** been **held**, and

  has a **Project Closure Report** been **generated** for this project | milestone release?

  - o   Recording such things as; the "**Degree of Compliance**" with the **Agreed Deliverables**, summation of the **Change Requests & Defect Reports**, list of the **Objective Quality Evidence**, summation of the **Book Keeping** issues (i.e. outstanding financial claims, financial accounts reconciliation, residual resources, Intellectual Property Rights), and those **outstanding Risks & Issues** and **associated Problem Reports**.

- ☐ Has the **Project Closure Report** been **signed off** (and dated) by representatives from both the customer & performing organizations, so as **to mark the formal end** of this project | milestone release?

- ☐ Has a "**Lessons Learnt**" session been held with the entire Project Team?

Post Completion Review

The **purpose of the "Post Completion Review (PCR)" is to reach a consensus among the various primary stakeholders that this project | milestone release has been completed satisfactorily in accordance with the agreed Acceptance Criteria**.

This **Post Completion Review would be held after this** project's | milestone release's **deliverables have been handed over to the appropriate recipients and** the deliverables have been **signed off**. Possibly, these deliverables have been operating in a steady state for some period of "in-situ evaluation".

Once the Post Completion Review is completed then the Project Closure Report would be created to record the outcomes of the review, then be examined for correctness of the facts before being **signed off by representatives from both the customer & performing organizations**. … And, **thereby sealing the closure of this project | milestone release**.

3.7.5. QUESTION: what did we learn from this?

Lessons Learnt – Introspective

Now for some self-assessment & reflection; aka "Lessons Learnt", "Sprint Retrospection".

- ☐ What were the project's **significant achievements**?
- ☐ What were **the positives** from each phase of the project lifecycle?
- ☐ What were **the negatives** from each phase of the project lifecycle?
- ☐ What were the **notable mistakes made** during each phase of this project's lifecycle?
- ☐ What were **the project's major failings** and the project's major successes?
- ☐ How could things be approached differently next time?
- ☐ Who were the project's **standout performers, what did they do** that worked so well, and **why did it work** so well?
- ☐ Who **did not perform as well as expected**, and **why** was this **under performance** so?

"This ain't meant to be a witch-hunt, just looking for actual reasons."

- ☐ What could be **done better next time** to ensure that any future projects (like this one just completed) are subsequently going to be a success?

…

And, some opinionated "in hindsight" questions:

- ☐ **Did this project meet the performing organization's Selection Criteria for acceptable projects?**
- ☐ **Did undertaking this project make sense by providing value to the performing organization and more specifically its owners / shareholders?**

And, if the project was unsuccessful then was this due to:

- ☐ **Mismanagement of the traditional project determinants** of [Scope], [Time], [Cost], and [Quality]?

- ☐ **Restraints imposed** on the availability of [People] and/or [Resources]?

- ☐ **Lack of understanding** of what the customer had intended, and/or **lack of relevant knowledge**?

- ☐ **Unforeseen technical difficulties**?

All of these are very important questions that need to be answered, honestly, ... And, not *"playing political games"*, nor in search of *"a scapegoat on whom to place blame"*.

Which brings us to the next rule of "Adaptive & Proactive" SDLC Project Management.

RULE 32: A "smart person" learns from their own mistakes. ... A "wise person" learns from the mistakes of others.

Lessons Learnt – Extrospective

While that previous lessons learnt – introspection considered this project | milestone release from the performing organization (and more specifically from the Project Implementation Teams) perspective; **what would be the perspective of the representatives of the customer organization and those involved third parties**?

- ☐ Would they consider this to have been **a mutually beneficial working relationship**, that was built on the principle of **fairly distributing the burden and the <Risk>**, when taking into consideration what they contractually signed on for? ... OR,

- ☐ Would they consider this to be **a lopsided self-interest deal** that evidently advanced another party's agenda **at their expense**, that they were **taken unfair advantage of**, and that they took on a **disproportionate amount of the burden and the <Risk>** when compared to what they "actually obtained"? ... And, not what they were "promised".

Qualitative Project Analysis & Review

Qualitative Project Analysis & Review **is about understanding the interrelationships that defined the project's outcomes.** That is, **anecdotal evidence & accounts of experiences** that **descriptively compare the project's "expected outcomes" against the "actual outcomes" based on those involved parties thoughts & opinions.**

This analysis & review would entail Project Autopsy and Project Reminiscing:

- **Project Autopsy** ... is a review that must be *"conducted in the spirit of learning"*, and not as a search for someone to blame, shame, and flame for the project's failures. That is, those **lessons learnt activities**.

- **Project Reminiscing** ... is those **stories told** long after this project | milestone release has concluded, and contains the essence of the major lessons to be learnt from that past experience.

Qualitative analysis & review is useful for establishing Quality Assurance guidelines, where the description of how things were done (and should have been done) can change the thinking of those persons involved with the **Process Improvement loop**.

Project Autopsy

While the title of Project Autopsy does sound like it would be a forensic analysis of the project's outcomes, the **Project Autopsy review must be "conducted in the spirit of learning"**, and **not as a search for someone to blame, shame, and flame for the project's failures.** If there is the slightest perception that the Project Autopsy will turn into a "blame-game" "witch-hunt" then the attendees of such a review will be averse to participating, knowing that they could end up being the one who is "targeted". In such a negative environment, the review's participants may decide that it is in their own self-interest to "save face" and either; hide those problems that they knew | know about, be reluctant to reveal anything that is less than perfect or redress the facts, or redirect the focus and eventual blame onto someone else's domain.

DO NOT turn the Project Autopsy review meeting into a search for someone to blame for the project's failures.

DO NOT turn the review **into a point scoring exercise** of overstating some groups / individuals to the detriment of others.

DO NOT turn the review **into a systematic & analytical appraisal** of the [Scope], [Time], [Cost], [Quality], [People], and [Resources] metrics.

DO NOT place the **emphasis** on **Earned Value Performance Measures**.

The **Project Autopsy** should be **carried out after** the project / milestone release has concluded (or the project has been halted because it has been considered as a failure). … And, **the associated emotions have subsided to a level where an open & honest conversation can be held** to discuss:

1.. **What went well** on the project?

- ☐ What processes & procedures contributed to a successful outcome?
- ☐ What activities worked to the benefit of each participant?

2.. **What didn't go so well** on the project?

- ☐ What processes & procedures hindered success or contributed to the project's unsuccessful outcomes?
- ☐ What activities & attitudes should not be encouraged in the future?

3.. **What** in each participant's opinion **were the mistakes made** with the project?

- ☐ What should be done and what changes should be made to ensure that such mistakes do not occur in the future?

4.. Were **there any** participants **who could have been more effectively utilized** in a different way? Alternatively, **were there participants who were utilized in an optimal manner**?

> 5.. **What** resultant **good ideas** should henceforth **be incorporated into** the performing organization's suite of **best practices**?
>
> 6.. **What** was the feedback received from the primary stakeholders?
>
> 7.. **What** was the feedback from the end-user customer?
>
> Notice how the previously listed points for the **Project Autopsy** were all **based around** "descriptive exploration" instead of an analytical analysis.
>
> **NOTE:** It is **highly beneficial to have** this Project Autopsy **review at the conclusion of each major milestone release**, as well as at the conclusion of the entire project. With an agile implementation, this self-assessment would involve the Project Implementation Team getting together at the end of each sprint for a round-table discussion. For a waterfall implementation, this discussion could be at the conclusion of each Phase Completion (Gate Review).
>
> Hence, the Project Autopsy needs to provide benefit to each participant so as to help them ensure that their next project (or the next phase of the current project) is a successful endeavour.

3.7.6. QUESTION: can everyone say good bye as friends?

To end the relationship amicably then those departing project participants should be:

- ☐ **Privately** & **publically thanked** for their efforts and their contributions.

- ☐ **Rewards** & **recognitions** distributed to those participants whose project performances have earned such gratitude (and potential financial bonuses).

- ☐ Ensure that **final payments are made within a reasonable amount of time** after each participant's departure (i.e. immediately when possible, but not several weeks or months later). Noting that, **the longer that it takes to pay them, then the greater will be the disintegration of** (and reduced potential for future) **working relationships**.

- ☐ Try to **resolve any outstanding grievances or underlying conflicts** with any of the departing project participants (and with those participants that remain).

"DON'T carelessly burn down relationship bridges, as you never know ..."

3.7.7. Closing Phase Completion

At this stage, the Closing Phase for this particular project | milestone release is completed.

Some Closing Things To Watch Out For

There are a few things the project's management needs to be on the lookout for:

- ☐ **Knowing when to call it quits** … there needs to be a **clearly defined set of Exit Criteria** that should have been agreed to (and signed off during the Initiating and Planning Phases) so that the project does not linger on from the stages of producing the deliverables, to Warranty Support of those deliverables, and then ongoing maintenance of those deliverables.

- ☐ **Know when tweaking & defect fixes transcends what was agreed** … that is, scope creep disguised as defect repairs, when project team members go beyond the bounds of the project's scope and give more than what was formally agreed to.

- ☐ **Know when & when not to give away work for "free"** … that is, need to resolve immediately whether questionable work is or is not within the project's [Scope].

- ☐ **DO NOT underestimate the impact of insufficient end-user training** … without appropriate training, the end-users can misunderstand, misinterpret, and not know what the deliverables are & are not capable of. Consequently, they could decree the deliverable as "broken" when in fact they are not using the deliverables as these were specified, designed, developed, and agreed as acceptable.

Well, that is the Closing part of the journey completed.

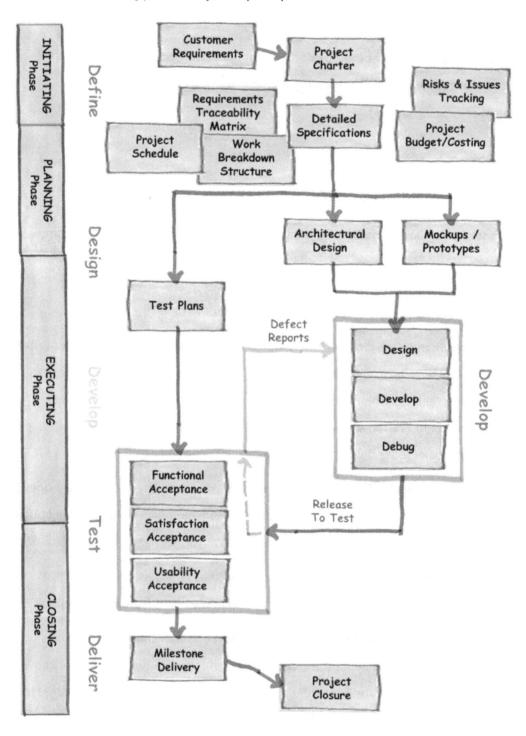

"IT IS THE COMPLETE PROJECT TEAM, AND THE INDIVIDUAL MEMBERS OF THAT PROJECT TEAM, WHICH MAKES OR BREAKS THE PROJECT, AND CONSEQUENTLY, WHAT MAKES OR BRAKES YOUR CAREER AS A PROJECT MANAGER."

...

"SO, DON'T TAKE THE LITTLE PEOPLE FOR GRANTED."

4. Your PM Techniques

4.1. Introduction

Now, that the foundation project management theory has been covered, let's push that all to the side, and start your project management from a *"clean sheet"*. … And, tailor a project management methodology specific to your leadership style and to your personality traits; as well as, compensating for your employer's (the performing organization's) processes & procedures, and taking into consideration your customer's expectations of, *"what a professional should do"*.

Remembering that, **"SUCCESS IS OPINIONATED"**, see [Section 1.2]. … And, for a project to be deemed a success is very much dependent on the primary stakeholders' judgment of the project as having been delivered; within the agreed [Time] frame, within the agreed [Cost] budget, with the agreed [Scope] of deliverables, with the agreed level of [Quality], using the agreed assignment of [People], and using the agreed allotment of [Resources]. And that is, irrespective of what project management methodology was | is utilized.

So, **DO NOT EXPECT the project to be deemed a SUCCESS, just because** you happened to have **followed some prescribed project management methodology to the nth degree**; as though, the written text on this methodology was gospel and not as a suggested guide.

Hence, I would recommend that, you start by taking some recognized project management methodology, which you feel comfortable with, and use this as the foundation on which to build your own project management techniques. … And then, **continually adapting** *"your techniques"* **to deal with the current real-world situation & prevailing circumstances**, by emphasizing those aspects which add value while deemphasizing other aspects which evidently aren't contributing to achieving your current project's successful outcome.

"As well as, blending in field-tested industrial best-practices, and incorporating those lessons learnt by yourself (and by others)."

4.2. A People First, Foundation

4.2.1. The Scenario

Well, the interview was a success, and today you start your new job at this smallish company that wants to build a bespoke product. ... Thinking back, the interviewing manager spent a noticeable amount of time talking about how this *"greenfield"* development project is going to produce a *"market leading" "innovative" "cutting edge"* product. ... And, when offering you the position, the hiring manager made a closing remark that, *"this project should be a walk in the park for you, given your extensive technical experience"*.

4.2.2. Building A Good First Impression

Your project (and all future projects, at that performing organization) starts the moment that you walk through the front door of the office, and are greeted at reception. Actually, it all started the moment that you were being interviewed for the role, and continued onto when you subsequently accepted the job offer.

"*First impressions count*", and these impressions are the foundations on which successful (or failed) working relationships are built. ... And, this first impression is not limited to how you look, how you speak, and how you act; but rather, by whom this is perceived. ... And, these perceivers are not limited to your bosses and the customer representatives, but also include; the receptionist, the office administrator, IT admins & support, the stores person, etc., i.e. those persons who form the underlying infrastructure on which your project operates (and hence your project management career resides). As these people's wilful cooperation at some point in the near future could be the difference between your project (and your career) continuing in the right direction or taking a nose dive.

Which brings us to the next rule of "Adaptive & Proactive" SDLC Project Management.

RULE 33: First impressions, like a first date, establish the foundations for either a good or bad future relationship.

RULE 33^A: A bad first impression take some amount of effort to recover from, and requires consistency of desirable behaviour (from the wrong-doer) to re-establish the potential for a good future relationship.

Building A Good First Impression, of PM you

Yes, building that first good impression.

(1) **Dress appropriately for the occasion.** ... unless specified otherwise, **dress ready to achieve the business's objectives** that are the purpose of your coming to work at this particular organization (and location). Though, **do conform to prescribed norms**, as how a person dresses, speaks, behaves, and their quirkiness traits **can influence the judgment of their (potential) performance and their perceived worth**.

(2) **Introduce yourself.** ... especially when you or they are new to the organization / project, then introduce yourself, **calmly & politely, confidently but not arrogantly,** as *"the Project Manager"*.

With a new Project Implementation Team (who have not worked with you previously), introduce yourself, give a <u>brief</u> background to your work history (try to make this relevant to "their project" that you will be managing). That is, **strive to build up their initial confidence that you know what you are doing**, and also make it clear that you intend to quickly come up to speed with the goings on with the project.

(3) **Confident, but not arrogant.** ... the Project Manager needs to **present a demeanour of confidence in one's own abilities.** However, **NOT** to be perceived as **arrogantly self-assured and overconfident**, as this type of demeanour can result in the manager being prejudged as someone who is prone to making rash decisions, as someone who will commit the Project Implementation Team to things that should never be agreed to, and as someone who will ignore the input | opinions | advice | recommendations from those *"little people"* who are subordinate to that manager's titled-position.

DO NOT expect to receive the respect of the Project Implementation Team just because you have been officially given the title of "Project Manager".

To obtain the respect of The Team will require you, as the project's manager, to earn and be worthy of such respect.

DO NOT consider the project hierarchy as a pyramid where (you as) the Project Manager **sit at the top issuing decrees** to your minions down below.

Rather, consider the Project Manager to Project Implementation Team arrangement as a **flat hierarchy** (possibly involving some form of mentoring) where the project's manager takes care of the non-technical matters (such as responding to stakeholder needs & wants, and dealing with inter-stakeholder politics), while **The Team takes care of the technical aspects.**

Which brings us to the next rule of "Adaptive & Proactive" SDLC Project Management.

RULE 34: True respect is earned by deeds, not bestowed by title.

(4) **Be a good listener, not a know-it-all** ... No one wants to work for someone who thinks they know everything and who doesn't listen to anyone (aka *"I'm the smartest person in this room"*). Because, after a while of having their input & past experiences continually ignored, many a project team member will start to question why they

should *"give a damn"* about the project, let alone spend the time to communicate with the Project Manager. ... And, once this **communications breakdown** occurs, then **risks & issues confronting the project can go unmentioned**, which could result in some *"arrogant know-it-all"* project manager being splattered by the *"proverbial bus"*; because, no one was willing & prepared to tell them the truth of the true state of things, nor bothered to provide meaningful input | opinions | recommendations.

(5) **Show interest in each person.** ... in small groups or one-on-one find out;

- o **What are each of The Team member's names?**
  Even if this is only their first name (not surname), and remember this by writing it down on a mud-map sketch of the office and where they usually sit.

- o **What are their roles** within the project and within the organization?

- o **What part** of the product / system / application architecture **are they working on and are responsible for?**

Partake in the **occasional** coffee room *"small talk"*, and get to know each one of them; their background, their interests, and pick up on their home life circumstances.

"Do they have children, and what did they do on the weekend?"

Hence, **show some interest in them as individual human beings**.

"At the end of the first day, and each day thereafter, when you get home from work and chill-out, can you mentally in your mind, walk around the work place and name each person, at least their first-name, and list something unique about each one of them?"

(6) **Know the stuff that they would expect the Project Manager to know.**

- o **Know the basic principles & concepts behind project management, and be familiar with the implementation methodology being used by the Project Implementation Team.** This does not necessarily mean that you have to be certified in this or that, but it does mean that you understand how & why the project should be undertaken in specific ways, and the "relevant" project management practices that should be considered for use.

- o **Know what is happening with the Project Implementation Team, around the project, and within the performing organization**; as The Team won't be too appreciative if any nasty surprises befall the project or them. Therefore, keep a watchful eye on the risks & issues confronting the project. This will mean that you as the project's manager will have to **learn to "*network*"** with other parties within the performing organization (and external to the organization).

Much can be achieved by **building up friendly working relationships with other people within the organization**; especially working relationships with your peer project managers, and with various people in different positions within the organization (as well as **within the customer organization** and **with 3rd Party** contractors / vendors / suppliers).

(7) Quickly come up to speed.

- o **Know the basic details of the project that you are to manage.** This means knowing; the overview of the project, the assumptions made about the project, the scope boundaries of the project, the constraints imposed on the project, the priorities associated with the project, the general requirements of the project (or at least where to quickly find these and extract the relevant information), what are the project's deliverables, and the relevant dates for the project's major milestones.

- o **Know the basic architectural details of the project's product / system / application that is being implemented.** This does not mean that you have to become a subject matter expert, but it does mean that, you are able to draw a relatively simple block-diagram explaining how it all fits together. … And, more specifically, you are able to point to what sections certain project team members (or sub-teams) are working on and are responsible for.

- Know the basic details of the technologies & processes underlying the project's product / system / application that is being delivered. This does not mean that you have to become a subject matter expert, but you will need to at least have an understanding of what the Project Team members (and the project stakeholders) are talking about.

Which brings us to the next rule of "Adaptive & Proactive" SDLC Project Management.

RULE 35: **You cannot hope to manage that which you do not know or reasonably understand what is going on.**

(8) **Be enthusiastic about the project**, about the Project Implementation Team, and about the project being a future success.   *... "Not a pessimistic grim-reaper."*

4.2.3.   Recruiting The Project Team

As the Project Manager, **it is very important** that you **strive to build a cooperative & collaborative Project Implementation Team**; as it is not the Project Manager, but rather "The Team" who transform the project from mental concepts into physical deliverables.

For each project undertaken, one of the following team construct scenarios will transpire:

(a) The Project Manager is new to the performing organization, and the members of the Project Implementation Team are also new (or have yet to be hired).

(b) The Project Manager is new to the performing organization, but some (or all) of the members of the Project Implementation Team are already established as a team at the organization. An extension to this scenario is when the new project manager is replacing a previous project manager who was either *"promoted internally"* or who *"decided to leave"*. ... Thus, leaving the new project manager with little to no handover of project related information.

(c) The Project Manager is established at the performing organization, and the members of the Project Implementation Team are new to the organization (or have yet to be hired). … Or, The Team members are "old hands" at the organization, but they have never really worked together before as an SDLC Project Implementation Team.

(d) The Project Manager is established at the performing organization, and the members of the Project Implementation Team are also well established at the organization. An extension to this scenario is that of a pre-established Project Implementation Team with its regular project manager.

For this textbook's scenario, it will be a combination of (a) and (b), where the Project Manager is new, some members of the Project Implementation Team will also be new, and the other members are old hands that are yet to join The Team in significant numbers.

> **NOTE:** There is a major caveat when it comes to the recruitment of personnel for the Project Implementation Team; as your performing organization most probably has defined processes & procedures related specifically to recruitment, and this could be undertaken by specific staff (possibly involving senior management) who do the entire process without consultation with the project's manager. Therefore, as the Project Manager, you may not be involved with any part of recruiting the Project Implementation Team. Thus, you may simply have to make do with the "resources" that have been assigned to your project.
>
> Hence, what is contained in this section could be constrained by pre-existing business guidelines & practices. However, what follows should provide you with food-for-thought, when it comes time for you to undertake the recruitment of "people". … And, especially when you are planning & budgeting for this aspect of your project.
>
> Though, this book will NOT be delving deeply into various recruitment methodologies & principles, as this would be a book inside a book.

Know Which Positions Need To Be Filled

If you were building your fantasy football league team, then you know that there are specific positions on The Team that have to be filled, else you won't have a complete team with which to compete.

Secondly, The Team only has a certain sized budget with which to "buy" its players and the associated coaching staff. To exceed The Team's allotted "salary-cap" could mean that The Club (aka, the performing organization) is driven into financial ruin; let alone the possibility that the Competition's Governing Body (aka, the customer organization) could fine The Club for breaching the assigned salary-cap, and/or they could exclude The Club from this year's competition, and/or they could ban The Club from competing in next year's competition (aka, no more contracted work for this performing organization).

Thus, **what positions on The Team do you need to field to be competitive**, while **remaining within the allocated budget?**

…

Unfortunately, there is seldom the availability of personnel within the performing organization to stack The Team with "star player" implementers. Hence, **financially & availability wise, the Project Implementation Team has to be composed of a mixture of senior, middle, and junior implementers who will have to fulfil a hybridization of various roles on The Team.** As the performing organization cannot (literally) afford to have one person filling only one role; rather, **each selected team member has to provide (partial or complete) coverage of several roles.** Hence, **the purpose of the project's manager is to build a team that doesn't have any field position gaps.** … While, remaining within the allocated budget.

Therefore, based on the skill levels & past experiences of the currently available project team members (including those recruited to join the organization), **compose a list of the necessary skills & desired experiences that any potential candidates would have to demonstrate proficiently, so as to be considered for a position on The Team.**

Sourcing Those Potential Position Candidates

Those potential position candidates can be recruited via various means, such as; *"networking"* with former work colleagues, directly advertising the role yourself, or indirectly through the services of a recruiting agent(s).

NETWORK RECRUITMENT ... involves recruiting via **personal contacts that you** (and other team members) **have worked with previously at other organizations.**

If using Network Recruitment then you have to go into it with the same perspective as you would "*if you were to be dropped behind enemy lines, then who would you want to be on your commando team, covering your back*".

Because, **by recommending a personal contact to your current employer then you are in affect going guarantor for that person's performance.** Hence, if they underperform or screw-up then this will reflect badly on you; however, if they do perform well then this will effectively bring positive credit to yourself.

Therefore, **when personally recommending someone for a position at your employer's organization, only do this because (in all honesty) you believe that your nominee is one of the best candidates that could apply for the role.** Conversely, do not recommend a personal contact, if the primary reason for doing so is as a favour to them (to *"get them a job"*). ... And, don't sell your personal contact a job that you know is not right for them.

There are **distinct advantages with engaging previous work colleagues**, especially if you can bring on-board a group of them:

- **The self-assuredness of knowing in advance that the personal contact is most probably going to perform positively as they did in the past**; whereas, a completely unknown person could turn out to be a performance disaster.
- **Some portion of The Team relationship dynamics** of Forming - Storming - Norming – Performing, see [Section 4.3.1], **will possibly carry over from the previous working relationship** and thereby **serve as a catalyst for team building** within the new team.

- [ ] By placing previous work colleagues in "strategic" team positions then **tried & tested work practices & work ethic expectations can carry over from the previous working relationship**. This can also reduce the combined efforts involved with "training" new team members as to the SDLC methodology that is to be used on the project.

- [ ] The **significant saving in [Time] & [Cost] that is involved with recruitment** when compared to direct self-advertisement and engaging recruiting agents.

"I was one of the first hires for a start-up company, and on my personal recommendations to the owners, they hired my former boss as the group manager, the systems architect, and several of the engineering staff (including juniors and seniors). These personal contacts once hired, in turn recommended other people, and so on, until in a relatively short time a well-rounded development group was built-up, with unknown persons fitting into already gelling project teams. Consequently, productivity rocketed sky-high to greatly exceed that which could have ever been achieved by composing the group entirely from unknown persons.

It was as though, specific parts of the development group were functioning telepathically, with the right-hand knowing in advance what the left-hand was doing. And, with many of the people knowing the processes & procedures to be followed, then those who didn't know the routine were shown how to by their peers."

However, there are **dangers with impregnating the Project Implementation Team(s) with former work colleagues**, as those persons could be perceived as having **"*too familiar a relationship*"** with the Project Manager, which could be perceived as disadvantaging those team members who didn't have that pre-existing working relationship. ... And, if The Team consists of noticeable proportions of both previous work colleagues and old-hands from the current organization, then **underlying tensions & conflicts can simmer away due to a "*them and us*" mentality**. ... *"which necessitates the building of a bridge between the two factions."*

SELF ADVERTISEMENT RECRUITMENT ... involves **the performing organization, by themselves, undertaking public recruitment for The Team position(s).**

To do this type of recruitment successfully requires:

(1) **Clearly defining in the advertisement what the performing organization is looking for and the range of experience that is desired.** ... **DO NOT cram the advertisement with "*a mile long*" wish-list** of skills & requirements; where realistically, no individual candidate is going to have most of those skills. As this bloated advertisement could scare away some of those best-fit potential candidates, and other best-fit candidates could be drowned out by the flood of half-fit candidates who possess some of those mass listed skills. Thus, **keep the "must haves" list to those absolutely essential skills and those highly desirable experiences.**

(2) **The advertisement should not be a marketing exercise** for the performing organization. For example; if existing staff read the advertisement and joke about the contents with comments like, "*yah, right, is that us*" or "*what a load of rubbish*", then you can expect some of those best-fit potential candidates to also see through this marketing dribble and then decide to skip applying for this role.

(3) **Once the advertisement is placed** onto a job-seekers website then **expect to be inundated with applications from variously skilled candidates.** Depending on the current economic conditions of the involved SDLC industry, there could be as many as 100 (or more) applicant resumes & cover-letters continually streaming into the recruiting manager's email inbox. Every one of these applications should be skimmed through, so as to construct a shortlist of candidates that should be evaluated further, and so on, until a shorter list is produced of candidates worthy of interviewing.

Remember that, [Time][Costs] money, and each candidate's resume & cover-letter is going to take some time to skim through, so as to "sort the wheat from the chaff". *For middle & senior positions this could take at a minimum 5 minutes per candidate document to be reviewed.* ... *Multiply 5 minutes review time by (say) 100 candidates* *equals 500 minutes which equates to 8+ hours, with an hourly rate of (say) $125 "bums on seat" cost of the reviewing manager's time equals $1000 per day to skim through these resumes. Then the next day is spent rereading through the shortlisted candidates' resumes with a bit more diligence, so as to produce a shorter list of potential candidates, and that is another $1000 per day. Then a senior manager (or two) will want to look at the shortlisted candidates' resumes, then discuss those candidates, so add a few more $1000s worth of [Time] to the tally. Then those shorter shortlisted candidates would be phone interviewed to get the list down to a few candidates that are worth face-to-face interviewing. ... So, that is somewhere between $5K to $10K in [Time][Cost], possibly more depending on the organization's recruiting procedures, to just get to the interview invitation stage (and not to the actual interviewing).*

"At this point, Network Recruitment is sounding like a real bargain."

A resume is like a second-hand car advertisement; what sounds great on paper could turn out to be a "lemon" when put to the test.

Hence, the recruiting manager needs to, in advance of placing the job advertisement, **create a scorecard of the "must haves" and the "desirable experiences", so that each appropriately sounding candidate's resume (and cover-letter) can be quantifiably evaluated.** Thereby, hopefully picking the best-fit candidates to be taken to the next round in the recruitment process.

(4) **The phone interview's purpose is to** confirm that the potential **candidate's spoken words does correspond with their submitted resume**, to establish that the candidate can **put together understandable sentences**, to establish that the candidate **understands the technical & industrial field that the project is based**, and to confirm that the candidate **has the necessary work rights** (i.e. citizenship, permanent residency, accreditations, clearances) that are required as a prerequisite for the position that is being recruited for. ... And, thereby **determining those candidates that are worth being invited to a face-to-face interview**.

> **A PERSONAL EXPERIENCE ...** *The "puppy dog fetch" test of a graduate candidate.*
>
> The reality is, most recent graduate candidates don't have the relevant work experience nor the practical skills to be judged for an SDLC implementer's role. They could be judged based on their academic record, but the appropriateness of this is very much dependent on the structure of the course(s) that they did; as their course could have been mostly theory with minimal "hands-on" practical assignment work, or a 50-50 balance of theory versus practical, or it could have been predominately practical.
>
> Fortunately for me, the local universities provide all of their graduates with the same generic resume template. Hence, for those graduate candidates who make it through the telephone interview round, when I call the candidate to invite them in for a face-to-face interview, I also request that they make specific changes to their resume, *"to make it more specific to engineering roles"*. Additionally, I ask them to put together a two page summation of their final year's project, where I list the sections and what to include. Similarly for a graphical user interface developer, I would ask them to prepare a few page portfolio of their work, where I suggest that they include a website design sketch, a wireframe mock-up, and the final graphics page(s). ... And, I emphasize the importance that they, *"email me the updated resume and attachment as soon as possible, so that I can present it to my bosses"*; then, I come to an agreement with the candidate as to when they will be able to provide the requested deliverables.
>
> In factuality, what I am interested in is; (1) how soon does the graduate candidate email me back their updated resume, (2) with the attached document, and (3) how well do these deliverables conform with my instructions & specifications.
>
> IMHO, the important ingredients of a graduate implementer is their enthusiasm, their willingness to learn, and their ability to follow instructions. Hence, this *"puppy dog fetch"* test allows them to demonstrate exactly this, and it also shows how much they really do want the implementer's role that is on offer.
>
> P.S. every recent graduate who has excelled at this test, when hired, has performed way beyond the expectations of senior management.

AGENT RECRUITMENT ... involves **engaging the services of a specialist company / person to take care of finding those potential position candidates**; after which, the performing organization does the final rounds of candidate interviewing.

The first thing to realize about engaging **recruitment agents** is that, **often they don't have the technical background to be able to spot those candidates who truly do have the right skillset and relevant work experience.** Rather, the agent is reliant on keyword searching through collections of resumes, targeting the "most ticks" matches. Hence, for this type of agent to be effective requires that **the performing organization provides the agent with a precise short list of "must have" technical skills & desired experiences, and then spend some time explaining what these mean.** ... Else, the recruiting process will end up being a "lucky dip" search of random chance selections; that, they put forward best-fit candidates.

Secondly, realize that **many an agent is paid based on commission**, and hence **their aim is to place as many hired candidates in as short a time as possible**. Thus, some agents who have not been in the local recruiting industry "for decades", **won't necessarily consider the compatibilities of the personalities & attitudes of both the potential candidate and the performing organization**, which would eventually have to work together.

Though, **using a recruiting agent does have the distinct advantage of not wasting the organization's management [Time] and effort** sifting through those piles of half-fit candidate resumes.

> Which brings us to the next rule of "Adaptive & Proactive" SDLC Project Management.

RULE 36: Recruitment is not a lucky dip; rather, it is a defined & logical search for the best-fit candidates from both a technical skills and past experiences perspective, as well as compatibility of personalities and attitudes.

Interviewing Those Chosen Position Candidates

Firstly, **interviewing is not an ad-hoc make-it-up as-you-go get together**; rather, it is serious business with specific objectives and desired outcomes.

The purpose of interviewing is to:

(1) **Give the candidate the opportunity to sell their capabilities & experiences** to the performing organization, via bi-directional communications with the interviewers.

(2) **Determine the best-fit candidate for the role** that has to be filled (right now).

(3) **Determine the candidate who will provide benefit & add value to the performing organization** (and to the project), **from both technical & interpersonal perspectives**.

(4) **Selling to the "desired candidate" the idea that joining the performing organization** (and the project) **is advantageous to themself**.

What the interview should never be, is:

- ☒ **NOT a power-trip for the interviewing manager to** belittle and/or intimidate the candidate (and those other interview participants) to let them know who is boss.

- ☒ **NOT self-justification** for the interviewing manager's position.

- ☒ **NOT a quest to find *"yes-men" "worker-drones"*** who, when hired, won't be potential threats to the interviewing manager's position. ... And, it is definitely

- ☒ **NOT making up the numbers** as cover for hiring a preferred preselected candidate.

> Which brings us to an addendum to "Adaptive & Proactive" SDLC Project Management.
>
> **RULE 36[A]:** The interviewer's goal is to find the candidate with the best fit capabilities & experiences, who would add value & provide benefit to the organization. ... And, to the current project.

When interviewing, there are a few things to watch out for with respect to your technique:

- ☒ **DO NOT do all of the talking**, as this makes the interview a complete waste of time; because, when is the candidate going to be given the opportunity to *"hero or zero"* themself as the best-fit for the role that is being recruited for.

- ☒ **DO NOT ask a question unless it has a definite purpose and a desired outcome to** reveal some information by which **to further evaluate the candidate**. ... And, this is not necessarily to find the candidate with the best-fit qualifications who also matches all of the predefined expectations; but rather, to **find the candidate who will provide benefit and add value** to your employer (the performing organization).

- ☒ **DO NOT waste time asking template interview questions** that the candidate has probably been asked umpteen times before; because, the candidate has by now formulated standardized (rehearsed) answers to such questions, and hence the answers may not be a true reflection of the candidate. ... *"Rather, their answers are reworded responses that they found via an internet search."*

    For example; "where do you see yourself in 5 years' time", "what are your personality strengths and weaknesses", "what do you see as your most significant achievement".

    Noting that, the HR interviewer and senior manager(s) will probably ask these very same template interview questions. But what they can't ask proficiently, are those SDLC related technical & practical questions, which are within the realm of your responsibility as the project's manager; so, concentrate on these questions instead.

- ☒ **DO NOT ask a question** (or word a question) **that could be perceived as prejudice**, in any form, whether this be in jest or not. That is, sexist, racist, religious, bigoted ... etc.

- ☒ **DO NOT tailor questions to wordsmith answers** (i.e. competency based interview questioning); because, this is an SDLC project for which the candidate is being interviewed. **What is important, is, ... determining the candidate's implementation capabilities, past experiences, and their technical & interpersonal value to The Team.**

COFFEE BREAK DISCUSSION ... *Competency Based Interviewing, is wrong for SDLC.*

Competency Based Interviewing – where the purpose of this interview style is to predict future performance based on past performances in similar situations. ... And, if the interview candidate has been pre-groomed properly, then all the candidate has to do is pick (what sounds like) a real-life example case and answer in the format of:

1.. Summarize the [S]ituation; including the where, the when, the whom was involved, and the potential consequence of inaction.
2.. Summarize the [T]ask / [O]bjective that was required of you.
3.. Summarize the [A]ction that YOU took, focusing on YOUR CONTRIBUTION and not that of the group; i.e. "I DID" and not "we did".
4.. Summarize the [R]esultant (positive) outcome of your actions, and what did you learn from the experience.

IMHO, competency based interviewing is inappropriate when it comes to engineers / coders / testers / implementers because these people are not wordsmiths; but rather, they are "doers" whose purpose is to get things done. ... And, where "WE DID" is more important to the Project Implementation Team getting things done, than the "I did".

"I was once going for this technical role and the recruiter prepared me with competency based interviewing questions, where the recruiter reorganized each of my answers into the desired S.T.A.R. format. The first time, the agent's reword of my answer was impressive; but, by the third time, it felt like a con-man's scam, as here was this agent who had absolutely zero technical skills & experience in that field, yet his answers were perfect enough that it would have been hard to pass him over for the role."

And, therein lies the problem with **competency based interviewing**, it **is not targeted at finding the "best doer" for the job**; rather, it gets the best answer from a "wordsmith".

"Oh, and these type of questions can be answered by relating what someone else did and then claiming those actions as your own."

COFFEE BREAK DISCUSSION ... *Aptitude Testing is racist, sexist, and wrong for SDLC.*

Aptitude Testing – where the interview candidate is given a series of numerical, verbal, and diagrammatic (spatial) reasoning questions. Sometimes this is combined with a sequence of written instructions where the process flow goes around & round & round (for half an hour per question) to eventually end up with some numerical answer.

IMHO, aptitude based testing is inappropriate for engineers / coders / testers / implementers because; firstly it is racist & anti-native language speaker prejudice, secondly it is sexist, and thirdly it is not related to the technical skills required of the person to get the implementation job done.

Racist & anti-native speaker prejudice – today's engineering community is ethnically diverse, and thus many of the interview candidates will potentially not have the interviewing language as their first-language; hence, such questions as *"what is the difference between affect and effect"*, *"lock is to key as pen is to"*, and "negative-not-case sentence constructs", does put these ethnic candidates at a linguistic disadvantage.

Sexist – males typically have better spatial abilities then women; hence, taking a two-dimensional image on the paper then converting it to a three-dimensional object in their mind then rotating this object in headspace then converting it back to two-dimensions can put female candidates at a spatial disadvantage. Similarly, males typically perform better at mathematical tasks and numerical sequences than do females.

In all honesty, what do these numerical, verbal, and diagrammatic (spatial) reasoning questions have to do with being a good engineer / coder / tester / implementer?

Why eliminate these potential candidates simply due to a series of questions that are completely unrelated to getting SDLC things done? And, therein lies the problem with **aptitude testing**, it **is not targeted at determining the "best doer" for the job**, let alone contributing to building the "best doer team".

"IMHO, aptitude testing is an academic theorist's con-job to prove how smart they are, and not how appropriate a DOER is the candidate."

A Recipe For An SDLC Interview ... *"show us what you can do"*

Let the HR interviewer and/or senior management open the interview process with their behavioural interview questions, and the introduction to the organization. After which it is your turn, as the technical interviewer, to do some proper SDLC interviewing.

1.. **Prepare the location.** ... In a quiet place with space for 4 to 5 participants, such as the meeting room (near as possible to the Project Implementation Team's area). Preferably, setup an open seating arrangement where all of the participants are not "shielded" behind a table, and that there is room for participants to stand-up and utilize a whiteboard. ... *"Make sure the whiteboard marker pens do work."*

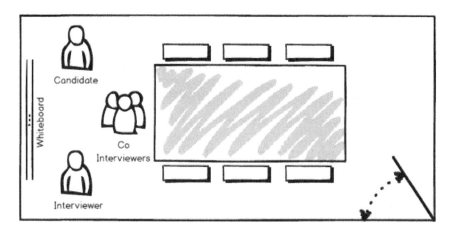

2.. **One-on-one opening.** ... With just the candidate and you (the interviewing manager) sit down in the interviewing place (without the other participants present) and have a short friendly social chat *"about the weather"*, *"their journey to the interview"*, and most importantly, *"thank you for coming in"* to be interviewed.

- ☐ Do a quick introduction to the organization, though ask if this has already been covered during the earlier HR Department / senior management interview.

- ☐ Ask the candidate if they have checked-out your company's website, and ask the candidate to, *"briefly describe what our company does"*.

- ☐ Ask the candidate to, *"briefly summarize"* their *"career thus far"*.

The purpose of these opening questions is to:

(1) **Establish** the interview location, provide time to get use to each other's spoken words (accents), and to provide **a baseline** as to their defensive body language. Knowing that, the candidate will be nervous and naturally defensive; so watch for their hands clenched, arms crossed, or their hands on their knees as though they are ready to sprint for the door. **Your objective as the interviewing manager is to make the candidate feel at ease; thereby, getting them to drop their defences**, and thus **exposing the implementer** that resides within them.

(2) **Determine their interest in the organization** by the candidate demonstrating that they have at least spent some time researching your company, and to see if they picked-up anything from the earlier HR / senior management interview.

(3) To check that the candidate's **resume does match with what they are saying**.

3.. **Zoning their skills**. ... For an SDLC project, start the practical part of the interview by quickly drawing on the whiteboard a simplified diagram of team positions. Then ask the candidate to mark on the board, *"what do you consider to be your 5 star areas"*.

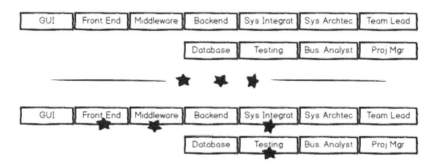

The purpose of this question is to confirm whether the candidate's primary areas of experience & skills (according to themself) corresponds with the role that is to be filled. For example, if this was a front-end developer's role, but they place their stars against the back-end & database, then they are probably not the best-fit candidate for this particular role; however, they may be better suited for that back-end role that you also have to fill, so be prepared to adapt the interview's focus accordingly.

If the candidate marks almost every area as their top skills, then potentially they could have an overinflated valuation of their own abilities; however, they may turn out to have good coverage of several roles. … And thus, be of great future benefit to the project, and to The Team (as well as to the performing organization).

Another purpose of this question is to make it evident that this is not going to be the typical question & answer interview.   … *"So better adapt to the situation."*

4.. **Meet some members of The Team.** … At this point, bring into the interview a couple | few of the senior implementers from The Team (or associated project teams) to participate as was pre-arranged. Preferably have a mix of ethnicities (as is representative of The Team's / organization's makeup), and briefly have each co-interviewer introduce themselves to the candidate.

<u>The purpose</u> of this phase of the interview is to see how the candidate interacts sociably with their potential fellow team members.

5.. **Opening design question.** … Without giving a verbal foundation to the next scenario, as the interviewing manager, clean the skills-star drawing off of the whiteboard (having given the co-interviewers sufficient time to see what the candidate considers to be their primary skills), and then launch into,

> *"I have this great idea, I don't think it has ever been done before."* … And, draw a starting sketch for a system in your company's field of endeavour.

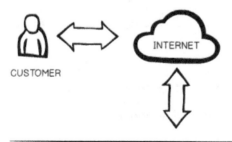

???

"*It's an online website where people can buy & sell anything for either a fixed price or as a bidding auction that is publicly viewable.*" ... Thus, having verbally described a very well-known global ecommerce auction site.

NOTE: For a hardware based role, this opening scenario could be a mobile phone or television device. For a systems engineer, a telephone or TV network. That is, pick a scenario that does not require specific technical knowledge of the subject matter; rather, something that can be extrapolated "on-the-fly" based on a general principle of, *"so how would I break that down"*.

Having completed your part of the drawing, hand the candidate the whiteboard marker and ask them to fill-in their design of the proposed system.

The outcome of this request is that, the candidate will either;

- ☒ remain fixed in their seat and try to verbally "*ummm & ahhhh*" an answer, or

- ☒ they will get up and chicken-scratch a couple of boxes & words, but not produce any interconnected thing (that would commit them to a possible wrong answer), nor will they be able to verbalize the formation of a solution. ... Or,

- ☑ they will draw a high-level system with summation blocks, and quickly build-up an overview map to their solution, or

- ☑ they will dive down into detailed boxes of sub-components that makeup portions of the system, but they won't necessarily have enough time to draw the entire solution.

The purpose of this question, other than to get the candidate to (literally) "*think on their feet*", is to provide the candidate with the opportunity to demonstrate that they can come up with an implementer's solution. Additionally, this question determines whether the candidate is a high-level or detailed thinker; as certain roles within The Team, such as business analyst and tester, will tend towards different styles of thinking. ... *"Don't build The Team with only one type of thinker."*

6.. **Expanding on their solution.** ... Invite the co-interviewers to ask questions about the candidate's whiteboard design. ... As the interviewing manager, watch for how the candidate interacts with their potential teammates' questioning.

7.. **Changing their design.** ... At this point, change the scenario by saying that,

> *"The website's operator also wants to be able to sell their own products and to be able to ship out customer purchased inventory, so how will this requirement change the design".*

The purpose of this question is to determine how (and if) the candidate adapts their solution to requested changes, and how do they react to the associated questions & suggestions from their potential teammates.

NOTE: The important thing is not the technical correctness of the candidate's solution; but rather, how they drew it, were they able to explain it, how did they interact with their potential teammates' questions & suggestions, and how adaptable were they to changing their design.

With that first design question answered, as the interviewing manager, wipe the board clean. While cleaning the board don't say anything, as the silence and time taken to clean the board will probably mean that the candidate returns to their seat, and thus to a certain seating body position. **Is the candidates seating position the same position as their earlier defensive (crossed arms / legs / hands) or are they sitting more openly?**

8.. **Project targeted design question.** ... Repeat the process of the previous question, but this time make the scenario as close as possible to the project's subject matter (taking into consideration any business confidentiality & security clearance issues).

The purpose of this question is to evaluate the candidate's expertise in the industry & technologies related to the project. This will also give the co-interviewers more opportunity to interact with their potential new team member. So, let your co-interviewers do most of the questioning & suggesting during this stage.

> **NOTE:** The technical viability of the candidate's solution is of importance here. And, especially of interest is how well that the candidate can integrate their potential teammates' questions & suggestions into the drawn solution.

9.. **Technical questions.** ... The targeted design question would then be followed up with specific technical questions from the participating implementers, preferably one question from each. For example, a coding question related to "*when would you use such-n-such and when would you use so-n-so*".

These questions should be of sufficient technical difficulty to test the candidate's knowledge & capabilities; however, these questions should not be intended to technically belittle and/or intimidate the candidate into feeling inferior to the questioning implementer (which could mean you, as the project's manager, could have a pending team problem with this interviewing questioner).

10.. **Difficult technical question.** ... If the co-interviewing implementers can pre-devise a question that candidates are likely to not know straight away, but have to figure out (and "dig themselves out") then this could be highly beneficial to determining whether the candidate is "a fighter" who during times of project difficulty will battle on through adversity.

11.. **Huddle voting.** ... Once all of the technical questions are complete, then ask the candidate to excuse you all, as the invited implementers "*need to leave and get back to working on the project*". The interviewing manager quickly departs with the implementers, and holds a very fast-huddle (possibly with the representatives of HR and senior management) out-of-sight of the candidate, and individually verbally vote **YES:NO on the "general acceptance" of that particular candidate progressing through to the next round of interviews**. After tallying the votes, if there were a combination of YES and NO votes, then any NO voters would give a very-very short summation as to why NOT. Keep this verbal voting very-very short, no longer than 5 minutes; and, it should never turn into a debate, as this debate can happen later.

241

While this YES:NO voting is only a "gut feel" evaluation of the candidate, often there will be a consistency of reasons as to why NOT. Hence, if the candidate has collected a majority of NO votes than most probably this candidate is not the best-fit addition to the Project Implementation Team (according to The Team member's & yourself).

"But, what happens when The Team's representatives vote NO, but it turns out that HR / senior management think YES, and vice versa?"

12.. **Candidate questions in reverse.** ... As the interviewing manager, return as quickly as possible to the interview location, and excuse the delay. This delay should have given the candidate time to think over the situation and come up with questions to ask. Inquire whether the candidate has any questions about the role, the project, and the organization at large.

13.. **Selling the organization.** ... If the candidate received predominately YES votes, then this is the time for the interviewing manager to sell the role, the project, and the organization to the "desirable" candidate. Conversely, if the candidate received a clear majority of NO votes, then keep the remaining conversation short & sweet.

Remember that, while your evaluating the candidate to hire, the candidate in turn is deciding if they really do want to work at this organization.

14.. **Explain what happens next.** ... Explain the process going forwards from here.

After the candidate has departed, then

15.. **Debate the YES candidates.** ... The interview participants (including HR / senior management) should come together to debate the merits of those YES candidates.

16.. **Job offer or the final decision.** ... The chosen YES candidate would then be either offered the job or invited to a final round interview with senior management.

17.. **Integration into The Team.** ... Sometime later, the chosen candidate will start work at the performing organization, and subsequently they need to be integrated into the Project Implementation Team.

4.3. Constructing A Team

4.3.1. Team Dynamics & Team Relationships

As with any group of individuals that comes together to form "a team", be it a Project Implementation Team, a sporting team, a social club, the resultant team will naturally go through five distinct stages of relationship development; [ref: Tuckman & Jensen].

1.. **FORMING** – **is when independent persons come together** as assigned team members and learn about the project, learn about their individual roles & responsibilities within The Team, and learn about each other. As with any relationship, **the Forming Stage is when people start to reveal themselves**; i.e. their values, beliefs, priorities, abilities, and their individual quirkiness.

   If a pre-established Project Implementation Team has lots of new people added (in a very short amount of time), or there is a marked change in the workplace location, then this Forming Stage may reoccur, in what was, a previously performing team.

    DO NOT expect a great deal of productivity in producing project deliverables during the Forming Stage, because The Team will be relatively dis-functional as they act as individuals trying to be accepted, and determining their position within the group; i.e. who is who within The Team, and the shaping of a "pecking order".

2.. **STORMING** – **is when The Team members start to "feel" their way around each other and around the project in general**.

   Differences in team roles & responsibilities will become clearer, and these will be formally & informally agreed upon by The Team members. Differences in opinions, perspectives, and ideas will be raised, considered, debated, (argued over), and the "better" ones will be acted upon.

This is also when the ground rules for expected & acceptable behaviour will formally & informally be established. Differences here could potentially result in highly detrimental conflict between team members, and (in a worst case) conflict with the Project Manager and with other project stakeholders. To resolve any such conflicts will require the guiding hand of the Project Manager, and the maturity of the Project Team members to communicate openly with each other.

 Unless The Team members' start to collaborate, compromise, and their self-interests converge, then the Storming Stage can be a very long dark miserable journey for some or all of the people that are involved with the project.

 DO NOT expect a great deal of productivity in producing project deliverables at this Storming Stage. Instead, concentrate on the constructive resolution of interpersonal conflicts.

The Storming Stage is when the Project Manager's "humanities skills" need to come to the fore as; a communicator, a listener, a negotiator, an influencer, a conflict resolver, a problem solver, a planner, an organizer, a delegator, a motivator, and a leader.

This Storming Stage is also when the Project Manager's leadership style (as authoritarian, participator, and laissez-faire, or combination thereof) will greatly influence the long-term resolution of any such conflicts.

3.. **NORMING** – **by this stage, each member of The Team should have established their (formal & informal) position within the group, and know what is their individual roles & responsibilities**.

Hopefully by now; internal conflicts have subsided to the occasional misunderstanding, productivity should have increased significantly when compared to the earlier stages, and cooperation between The Team members should have become the norm.

4.. **PERFORMING** – by this stage, **The Team should have established (formal & informal) cooperative working relationships between the various team members**, as they work their way through the Executing Phase of the project's life cycle.

The Team should now be functioning as a cooperative & collaborative team; resolving their own internal problems & conflicts, having understood each other's strengths – weakness – values – beliefs – priorities, and with the common goal of successfully completing "their project".

An effective / high-performing team would spend the majority of their time in the Performing Stage; whereas, an ineffectual / under-performing team could spend a large proportion of their time bouncing back & forth between the Storming, Norming, and Performing Stages.

Then of course, at some point in the future.

5.. **ADJOURNING** – by this stage, the project is (hopefully) **concluding and the Project Implementation Team starts to be disbanded.** The Team members come to the realization that they will not be working together with some or all of these people again, which could result in a feeling of lost mateship, and insecurity about what is next for them (as a group and as individuals).

> Which brings us to the next rule of "Adaptive & Proactive" SDLC Project Management.
>
> **RULE 37: Forming – Storming – Norming – Performing – Adjourning is the natural sequence of team life. Where one cannot jump from infancy to adulthood without going through the awkwardness and traumas of adolescence.**

4.3.2. Team Work and Team Cohesion

When you were a kid and played Saturday team sports, what made your side a winner, or more specifically, what made your team enjoyable to play with?

So, what is different now that you are all grown-up as an adult at work, instead of a kid playing in the park?

Though, you as the project's manager will now play the role of coach, not captain.

Transforming "a team" into an "A-Team"

For "a team "of individuals to be transformed into an effective & efficient "A-Team", then the project' manager needs to **facilitate a working environment that**:

(1) **Gives a common sense of purpose, and emphasizes a commitment & dedication to the common objectives**. ... Of, delivering the project with the agreed [Scope], by the agreed [Time], within the agreed [Cost] budget, and with the agreed level of [Quality].

(2) **Encourages open, affective, and efficient communications** between The Team members, with the Project Manager, and with the project's other stakeholders.

(3) **Instils trust & mutual respect** between The Team members, with the Project Manager, and with the project's other stakeholders.

(4) **Formulates a set of formal & informal ground rules defining the expected & acceptable behaviour** for The Team members, for the Project Manager, and for the project's other stakeholders.

(5) **Establishes informal & formally accepted ways to constructively resolve internal & external conflicts** between the Project Team members, with the Project Manager, and with the project's other stakeholders.

(6) **Outlines informal & formally accepted ways that decisions will be made** by the Project Team members (technical) and by the Project Manager (interpersonal).

(7) **Encourages The Team to solve their own problems in a collaborative manner**, before escalating these issues to higher management.

(8) **Tries to inspire & motivate** The Team **to achieve higher performances**.

(9) Organizes and arranges for **guidance & support from the senior management** at the performing organization.

(10) **Provides timely feedback & constructive criticism** on The Team's performance, as well as relaying feedback from the primary stakeholders.

(11) **Ensures that there is a common identity for The Team**, even giving the project an identifying name (if one is not already being used), not an acronym-ized number.

(12) **Encourages a sense of belonging.** Ensures that **all team members participate**, are given "a fair go", have their individual input sort, heard, and where relevant accepted.

(13) **Tries to co-locate the project's team members.** If this is not **physically** possible then tries to arrange for **virtualized** co-location to be established via the utilization of telecommunications technologies. *Such as; Virtual Private Network (VPN), shared screen desktops, shared digital whiteboards, Voice Over The Internet (VOIP), video conferencing, instant messaging – text chat, shared cloud storage of files ... etc.*

(14) **Ensures that appropriate rewards & recognition** are obtained **for good performance**, as a whole team and also for those individual standout doers.

(15) Tries to provide The Team (and its individual members) with work that is challenging, and provides **opportunities for each team member's career advancement & knowledge growth.** ... *"And, not just the advancement of the manager at The Team's expense. ... to rise upon the crushed backs of others."*

(16) Tries to maintain a **friendly, healthy, & sociable atmosphere** for all members of The Team. ... *"Including the quiet, the withdrawn, and the newbies."*

(17) **Removes those barriers that prevent The Team from achieving the project's objectives.**

SDLC Team Building … *"welcome to the legion"*

Now that the individual members of The Team have been recruited, the next concern is how to mould this body of people into a collaborative & cooperative team? … While at the same time, resulting in productivity returns for the performing organization.

This sort of problem should not be prevalent for a pre-existing Project Implementation Team that has been working together in its current form for some time, nor should this be much of a problem for a pre-existing team that has to absorb a couple of new team members. Because there should, by now, be well established team dynamics as to the implementation processes & procedures, and the work practices & behavioural expectations that are carried over from The Team's previous projects.

However, what about when The Team is made-up of persons who have never worked together before, or a pre-existing team that is performing way below expectations?

Break Them Down and Build Them Up

If you were building a fighting force which is composed of recruits with different nationalities, different ethnicities, different previous military training, and different levels of combat experience, then the first thing that you would do is to break each individual recruit down to a common baseline, and then build them up together as a cooperative & collaborative unit (aka, "A-Team").

So, how to do this for an SDLC Project Implementation Team?

1.. **Co-locate The Team.** … Based on geographical, logistical, and political limitations, try to have as many members of the Project Implementation Team physically located together in the same office | floor space. The **primary advantage of co-location** is that, it's **so much easier to get people to start communicating with each other** if they can simply turn around in their chair and **"talk face-to-face" with the person that they need to exchange information with**. Additionally, try to provide them with a few portable whiteboards so that they can collaboratively discuss their ideas.

2.. **Give them a common problem to resolve.** ... When a Project Implementation Team is newly put together, then most probably the project has just started; hence, the project could still be in the SDLC phases of either Initiation or Planning. With the project having not long been underway, this provides an excellent opportunity to have the individual members of The Team work together in small groups to understand the project's objectives, requirements, and deliverables. ... And, to learn what are the "needs & wants" of the project's primary stakeholders.

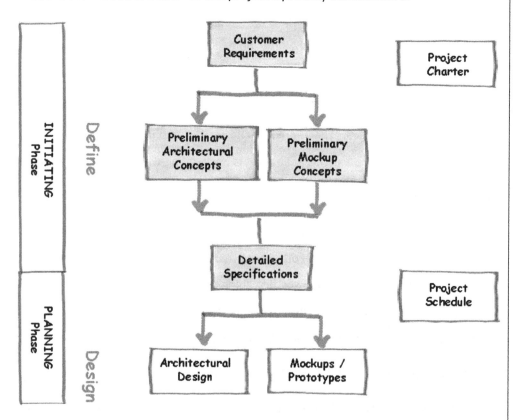

With a Project Implementation Team, ideally composed of 6 to 7 members, amongst sub-groups of pairs | trios of these individual team members, breakup the following activities; (1) Customer Requirements updates / draft of the Detailed Specifications, (2) preliminary architectural design concepts, and (3) preliminary mock-up concepts. Get each sub-group to work on their given part for exactly ONE DAY, before they have to handover their work to another of the sub-groups to continue on with.

Remember that this is still very early in the project's lifecycle, possibly this is only the first week of The Team being put together as a group of individuals. ... And, various persons composing this group will have differences in skill levels & past experiences (as either senior, middle, junior, or a recent graduate implementer).

<u>The purpose</u> of the ONE DAY swap over and the parting up of Project Initiation is to:

- ☐ Get individual team members to work together as pairs or trios, so as to cooperatively build a common understanding of what the project is all about.

- ☐ For each sub-group to communicate with each other at the handover.

- ☐ For every person in The Team to obtain a common exploratory introduction to the project's technological basis, and to get communal design ideas flowing.

- ☐ For every person in The Team to, hopefully, interact with each other.

NOTE: It is very important that the swap over occurs after only ONE DAY's duration, ideally swapping straight after lunch. The reason being that, in one day a fair amount of work can be done by the pair | trio, but not too much work being completed, so that other sub-groups are able to keep up and make changes & suggestions to each partition of work. Thereby, providing a catalyst to the Forming Stage of learning about the project and learning about each other.

3.. **Establish points of contact.** ... During this first week of the project, if possible, don't declare someone as the "team leader"; instead, ask for people within the group to volunteer to be a "Point Of Contact" (POC) for specific partitions of understanding the project. *For example, someone to take responsibility for updating the Customer Requirements, someone to take responsibility for the preliminary architectural design concepts, and someone to take responsibility for the GUI mock-up concepts.*

Let the natural leader arise, and then appoint the most appropriate POC as The Team leader. By doing it this way, team members will have a good feeling as to who is better suited to technically lead this particular project. ... And, this can greatly reduce resentment within The Team because "so-n-so" was appointed team leader.

NOTE: Up to this point, the **(hidden) objective has been to convert each member of The Team from thinking & speaking in terms of, "*I can do this, you can do that*", across to a mindset of "*WE CAN do it'*.** ... And, this literally does mean how they use the words, "*we*", "*you*", "*they*", and "*I*"; because, the moment that each one of them starts using the term "*we*" when referring to the project, then the more likely it is that they are seeing themselves as "A-Team" and not as individual members of a body of people (which happens to be called, "*a team*").

Thereby, **encouraging "team buy-in" which builds team ownership** of the project, which **promotes self-organization** within The Team; instead of, the perceived drudgery of being assigned to work on "*someone else's project*". This type of "team buy-in" will potentially give the project's manager more time to concentrate on the stakeholder's needs, the risks confronting the project, and the necessitated changes to the project. ... And, allow the Project Manager to deal with other projects that they happen to be simultaneously allocated.

WARNING: watch out for The Team member who uses their senior implementer's job title and/or says things like, "*LISTEN HERE, I'VE GOT ## years of experience in this field*" to push their point of view, so as to win a team discussion; as this person is potentially not going to be much of a team player. This person could in fact be very detrimental to The Team's cohesion & morale, because they stifle others willingness to contribute.

As the project's manager, I would recommend that you privately take this person to the side, and "have a talk" about the impact that their actions are having on The Team as a cohesive unit. ... And inform them that, "*at this stage, a perfectly correct technical solution is not as important as everyone understanding what the project is about, and team participation.*"

A Common Foundation For Daily Stand-up Meetings ... for Agile, Iterative, and Waterfall based projects.

During this first week of The Team being together, I would recommend that the Project Manager take on the role of the **Daily Stand-up Meeting Master**; so as to establish team oriented work practices & behavioural expectations, see [Section 3.5.6].

It's our scrum, and we will run it how? ... For the first daily stand-up meeting (henceforth to be held each work day), as the project's manager, establish by demonstration the format for all future stand-up meetings, that in a circular fashion, each team member will summarize:

1.. **What did I achieve during the last 12 working hours** (i.e. yesterday and the time prior to holding today's stand-up meeting).
   ... Then, it is the next team member's turn, until the circle is complete.

2.. **What is "blocking" me from doing my work in the coming 12 working hours, have I identified any <Risks>** to the sprint's timely completion, and **do I need help?**
   ... Then, it is the next team member's turn, until the circle is complete.

3.. **What do I plan to do in the coming 12 working hours** (a work day and a bit)?
   ... Then, it is the next team member's turn, until the circle is complete.

The project's manager, as the (first day's) Daily Stand-up Meeting Master, should after each team member has stated a blocking issue / raised a concern / asked for help, then bullet-point this on the whiteboard that is present in the meeting area; but, DO NOT discuss / resolve this point now, just keep the circular flow going. If a team member(s) diverges off topic or digresses into a discussion then, as the Daily Stand-up Meeting Master, stop them there and remind the entire team that such discussions are to be done independently with only those people who really need to be involved.

At the (second day's) Stand-up Meeting, as the project's manager, select one of the junior team members (*e.g. the graduate*) to be today's Stand-up Meeting Master, and when they happen to miss a step (in the previous day's demonstrated process) then hint them through that bit of the process.

Do the same thing for the next day's stand-up meeting, until everyone in The Team should have had a go at being the Daily Stand-up Meeting Master. ... Then, the project's manager should step away from being actively involved / present at each day's stand-up meeting; so that, The Team is effectively running their own stand-up meetings and self-coordinating their work activities.

The purpose of the demonstration of the Daily Stand-up Meeting process is to:

☐ **Establish a common process for all members of The Team to follow**, given that, some members will have done it differently at other organizations.

☐ By giving each team member (including the graduates & juniors) a go at being the Daily Stand-up Meeting Master, this will help **flatten The Team's hierarchy** from a pyramid structure with the senior members at the top, to something more conducive of a cooperative & collaborative "fellowship".

☐ **Establish a mindset of**; ... ... (1) **Recognize & Reassess** what happened yesterday,

(2) **Reassess & Revise** if something different needs to be done,

(3) **Plan to Apply** what needs to happen for the rest of today and for tomorrow. See [Section 2.5.1] on Project Rescue silently meshing with the project's routine 'PLAN – DO – CHECK – ACT' cycle.

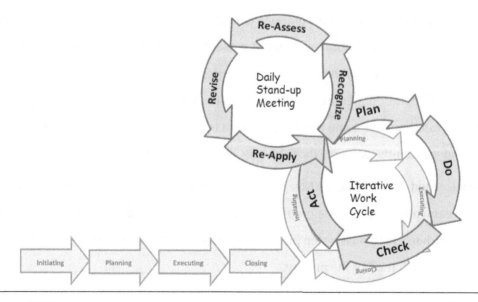

"This technique works particularly well for a team of 6 to 7 members, though it works just as well for teams as large as 12 to 14 members. It especially works well, when The Team has a good mix of seniors, middles, juniors, and graduate implementers. ... And, it's highly beneficial when The Team is ethnically diverse, as everyone gets to participate as a meeting input provider and as the occasional meeting chairperson.

However, it may not work out so well with a team that consists predominately of senior implementers, because they could feel that they are being patronized (i.e. treated as though they know nothing)."

NOTE: In the following weeks & months, The Team may decide to revise the format of their Daily Stand-up Meetings; with, for example, The Team leader acting as the meeting's bullet-point taker and functioning as chairperson who decides whether (The Team agreed) meeting processes & rules are being abided by.

However, what the project's manager has to watch out for, is if The Team leader starts remoulding The Team structure into a pyramid with themselves at the apex (as the self-appointed "Project Leader") and not just as the technical lead.

> *Such restructuring is exemplified by the Daily Stand-up Meeting becoming a platform by which The Team leader does the majority of the talking and doles out each team members work assignments (i.e. micromanaging); hence, voiding the Agile practice of The Team deciding the 'To Do Items' for the Sprint and that it was supposed to be The Team who choose for themselves what they should work on next.*

Subsequently, as a result of pyramid-izing The Team's structure (combined with the possible formation of a "clique group" within The Team), this can cripple the cohesiveness, cooperativeness, and the collaborative nature of The Team.

"Effectively wiping out all of that effort which went into building an A-Team in the first place. ... What you've now got is a D-Team that puts in a commendable effort, but finishes last."

EXAMPLE CASE … *Miss reading a "Clique Group" as a strong project team.*

 WARNING: watch out for the "clique group" | "in-crowd" forming amongst the members of the (large) Project Implementation Team.

Comedy movies portray this type of clique group at its extreme, with the 3 to 5 trendy teens who strut their stuff, and make life a living hell for their "lesser" classmates.

In the SDLC world, this "in group" would be composed of 3 to 5 developers / engineers, probably with The Team leader / senior implementer as the (male or female) equivalent of the "queen bee". Where this "queen bee" (unofficially) structures The Team's operations so that he/she is at the centre, with the "in" team members as the next layer, and those "outer" team members on the fringes. This "queen bee" will also make themselves the sole external representative of the Project Implementation Team, having no immediate second-in-command (else a second with limited power & responsibilities).

In the worst cases, the clique group (more specifically, the "queen bee") can:

- ☒ Push out (by isolating / ostracising) those existing project team members that don't fit-in with the "in" group's perceived ideals. … *"Bye-bye to whoever asks why."*

- ☒ Pull in by enticing those implementers that they want for their group, even going as far as directly and/or indirectly undermining other project teams to get that "right" person aboard their team.   … *"Mini empire building at others' expense."*

- ☒ Reserving the "interesting" / "important" tasks for their fellow "in" team members, while distributing the "crappy" / "boring" tasks to those "outer" team members.

- ☒ At the extreme; considering the project's manager (and senior management) as outside of their "in" group, and hence (in an attempt to protect their "in" group) are prepared to report mistruths when the project's scheduled tasks are not going as well as was planned. Though, this "in" group is quite willing to place blame for the project's failings on those "outer" team members. … *"Throw 'em under a bus."*

Initially, the morale & work ethic of this "clique group" structured team will appear to be high, and they may for all available evidence be progressing excellently; but, this is a falsehood that will eventually fall apart (the longer that the project goes on), because:

(1) Those "outer" team members will silently "want out" as they feel that;

- ☒ their potential constructive contribution to the project (and to The Team) is being squandered, and their skills aren't given the opportunity to be exercised.
- ☒ they feel discriminated against by only being given the "crappy" tasks to do,
- ☒ their future career advancement & knowledge growth with this project (and at the performing organization) is greatly limited to non-existent.

(2) Those "in" team members while receiving the benefits of getting the "interesting" work to do, will become overloaded & overworked if / when the project falls behind schedule, as those "outer" team members can't take on more of the heavy burden because they weren't "trained for" (previously allocated) those important tasks.

(3) The Project Team's cohesion is predominately via an imposed pecking order, and not via cooperative & collaborative fellowship (aka "mateship").

(4) The Team is being driven forwards by a micromanaging team leader "queen bee" who does not necessarily have access to the strategic understanding of how each project release fits into the "Big Picture" of the performing organization's plans.

If all does proceed along well with this Project Implementation Team being productive; what happens when the "queen bee" team leader gets promoted into another role, moves to another project, or leaves the performing organization for "greener pastures"?

When the "queen bee" departs (or is promoted), then most probably the cohesion of this "clique group" will disintegrate, The Team will quickly unravel as the "in" team members jostle for position in the power vacuum, and the "outer" team members will become under-utilized as there is no "queen bee" to allocate them work and issue new instructions. ... And thus, The Team would be back to an unproductive Storming Stage.

4.4. A People First Manager

4.4.1. Being A "Humane" Project Manager

For the Storming Stage it was stated that, *"it is during this stage when the Project Manager's humanities skills need to come to the fore as; a communicator, a listener, a negotiator, a influencer, a conflict resolver, a problem solver, a planner & organizer, a delegator, a motivator, and as a leader."* ... *"But, no longer as an implementer."*

A Communicator

As stated previously in [Section 3.4.8], *"**a project manager will spend a lot of their time communicating in both written and oral forms**"*. Hence, the Project Manager has to be able to communicate effectively & efficiently with all of the project's stakeholders, and especially with each member of the Project Implementation Team.

"What did you think the Project Manager was going to do, just sit in their corner office and manage by telepathy? Unfortunately, during mine (and most probably your) career, you will encounter managers, even at very senior levels, who evidently believe in management by telepathy, because they don't TALK to any of their ~~people~~ subordinates, other than to bark-out the occasional order and (at the top of their voice) verbally reprimand someone for something they did wrong. ... Which is often due to that manager having not clearly stated what they wanted them to do."

A Listener

Communications is bi-directional, where **someone is transmitting** (telling) **while the other party is receiving** (listening), then vice versa, and versa vice. While the speaker's role is to effectively & efficiently transfer the information, the listener's role is to **actively take in the information, and then provide some form of feedback to indicate that the information was received & understood.**

Hence, the Project Manager needs to demonstrate that they are an **"active listener" who takes in what was said and then (evidently) acts upon this information**.

A Negotiator

The reality is that, not every situation can be "win-win" for all of the involved parties; therefore, the Project Manager has to be able to **work towards a "compromise solution" that is the best possible outcome** for (most if not all of) **the involved parties**. … And, not just "winner take all" for the limited few.

To be a good negotiator involves:

(1) **Being a good Listener & Communicator**.

(2) **Trying to understand the situation** that is causing the problem. This means trying to "see the situation" **from each party's perspective**. … And, not just your own, nor solely that of the performing organization (which can result in some ethical dilemmas).

> *"You will never understand a person until you have walked a mile in their shoes", and "carried some of their burden upon your own back", or sung along with them at a karaoke bar."*

(3) **Determining each party's (real) "wants" and (real) "needs"**.

> NOTE
> - **"Wants"** is the high point of what the party is after.
> - **"Needs"** is the low point of what that party will accept to be satisfied.
>
>   If all involved parties "needs" are at least being met (in some way) then a "compromise solution" is potentially achievable.
>
> - **"Greed"** is to want more than one could possibly need, solely to satisfy one's self-interest, irrespective of the detriment to others.
>
>   DO NOT discount on Greed from (inexplicably) hindering the negotiation towards an amicable resolution.

(4) **Concentrating on the issues & concerns** of the involved parties, and **DO NOT focus on the emotional position that these involved parties are taking**; because, by focusing on their issues & concerns, one can obtain a better understanding of their actual "needs" rather than being diverted by their perceived "wants".

(5) **Realizing that the involved parties personal characteristics, ethnicity, and cultural background will influence how they will approach any negotiations**. ... And more importantly, how far will they be prepared to compromise; *e.g. "a need to save face"*.

(6) **Ensuring that all involved parties are receiving a "fair deal"**; because, if one or more of the involved parties feels that they have been taken unfair advantage of, then the negotiated agreement / arrangement will not last very long. Subsequently, everyone will be dragged back into more negotiations; however this time round, everyone will probably be in a lot less malleable mood because their hands (i.e. advantages & disadvantages, strengths & weaknesses) will have already been revealed during the previous rounds of negotiations.

(7) **DO NOT make a concession without at least appearing** (in the other party's opinion) **to be giving up something of value**; because, this will help form an unwritten obligation for the other party to also give up something of value in return. That is, all negotiations involves some form of bargaining (aka "haggling").

(8) **DO NOT be the party who is continually surrendering their "needs & wants" in order to try to "reach a compromise"**; because, the other party could soon perceive you as a "pushover", and thus they are highly likely to continue taking until there is nothing of value left to give.

(9) **DO NOT over sell your position**; because, if you **are caught out then** you could be **perceived as being in a weaker position** (than you actually are). As a result of this perception, the other party may think that they themselves are in a stronger position, and consequently they may be a lot less willing to compromise, because they are confident that you will be forced to give in first.

(10) **Do your homework**. ... If you're able to **"find out what's going on behind the scenes" then you will be in a better position to understand the why and the how come certain involved parties are negotiating**; i.e. their true objectives, their real "needs & wants", their actual advantages & disadvantages, and their strengths & weaknesses.

An Influencer

A good project manager is able to affect people & events without necessarily being directly involved (hands-on), as though using some magical force. Though, the reality is that, while we would all like to have "super-natural influencing power", most of us mere mundanes have to make do with the following techniques.

(1) **Know what you want to accomplish**. ... You cannot redirect them towards your objectives unless you understand what you are really trying to achieve.

- **See your perspective** – that is, getting them to understand & acknowledge your view of reality, and that they are going to give their best while working towards your stated objectives.

- **Modifying their behaviour** – that is, getting them to change their behaviour, **based on your perception of its desirability or undesirability**. This is done by using positive enforcement to encourage those desirable behaviours and ~~negative reinforcement to~~ discourage those undesirable behaviours.

    "IMHO, if you have to resort to negative reinforcement then you have lost the capability to sustain long-term influence over them."

(2) **Know what they "need" and what they "want"**. ... This is similar to negotiation, you cannot redirect them towards helping you to achieve your objectives unless you understand **"what's in it for" them** (i.e. what is in their self-interest).

- **Rewards**; i.e. self-esteem, self-confidence, a sense of achievement, praise & gratitude from others, recognition & respect.

- **Gains**; i.e. financial, material, experience, knowledge.

- **Safety**; i.e. job security, stability of employment, belonging to a group.

Which all relates to **Maslow's Hierarchy Of Needs**.

Having determined what their "needs & wants" are, your **message can** then **be better tailored towards matching these**; thereby, being able to emphasize "what is in it for them", rather than "what is in it for me". In addition, by helping them accomplish their "needs & wants" they are more likely to reciprocate and help you achieve yours.

(3) **Look, listen, learn, and ask**. ... How else can you know their "needs & wants" unless you pay attention to what they are saying, and more specifically, what they are not saying? You will also have to observe **what work & home life constraints would prevent them from being able to do what you require them to do**.

Hence, **to be a good influencer requires being a good observer, communicator, and listener**.

(4) **Listen to and be open to counter proposals**. ... Sometimes to get what you "want", you will have to **re-evaluate & re-adjust your proposal based on their responses, and** more specifically, **their counter-proposals**. Noting that, it is easier to get someone to do what you "want" when they are involved in the decision making. ... And, this is especially true when they feel that it was them that made the decision; *e.g. the 'To Do Items' that are to go into the Sprint Backlog that is to be done by them in the upcoming Agile sprint is selected by The Team members and not the Project Manager*.

(5) **Keep focus**. ... That is, **keep their attention on what you "want" them to do**, and get them to **commit to a date** by when they will do that which they have agreed to do.

However, this is a "two-way-street", so if you have made a commitment to do something for them then it is advisable that you do exactly that (by when you said it would be done), and thereby creating reciprocal expectations on both parties.

(6) **Communicate openly, clearly, and confidently**. ... No one will bend to your will if they do not clearly understand what you are wanting of them; so, confidently inform them of exactly **what you "want" (in terms that they will understand)**.

Be consistent & persistent with the definition of what you "want"; but, choose an appropriate time to reiterate the definition (i.e. not too often to cause annoyance). And, state it **without ambiguity and multiple interpretations**. *For example; "was that delivery on the 1st of January, or was the delivery on the first working day in January, after the Christmas & New Year holiday break".*

(7) **Time your criticism**. ... Occasionally the person you have influenced may not have done exactly what you requested of them, or they may have done it poorly. In such cases, negative feedback will be necessary; however, **how this negative feedback is administered will greatly affect how much influence you will be able to exert on them in the future**.

A misdirected or inappropriately timed negative feedback (given their situation and their prevailing circumstances) **is a sure-fire way to lose** their respect and reduce their willingness to give you (the criticizer) that extra bit of effort in the future.

(8) **Criticise as soon as possible after the event that initiated the criticism.**
Though, limit the amount of criticism that is dispensed at any one time or is dispensed over a relatively short period of time, else you will be perceived as someone who cannot be pleased (or worse, you will be perceived as "a bully").

(9) **Get straight to the point**. ... be very clear, precise, and honest about the criticism.
DO NOT waste time "buttering them up", and
DO NOT skirt around the crux of the issue, "afraid of saying the truth".

(10) **Let them respond to the criticism**, so that they can "tell their side of the story", as there may be mitigating circumstances, key information may have been missing, or the situation at the time and the prevailing circumstances may justify why they did what they did.

(11) **Be calm & clear,** because being frustrated, angry, or out-right hostile will only make the criticism's recipient defensive to whatever you have to say.

(12) **Be serious the entire time** that the criticism is being given. ... Because, **criticism is NOT a joke**; to treat it as such will be interpreted as not showing real concern.

(13) **Never criticize something that cannot perceivably be changed** (i.e. you cannot criticize a cat for not being a dog), or because they didn't perform as well as a certain other person would have / did.

(14) **Criticism that can be invested**. ... That is, if you must criticize someone then do it in a **constructive manner**, so that the recipient will **learn from the experience** and not be resentful / defensive.

- **Always make sure that you are "actually correct" before unleashing your criticism on someone.** Make sure that you have all of the facts and that you **understand the particulars of the situation & prevailing circumstances** at the time of their supposed wrong doing.

- **DO NOT criticize them for** dealing with an issue / problem the way that they did if that issue / problem resulted from **your own failings** to do something about it.

    If they made multiple requests of you (as their manager) to perform some action and/or to provide them with the appropriate resources, but **you failed to deliver in a timely manner, then it is not appropriate** for you **to reprimand them for taking the initiative to do what needed to be done** to complete their assigned tasks. ... And, especially if you told them to, "*sort it out for yourself*".

- **Never "badmouth" theirs or your predecessor's performance** (i.e. the person who previously did theirs or your role) as this will raise suspicions about what you are currently saying "behind their back" about theirs and others performances.

    "Let alone what negative things you'll be saying about them in the future. And, bad mouthing the departed is a sure-way to lose the respect of those who remain and marks you as a poor manager."

- **Never turn the criticism into a personal attack** on their character. Stay with the facts at hand, and be precise with the criticism.
  DO NOT deviate off-course into an attack on their personality (or worse their **culture, ethnicity, religion, believes, sexuality, or gender**).
- Having dispensed the criticism, now **make it clear what is expected of them**.
- If **punishment** must be meted out then ensure that this punishment is **proportional to the crime** that they have committed.

(15) **Do unto others, as you want them to do unto you**. … That is, before you can get someone to do what you "want", they must at least "tolerate" you and may even "like" you. Because, **if you treat them badly,** or are unfriendly towards them, **then it is highly unlikely that you will get them to voluntarily do what you want them to do** (let alone to the standard that you expect).

"Agreed, if they really dislike you then they will keep putting off doing what you want for as long as they possibly can; and, when they do get around to doing it, they will probably, only put in as much effort as they believe is necessary to make you go away.

There is an old adage that "you catch more flies with honey than with vinegar", which translates to, it is the tone & attitude of what you say, how you word it, and the use of the words 'please' and 'thank you' that win your way."

So the next time that you want to influence others then remember to be:

- ✓ polite & diplomatic, courteous & tactful, sociable & friendly, generous & kind,
- ✓ concerned with the plight of others,
- ✓ show respect & gratitude,
- ✓ give others your undivided attention,
- ✓ show appreciation for others efforts,

- ✓ strive to resolve the problem at hand,
- ✓ compliment people and praise them when they do a good job,
- ✓ when possible take the time to give them a helping hand,
- ✓ when necessary go out of your way to help them out.

But, don't expect to be a successful long-term influencer, if you are (perceived as):

- ✗ anti-social, snobbish, arrogant, know-it-all,
- ✗ mean, nasty, unkind, insensitive to other people's predicaments,
- ✗ put people down, belittling, degrading,
- ✗ intimidating, bullying, physically threatening, verbally abusive,
- ✗ back-stabbing, bitchy, rumour mongering,
- ✗ steals the credit from others, or passes others' work off as your own,
- ✗ passes the blame onto others, and is prepared to "throw them under a bus",
- ✗ selfish, show-off, look-at-me attitude,
- ✗ ignorant to others problems,
- ✗ ignorant to the fact that a problem exists, to "bury your head in the sand",
- ✗ manage situations by avoidance, and in the worse cases, actively going out of the way to avoid a confronting issue.

 DO NOT manipulate others by devious means, because when a person feels that they are being wangled into doing something that they would not normally want to do then they will eventually resent doing such a thing, and they will subsequently be negative | hostile to doing other future things.

DO NOT dictate and reframe from micro-management, because a bit of freedom will help in satisfying their own "needs & wants".

THEORY ... *Maslow's Hierarchy Of Needs.* [Ref: Maslow]

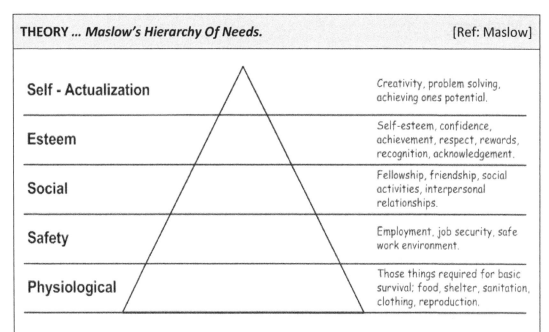

Figure 82: Maslow's Hierarchy Of Needs.

The techniques that the Project Manager could use, include:

1) **Be aware of** (cognizant to) each team member's **efforts & contributions** towards the project's objectives; i.e. to know what they are doing and what they have done.

2) **Give praise** to the relevant team members as soon as possible after they have done a good job.   ... *"The proverbial 'pat on the back' can do wonders for a person's self-esteem and the sense of their work being appreciated."*

3) Provide **timely feedback** on their performance (and if necessary, **constructive criticism**).

4) Provide **challenging tasks** and not just the repetitive boring / "crappy" bits.

5) Provide **new, expanded, or different roles & responsibilities**.

6) Provide **opportunities for individual growth and career advancement**.

7) Provide **acceptance of their proposals and consideration of their ideas**.

8) Ensure that they have a **balance between their work life and their home life**.

A Motivator

The Project Manager has to be able to get the Project Team members personally committed to delivering the project in accordance with the agreed Acceptance Criteria for [Scope] & [Quality], within the agreed [Time] & [Cost].

To achieve these goals, the Project Manager needs to "motivate" the Project Team members' individual & combined efforts. So, how does the project's manager do this?

Well, ... it is, the exact same things that motivate you each workday.

> To wake up before the dawn to that incessant alarm clock. ... To drag yourself out of that nice warm comfy bed. ... To have a shower & groom while your eyes are still welded shut. ... To apply that sociable face. ... To put on that not so comfortable outfit & shoes. ... To jealously kiss your partner on the forehead while they get a few minutes more sleep. ... To say goodbye, but not be heard as your youngest one plonks their butt down in front of that big-ass TV screen that you never have the time to watch, as the older one giggles away at something indecipherable that their invisible friend responded with.
>
> To walk to the bus stop or train station in dark miserable weather, to stand like emperor penguins for an hour or more with people squished into your personal space. Alternatively, to listen to advertisement-interrupted senseless breakfast radio as you stare at the stream of glowing red taillights snaking off into the distance as you crawl in traffic away from your massively mortgaged little kingdom in the burbs.
>
> To enter that open planned cubical farm of half-height padded grey walls where everyone can see everything you do and hear everything you say. ... To sit down in that less than ergonomic chair, surrounded by an anaglyphy of post-it notes and faded photos of when your kids thought you had a relevant opinion.
>
> To finally, sip on that re-boiled pick-me-up in your brown stained "world's greatest" coffee cup, while having breakfast at your assigned desk, as you are confronted by the daily barrage of unanswered emails.
>
> *"That is so depressing. ... I think I'll go get another coffee."*

So, ... what motivates you to come to work each day, and give your all for the project's success? Then repeat this cycle again & again, day after day, for weeks – months – years to come?

Motivating Myth: Financial & Material Gains

Does the hourly thought of coins falling into your piggy bank really motivate you to give your best efforts at work?

Nope, ... **money, bonuses, and share options are NOT motivators; rather, these are "initiators to action"**. That is, if you are currently unemployed or in (what you know is) a grossly underpaying job then money will be a good motivator for you to find a new job. However, when you currently have a relatively secure job that pays you at an acceptable level then money is no longer a motivator; in factuality, payment is an expectation, an inane right (given that daily drudgery of coming to work).

Also, you would probably be paid weekly / fortnightly / monthly, and those bonuses are only handed out at the end of the financial year. Hence, these gains are not directly associated with your daily work efforts; rather, these incentives are determined by the annual / biannual review of your past performance (if and when such reviews do occur). Alas, such gains do not drive you (nor the Project Team members) to give all for the project, let alone give every day your best efforts to the performing organization.

"Oh, and if your project team consists of contractors then they probably won't get any performance bonus let alone performance reviews. Instead, they are often shown the door when the project is done."

And, ... the Project Manager probably won't have a direct say over what the Project Team members are individually paid, as these people were probably, just assigned to the project.

Hence, **financial & material gains are NOT motivating tools in the Project Manager's arsenal.**

"Oh, by the way, word is that there won't be any pay rises this year (for the majority if not all of the staff), not even an inflation matching CPI increase. Subsequently, the work place will be demotivated for some time to come; and, there will be a few disgruntled employees who will be contemplating looking for work elsewhere, so as to find an employer who is prepared to compensate them appropriately for their efforts."

269

Motivating Myth: Rewards & Recognition

Does being rewarded & recognized for yesterday's performance really spur you (and the Project Team members) to perform today, and tomorrow?

Some companies try to use rewards & recognition, such as "employee of the month" or "most valuable employee", to stimulate performance. While these maybe acceptable in a competitive work environment such as sales; in an SDLC project environment these **"one winner takes all" motivators** would in fact be **detrimental to the Project Team's cohesion & morale, because these supposed motivators encourage individual self-interest over teamwork.**

Also, with rewards & recognition, the point of application of these supposed motivators is greatly removed from the moment when the performance occurred.

"Oh, and the company rules probably dictate that contractors are not eligible for such rewards & recognition. So, there goes that as a motivating tool for a portion of the Project Team."

And, ... as the Project Manager, you probably can only make recommendations about which project team members should be considered for such rewards & recognition (which are then doled out later on at some ceremonial get together that is completely disassociated from when that performance actually contributed to the achievement of project success).

Hence, **rewards & recognition are NOT motivating tools in the Project Manager's arsenal;** rather, **rewards & recognition are "anti-demotivating" "thank you" tools, for the performing organization to deploy (not for the Project Manager to utilize).**

Motivating Myth: Threats, Intimidation, and Punishment

"Seriously ?? ... Threats, intimidation, and punishment, ... as a motivator ??"

Threats, intimidation, and punishment (including across-the-board harsh / vindictive performance reviews) will **only work as motivating tools for an SDLC environment where the recipient implementers "have nowhere else to go"**, due to:

- ☐ **Job insecurity** – of high unemployment rates, incumbent age discrimination, and the financial realities of *"having this crappy job, is better than having no job at all"*.

- ☐ **Lifetime commitment** – the recipient has been with that particular performing organization for so many years, that to leave now would be tantamount to getting a divorce (having worked at this place for longer than they have actually been married).

- ☐ **Home life priority** – work strategically positioned for ease of home life circumstances.

- ☐ **Personality type** – the recipient has a subservient "drone like" demeanour.

Also, ... **threats, intimidation, and punishment are only viable in response to already existing poor performance**; to use such "motivators" prior to the actual performance (other than in response to pre-existing poor results) is equivalent to a pre-emptive strike.

While the contracted business arrangement between the customer organization and the performing organization may include penalty clauses related to liquidated damages and performance bonds as guarantee; the performing organization should have incorporated into the contracted price a contingency for such penalizing situations, and potentially the performing organization would be entitled to performance bonuses.

However, **for mere implementers then such threats, intimidation and punishment only serve as encouragement to leave as soon as possible**. ... And, if such management behaviour results in work related stress being expressed at home, then that implementer's (misery affected) partner may also be encouraging them to leave. Subsequently, there could be a contagion of staff resignations, as an ever shrinking core of SDLC implementers become more & more demoralized; which can lead to the constant "churning" of staff.

"IMHO, if there is a Yearly Resignation Turnover Rate exceeding 15% for an SDLC group, then the performing organization has got real (pending) problems with technical knowledge and intellectual property drain.

If the SDLC personnel's Resignation Turnover Rate is at 25 - 33% levels (i.e. a quarter to a third of the implementers are changing each year), and given that it takes at least 6 months for their replacement to get up to speed to be a truly self-sufficient implementer, then productivity is going to take a major down turn or remain appalling. Any Resignation Turnover Rate greater than these levels, is organizational suicide.

Mathematically, this SDLC group could have a half-life of only a couple of years (where more than 50% of the staff have changed), and people only stay for a year and a bit, as though serving a Tour Of Duty in a war zone. Subsequently, the "old timer" implementers won't bother properly training and mentoring the newbie because they or them won't be around long enough to justify the effort involved.

Consequently over time, those implementers who do remain will progressively be transformed into involuntary "yes men" "drones" who no longer possess the creative intellectual spark to drive the SDLC group forwards to higher performance. As a result, the performance levels and morale of this group will baseline to a plateau of stagnancy."

Such **high Yearly Resignation Turnover Rates of SDLC implementers is an indicator that the performing organization has been inflicted by a sociopath manager(s)** somewhere up the management hierarchy; **subsequently, the SDLC group won't be able to reach anywhere near full potential of effective & efficient productivity.** … And, morale could be so low that, most probably, past & present staff will be "bad mouthing" the company; hence, the organization will garner the reputation as *"a place to steer clear of, unless you're really desperate to get work, and then, only as a short term stay until you find a better job".* … *"At which point, try delivering quality of projects on time."*

Hence **threats, intimidation, and punishment are motivating tools that SHOULD NEVER be in the Project Manager's arsenal.** … *"Unless you're a narcissistic psychopath."*

Motivating Tools

To motivate the Project Team, the project's manager needs to create a work environment that is conducive to each project team member's intrinsic & extrinsic motivations; while simultaneously, minimizing those demotivating factors. See [ref: Herzberg].

- **Intrinsic motivation – is generated within the individual person due to the type of work that they are performing;** i.e. a sense of achievement & worthwhileness, plus a feeling of personal growth.

- **Extrinsic motivation – is generated externally to the individual person** and includes such motivators as; financial & material gains, rewards & recognition, competition, as well as threats, intimidation, and punishment.

- **Hygiene factors** – are not motivators as such, but if these factors are not present then the work environment would be less "healthy"; i.e. a less enjoyable place to work. Also referred to as "**demotivating factors**". ... *"I hate this job."*

As described previously, these **extrinsic motivators** are not tools that the Project Manager can effectively utilize; because, these **are administered after the Project Team** member's **performance has occurred**. Hence, the Project Manager needs to concentrate more on those **intrinsic motivators** that **are closer to the time when the work activities are performed, or** (even better), **prior to such work activities being performed**; thereby, being beneficial to the project's quality assurance processes.

To intrinsically motivate the Project Team, the project's management needs to establish a work environment that encourages (and is a catalyst for); job satisfaction & self-esteem, a sense of accomplishment & achievement, a sense of belonging, and a feeling of kinship.

While the Project Manager **deals predominately with the intrinsic motivators**, the **project's primary stakeholders should be dealing with those extrinsic motivators**. **Whereas everyone**; including the Project Manager, the primary stakeholders, and the Project Team **needs to deal with those hygiene factors**.

Motivation … *"Getting More Without Punishment & Pay"*

So how to motivate the members of the Project Team without resorting to bringing out the punishment stick or bribing them to work harder?

- ☐ **Bi-directional Trust & Respect** … A manager who **trust**s their people to get things done **and respect**s that (when possible) their people will do what they think is the right thing to do; while these people can trust that their manager will "protect their backs" if & when things do go wrong.

- ☐ **Boundless Contribution** … A manager who allows their people to **contribute** beyond the predefined boundaries of their job roles; which means **empowering** them with varied – interesting – challenging work and giving them the **authorization & responsibility** to perform such work, thereby providing them with **opportunities** for their own **personal growth**, a **sense of achievement**, and **career advancement**.

- ☐ **Recognition – Timely Feedback – Constructive Criticism** … A manager who is aware of and actually **recognizes** their people's **efforts & contributions**, gives **public praise** for a job well done, provides **timely feedback** on their performance, and provides **constructive criticism** on that which they actually have the **opportunity to improve**.

- ☐ **Team Participation & Involvement** … A manager who is conducive to a **participatory team environment** where all members are given **equal opportunity to contribute** to the resolution of a **common goal** / problem / the objectives at hand.

However, **which motivators will produce the best results will depend on the characteristics of the individual** that the motivator is being applied to; because, their **personality traits, ethnical & cultural background**, their age (i.e. maturity), and where they currently are at that **stage in their life** will greatly influence the effectiveness. Consider "Maslow's Hierarchy Of Needs" of **Physiological – Safety – Social – Esteem – Self Actualization**, in [Figure 82], … now consider your **own personal situation & prevailing circumstances**. … So, what arrangement and weighting of these will influence your personal motivation to give your all for the project and for The Team, right now ??

A Team Builder

As stated previously in [Section 4.2.3] for recruiting the Project Team, … *"it is very important that the Project Manager strive to build a cooperative & collaborative Project Implementation Team; as it is not the Project Manager, but rather The Team who transform the project from mental concepts into physical deliverables"*.

Hence, the success of the project is very much dependent on The Teamwork that can be (and is) generated within The Team; and therefore, how well the project's manager encourages & fosters an environment that emphasizes team-building & team-work over individual self-interest & self-gratification.

While the project can have a pedantically detailed plan, lots of [Time] on hand, cash in the bank, an abundance of [Resources], and the best available [People] as team members; nevertheless, without the "right blend" of people to execute The Plan, then the project has at best a 50-50 chance of success (i.e. to not perform optimally as is desired / expected).

At this point, this book could detail "team inventory" [ref: Meredith Belbin] and describe the nine team roles; i.e. plant, resource investigator, coordinator, shaper, monitor evaluator, team worker, implementer, completer | finisher, and specialist. … But, for the majority of cases, as the project's manager, you will most probably "*just have to make do*" with those persons who have been assigned to the project by the performing organization's senior management (possibly based on who was / is available at the time).

Additionally, if you were involved with the recruitment of new people to work on the project, then, … unless you did some "stand-up & show us what you can do" SDLC Interviewing, see [Section 4.2.3], … then most probably you won't have encountered the recruit's true character; but rather, their "spit-n-polish" persona that they presented during the job interview. Therefore, let's forgo this team inventory theory, and instead concentrate on the apparent personality of The Team members that are at hand. That is, "*just make do with what you've got*", not who is desirable to have.

Ideally the Project Team should be built with people that have different personalities types; because, while a team of highly intellectual clones (or malleable worker drones) may be successful in the short term, this grouping will lack the spark to continually generate new ideas & initiatives (nor continue to be productive). Also, after a while this work place could feel stale, monotonous, and void of recognizable human life.

"Working in a cathedral of monotone faced zombies and drones."

The successful project team needs a blend of both technical minded and socially adept individuals; as a good mix of both types will result in a project team that has good morale, spirit, work ethic, mateship, and commitment to each other (to finish "their project").

Which brings us to the next rule of "Adaptive & Proactive" SDLC Project Management.

RULE 38: **It is not who has the most talented individuals in their team; but rather, who is able to get these individuals to function as a cohesive unit ... as an A-Team who strives to win.**

RULE 38[A]: **Project team morale is that elusive ingredient that can give the project continued drive (and faith in eventual success) when the current situation and prevailing circumstances have turned bleak.**

Motivation & Team Building ... *"a poor man's bonding sessions"*

It is a simple matter of economic reality that most small (to medium) sized organizations don't have the budget (nor the time) available for formalized team building activities that are targeted specifically at "team bonding" and improving "group dynamics". Hence, it is not feasible for members of staff to go offsite for sessions with "Human Relationship Consultants" to "workshop" The Team building experience.

Instead, The Team building will come down to the Project Manager, the Project Team members, and senior management taking it upon themselves to encourage & foster a friendly – sociable – cooperative – collaborative work environment.

This makeshift team building could be as simple as:

- Lunch time chats (away from ones' assigned desks) talking about anything other than work related matters. Even better, getting out of the office, in small groups for a sit down lunch at the local diner / cafe / park bench.
- A coffee break discussion about the latest goings on of some TV show, that movie which some have and have not seen, or that internet video cat sensation.
- Stirring each other over one's obsessive devotion to a make of car / sporting team / operating system / mobile phone brand; i.e. in jest stirring of each other.
- Banter about the kids, what they did, and the last in-thing which you just don't get.
- Friday afternoon (non-monetary) betting on how regional / national sporting teams will fair that coming weekend, combined with the obligatory Monday morning expert analysis of the results and the ribbing thereof (until next weekend's round).
- Going off for a walk, playing some sports, ... just having fun together as a group.
- It could be (after work – off of the corporate computer network) kicking back as a "clan" clashing swords with Orcs, launching that counter-strike to recapture the battlefield's high ground, or joining forces to holdout against suicide waves of rampaging alien hordes (or those irks from accounting).
- And, encouraging others (especially the newbie and the shy ones) to join in.

A Conflict Resolver

Whenever two or more people are involved with something, there will always be some form of conflict. Such conflict will generally stem from the involved parties "needs & wants" (for resources / territory / priority), for acceptance of their opinions, over principles or believes, due to poor communications, misinterpretations, or misunderstandings.

Surprisingly, often the simplest solution to conflict resolution is to **"listen" to the involved parties "airing their grievances", then "communicating" back** to them your interpretation of what they have said, and **asking "questions"** about the situation & circumstances of the issues. Sometimes, this **is sufficient to initiate the Conflict Resolution Process**.

- ☐ **Let each involved party state their case without being interrupted**, unless it is to politely tap them back on course because they have gone off topic or they are going over the same point again & again.

- ☐ Ensure that the "discussion" concentrates specifically on the **current issues** at hand and **does not degenerate into personal attacks, nor encompasses ancient history, nor misdeeds from the past**.

- ☐ **Show some indication to all of the involved parties that you understand what they have said** (even if you happen to disagree with their point of view); e.g., *the occasional head nod, and "hmm, yes, I see"*. Then **ask them how this issue can be resolved to their satisfaction; thereby, establishing a starting point for negotiations**.

The solution for how to satisfactorily resolve the issue may now be apparent, **given that all parties have had the opportunity to express their viewpoints, and state their case**.

Thus, **conflict resolution has to be undertaken in an open constructive manner**, using those techniques that were previously laid out for influencing & negotiating; as well as, listening & communicating.

| **THEORY ... *Conflict Resolution Techniques*.** | [Ref: Thomas & Kilmann] |

(1) **Withdrawing | Avoiding – is where one of the conflicting parties retreats from an existing or potential conflict / disagreement**. This technique deploys tactics such as; not bringing up the issue, sweeping the issue aside by changing the subject, or saying it will be dealt with later on (though fully intending not to do so). These tactics can thus be used to; postpone the conflict until one is better prepared to handle the situation, cool-off the involved parties (especially the aggressor), or simply to not have to deal with such a (trivial) issue.

If the Withdrawing | Avoiding technique is used inappropriately or continually then this can result in bottled up tensions, and conflicts simmering away beneath the surface of the relationships. ... And, this could very well boilover into a larger conflict than that which was being avoided in the first place.

(2) **Smoothing | Accommodating – is where one of the conflicting parties forgoes their own "needs & wants" so as to oblige the other conflicting parties.** This technique deploys the tactics of modifying or completely changing ones position (or opinion) to coincide with the other parties; possibly aligning with the consensus, the subject matter expert, the superior rank, or the dominant personality.

If the Smoothing | Accommodating technique is used inappropriately or continually then this can result in the obliging party being seen as **"a weak pushover"**.
In addition, this technique may conceal the underlying cause of the conflict between the involved parties. This technique could also be used as an extension of the Withdrawing | Avoiding technique.

(3) **Collaborating – is where the conflicting parties work together to find a "win - win" resolution that is "mutually beneficial" to the involved parties.** This technique involves; having open discussions about the conflict and its impacts on their respective objectives, with the involved parties contributing to the discussion, the proposing of alternative solutions, and finally agreement on the chosen solution.

If the Collaborating technique is used inappropriately or continually then this can result in the involved parties being bogged down in "design by committee" which may not be able to achieve much relative productivity in the available time. … And, this could result in a solution that is a compromise of political diplomacy versus practical viability.

(4) **Compromising** – **is where the conflicting parties work together to find a "no lose – no win" resolution that is "mutually satisfactory" to all concerned.** This technique involves; discussing the problem (possibly with the involvement of a neutral third party acting as mediator) where each party gives up some portion of their "needs & wants", bargaining & negotiating occurs between the involved parties, and finally the parties come to an amicable agreement / arrangement / understanding.

This compromising technique should be used when the conflict is escalating but there is still time to come to a mutually satisfactory resolution.

(5) **Forcing | Dominating | Competition** – **is where one or more of the conflicting parties adopts an assertive posture where their "needs & wants" are placed ahead of the other parties**; potentially to the other parties' detriment due to **a "win - lose" resolution.** This technique is based on capitalizing one's; authoritarian position within the management hierarchy (or within the Project Team), influence within the organization, dominance in the field/market, superior subject-matter knowledge, or simply by aggressive behaviour.

If the Forcing | Dominating | Competition technique is used inappropriately or continually then this can result in animosity between the conflicting parties, resentment by the "losing" parties, a feeling of superiority & dominance by the "winning" parties, a feeling of disempowerment by the "losing" parties, reduction in morale & motivation, and an antagonistic atmosphere amongst the involved parties.

The Forcing | Dominating | Competition technique should only be used to get things moving when a quick decision is essential, such as during a crisis event.

A Problem Solver

The Project Manager should live to solve problems, either by themselves or via the intermediaries of other people.

If you've come from a technical implementer's background then, most probably you're already adept at problem solving, and hence you don't have much to learn here. … How, ever, there is a major difference now that you are the project's manager; whereas, previously you would solve inanimate technical problems using your copious amounts of technical **"hard skills"** (i.e. analytical, mathematical, mechanical, logical experience), instead you will have to solve issues that are related to human beings and their inter-relationships. Thus, utilizing your humanities **"soft skills"** as a communicator, listener, negotiator, influencer, motivator, and conflict resolver.

As with the solving of technical problems, **an ad-hoc solution for people problems will often generate other unforeseen problems**, in some kind of ripple effect. A way to alleviate this is, to employ a **"PLAN – DO – CHECK – ACT" problem solving strategy** of:

1.. **Determine the real cause of the problem** (and not just the evidential side effects).

2.. **Break the problem up into its constituent parts**, come up with alternative solutions to each part, and then accumulate a combined solution(s) to solve these problems.

3.. **Choose the optimal solution** (or collective solutions) to be implemented.

4.. **Determine whether the implemented solution**(s) has produced the **desired results**.

5.. **Go back to step 1 and repeat** the sequence until the problem & associated problems are either resolved to a satisfactory resolution, or are no longer a pressing concern.

Though, often the **best problem solving result is via a collaborative approach of involving other** (appropriate) **parties** in the **PLAN – DO – CHECK – ACT** resolution; see [Section 2.4]. And, the **Recognize – Reassess – Revise – Reapply** revolution; see [Section 2.5].

A Planner & Organizer

The Project Manager has to be able to coordinate with the project's stakeholders so as to ensure that the necessary [People] & [Resources] are made available and are kitted-out ready to perform the agreed [Scope] with the agreed [Quality] in the agreed [Time], and to ensure that the agreed [Cost] budget is available so that there is sufficient cash flow to pay for the implementation work to be done.

That is, **to orchestrate others to transform "The Plan" into the reality of deliverables**.

If the project's manager is a poor organizer, then no matter how diligently the planning was conducted, the project will be "spinning its wheels going nowhere fast" or come to a grinding halt. This is because, someone [People] or something [Resources] will not be available when required; consequently, other [People] and/or [Resources] will have to "hurry up and wait". Hence, there will be [People] standing around wondering what they should do in the meantime, there will be [Resources] gathering dust & going to waste. And, all the while the project's [Time] clock keeps ticking away, as the project's [Cost] expenditure keeps tallying up. Subsequently, the project could operationally turn into a complete debacle, irrespective of what "The Plan" had dictated.

Thus to succeed, **the project's management has to be able to "compartmentalize" sections of the project into "work packages" that when necessary can be delegated (handed-over) to others to complete** within a given [Time], within a given [Cost], using given [People] & [Resources].

"And you thought herding pre-schoolers (or soggy cats) was hard. Wait until you're monitoring & controlling implementers, 3rd parties, and project stakeholders; as this is just as challenging, if not more difficult because you can't just pick them up and dump them wherever you need them to be."

A Delegator

To succeed as a planner & organizer; the Project Manager has to be able to (and willing to) designate & allocate activities to other project stakeholders to undertake. This delegation could be downwards to individual Project Implementation Team members, sidewards to other project managers (or equivalent), and when the situation necessitates, then upwards towards the project's primary stakeholders (i.e. to senior management).

However, **delegation does not mean handing over to the delegation-receiver the final responsibility for the activity's outcome, as this remains with the delegator; i.e. the responsibility remains with the project's manager.**

 Be wary of delegating upwards to the project's primary stakeholders, because doing this too often could start them wondering what is the point of you, the Project Manager. Therefore, **only delegate upwards those essential decisions related to sensitive risks & issues, and changes to the agreed & approved [Baselines]**.

By not delegating and keeping all of the work to oneself, the Project Manager is burdening themselves with a potentially excessive workload. A workload which will be greatly increased when senior management decide to (reward you for the great job that you are doing by) assigning you other projects to simultaneously manage.

In addition to reducing the workload on the Project Manager, delegation has another useful benefit; by **delegating** some of the work **to subordinates**, the Project Manager is **training their potential replacements**.

Why delegate, and thereby make myself replaceable?

Well, **one way to make yourself un-promotable is to make yourself irreplaceable** in your current role. ... And, if no one else can do your job, **then you are "a single point of failure"**, **where there is no contingency for your unavailability** for some period of time.

"Also, its highly motivational if correctly executed with empowerment."

Additionally, not only is the Project Manager limiting their own promotional opportunities but they are also curtailing that of their subordinates who never get the opportunity to really expand their skillsets & responsibilities. ... And subsequently, this can have negative effects on the long-term motivation of the Project Implementation Team members.

"So, not only are you curtailing your promotional opportunities, but you are also impacting on your time availability to your family & friends. Oh, and you can forget about getting a swag of time off for that couple months European Vacation, ... else, when you get back from your holiday then expect bucket-loads of work and even recriminations."

Hence, **every member of the Project Team should be mentoring his or her own replacement, including the implementers and yourself, as the Project Manager**.

4.4.2. Being a "Leader" Project Manager

How diminished are the project's chances of success if the project's manager cannot **focus The Team's efforts towards the common goal** of producing the project deliverables within the agreed [Time] & [Costs] while providing the agreed [Scope] coverage and conforming to the agreed [Quality]? That is, being able to "lead" The Team to achieve or exceed those judgements of "Customer Satisfaction" and "Performance Measurement", [Section 2.8.2].

What Is Leadership?

"Well, one thing for sure ... leadership is NOT a popularity contest, nor who is the most charismatic, nor who thinks they are the smartest person in the room (just because of their subject matter expertise), nor who has the most academic degrees framed upon their office wall."

Leadership is the ability to communicate a common vision of the objectives then inspiring others to be driven to achieving those objectives (as effectively & efficiently as possible).

Leadership Styles

The Project Manager needs to realize (and understand) **what leadership styles** are compatible with their own personality traits; because, the project's current situation & prevailing circumstances will necessitate that the project's management apply differing portions of the following leadership styles:

- **AUTHORITARIAN** – would be **beneficial during a "crisis event" where quick steadfast decisions are paramount** to success.

- **DEMOCRATIC** – would be **beneficial** during the project's Planning & Executing Phases **where the input of different people's perspectives & opinions could produce better options and more realistic solutions** (than could be perceived solely by an individual).

- **LAISSEZ-FAIRE** – would be **beneficial for the day-to-day functioning of The Team**, while the Project Manager deals with those external issues confronting the project.

When **selecting the appropriate style of leadership to be applied to the current situation & prevailing circumstances**, the Project Manager **will also need to consider** what is more important to the performing organization at that moment in time;

(a) **being task oriented** – of completing the project's objectives asap,

(b) **being relationship oriented** – of maintaining & sustaining working relationships with the project's team members (and other parties).

These considerations will mean that **sometimes the Project Manager will have to "make hard decisions"** that put achieving the project's objectives ahead of the interests, concerns, and the individual outcomes for The Team members.

> Which brings us to the next rule of "Adaptive & Proactive" SDLC Project Management.
>
> **RULE 39:** Leadership is not the wielding of power; but rather, orchestrating & influencing others towards achieving defined & tangible objectives.

| THEORY ... *Leadership Styles.* | [Ref: Lewin, Lippitt, White] |

(1) **Authoritarian / autocratic** – **is where the leader is central to all decision making**; i.e. an "absolute ruler" come dictator.

While this style of leadership does result in quick decision making, it does not necessarily result in the correct or optimal decision; because, this type of leader does not always listen to or accept the input & opinions from those whose station is subordinate (or they consider as subordinate) to their own hierarchical rank.

Team motivation & participation, as well as their supposed respect for the leader, are often induced via negative reinforcement and threats of punishment.

However, some form of authoritarian leadership style maybe beneficial during the project's Initiating Phase and especially during the project's Closing Phase when The Team could be inflicted with "perfection procrastination" (of aspiring to be 110% complete) before moving onto the next activity.

(2) **Democratic / participative** – **is where the leader consults with the group before making the decision, or the decision is based on the consensus of the group.**

Team motivation & participation is via self-interest to fulfil the common vision and objectives.

The democratic leadership style is most beneficial during the Planning Phase and the Executing Phase when "team buy-in" is essential for the project's potential success; hence, why it is the basis for agile techniques.

(3) **Laissez-faire / free rein** – **is where the leader removes oneself from the decision making process** (almost in entirety), **instead entrusting the group or specific members of the group to handle certain domains from thereon.**

Team motivation & participation is via the empowered self-interest to fulfil the common vision and objectives. Thought, this leadership style can leave The Team members wondering what is the purpose of the Project Manager.

> The **laissez-faire leadership style is beneficial when the Project Manager is predominately outwards facing** so as to deal with the primary stakeholders' needs & wants – expectations – perceptions – concerns (i.e. external issues); **while the senior project team members (such as The Team leader) are inwards facing** so as to deal with the technical issues and the day-to-day tasks of continuing to move the project forwards. Whereby, it is the Project Manager's role to remove those barriers preventing The Team from advancing towards those stated objectives. Thus, why this leadership style is an essential ingredient to agile techniques.

What Makes A Good Leader?

Before listing what does and does not make a good leader, it must be realized that, the **current situation & prevailing circumstances are often determining factors as to whether a person is** (in hindsight) **a good or bad leader**.

For example; there has been many a political leader (who according to the opinion polls) were consider by many of their constituents to be incompetent to say the least, yet when war or natural disasters befell the nation / state then they stood-up and proved to be the right person to do the job. Similarly, there has been many a great crisis-leader who has fallen from grace during civil times.

What are the traits of good leaders?

- ✓ The presence of self-confidence, high self-esteem, and a belief in themselves.
- ✓ They provide their followers with a feeling of purpose, well-being, safety in numbers, security, and stability.
- ✓ They are good communicators of their vision, and they are focused on precisely communicating those objectives that they see as achievable in the short to medium term (with some conceivable connection to a long term plan | future).

- ✓ They are able to persuade others to follow them, and they are able to drive others towards achieving those short to medium term objectives.

- ✓ They earn the respect & trust of their followers, and they respect & trust their followers to do what is required / requested of them. ... Not micromanaging every aspect of The Plan.

- ✓ They empower their (in the field) subordinate leaders with the authorization and authority to modify & improvise on The Plan as they believe is necessary to adapt to the current situation & prevailing circumstances that are in affect at the point of implementation.

- ✓ They (themself) as leader adapt to the ever-changing situations & circumstances. And, not locking themselves away in their office bunker, as they hope to ride out the consequences of their plan being different to the unfolding reality.

- ✓ When necessary (for the greater good of all) they are prepared to make those hard decisions, irrespective of the negative consequence to themselves and their followers.

- ✓ They take personal responsibility for the outcomes of their decisions, whether that be good or bad. ... And, if it does go all wrong then a leader does not look for the closest scapegoats on whom to lay the blame; instead, the real leader is more concerned with getting back-on-track and determining how best to resolve the current issues.

What a good leader is NOT:

- ✗ Narcissistic – arrogant, self-absorbed, egotists, with an unassailable appetite for power.
- ✗ Toxic – leaving their followers (i.e. the Project Team and the performing organization) in a worst state then when they first started to lead the group.

Which brings us to an addendum to "Adaptive & Proactive" SDLC Project Management..

RULE 39[A]: What makes a leader is not the willingness to charge into adversity; but rather, the sense to know when & when not to charge.

| THEORY ... *Management Styles.* | [Ref: McGregor] |

The Project Manager needs to determine **what style of manager** they are likely to be.

- **THEORY X MANAGER – the "no you can't" manager will intrinsically need to micromanage every little thing that the Project Team members do**.

This style of manager will often be driven by some innate mistrust of The Team members that they will not do their jobs, because they are lazy unmotivated slackers who will (whenever possible) go out of their way to avoid work and they will cheat & lie about the work that they have & have not done. And/or, this Theory X manager believes that The Team members don't have the sufficient skills to advance the project forwards without that manager's continual guidance & instructions.

Hence, the **Theory X manager will be highly restrictive with the delegation of work and the allocation of responsibility** to the Project Team members, even going as far as to **silo team members in discretely demarcated job roles**. This will all be based on the manager's belief that the only way to get these "workers" to do their jobs (other than by the pay-cheque) is by the **use of rigidly defined hierarchical work structures and draconian work processes & procedures** (which includes baseline quantifiable statistics for performance goals and the punishments for non-achievement).

In a performing organization where there is a desire to be people oriented instead of purely procedural (e.g. agile instead of production line), the **Theory X manager can have a devastating effect on the Project Team's morale.** This is because, the Theory X manager often believes that whenever something goes wrong, it is the workers' fault *"because they must have not abided by the details of The Plan and/or they did not follow the stipulated processes & procedures"*; instead of considering the possibility that the failure was due to mitigating considerations of the current situation & prevailing circumstances. *... "Or due to the manager's own failings."*

For a performing organization that is using an agile implementation methodology then the **Theory X manager is completely at odds with the underlining principles of agile**, that it is The Team and not the manager that drives the Executing Phase of the project.

The **Theory X manager is more suited for the traditional waterfall** (and some iterative) implementation projects where micromanagement is the expectation.

- **THEORY Y MANAGER** – the "yes we can" manager will trust the Project Team members to do their jobs, and feels confident that each member will not let The Team down (and for that matter, let the project's management down).

This style of manager will be driven by the belief that The Team members will be self-motivated to do their job as best that they can, given the current situation & prevailing circumstances, and based on the training & experience that they have.

Hence, the **Theory Y manager will be prepared to delegate areas of work and allocate responsibility to The Team members**, even going as far as **encouraging cross-functional team arrangements**. This will be based on the manager's belief that to motivate The Team members (to perform at their best) then they have to be presented with challenging work, problems to solve, opportunities for growth & advancement, and diversity in their roles & responsibilities.

In a people orientated performing organization, the Theory Y manager will have a beneficial effect on the Project Team's morale. Hence, the **Theory Y manager** is what **is required for a project being implemented using an "adaptive" agile methodology.**

However, in a procedural (waterfall) performing organization, the Theory Y manager could be driven mad by the bureaucracy & inflexibility. … And, downright frustrated by the inability of the organization's senior management to (dynamically) adapt to the changing situation & prevailing circumstances that now confront the project.

Though, a project manager is not all Theory X : Theory Y, or a sliding scale with the two Theories at opposite ends; rather, a spectrum of both styles amalgamated into one.

4.4.3. Your Personal PM Credo

When you get up in the morning and look at yourself in the mirror, what defines the person that you see reflected back at you?

At some point in your project management career (especially when the economy is bad and your employer is not doing so well), **you could be requested to do things that are at odds with your sense of business ethics and your personal morals.**

So, how will you handle such requests; yet, still be able to look at yourself in the mirror?

P.E.R.I.O.D.

Patience – Effort – Respect – Integrity – hOnesty – Dignity.

- **PATIENCE** – to give reasonable & sufficient time for individuals & groups (and yourself) to get things done to an acceptable level of quality and scope coverage. ... And, not rush things through, so that, all that can realistically be delivered is a makeshift resolution that is riddled with unsatisfactory compromises.

- **EFFORT** – to give (and expect) reasonable & sufficient effort to be put in by individuals & groups (and yourself) to get things done to an acceptable level of quality and scope coverage. ... And, not to over-work / under-work people, so that, all that can realistically be delivered is burnt out individuals and/or a half-arsed resolution that is riddled with unsatisfactory compromises.

- **RESPECT** – to give due regards to individuals & groups (and your own) efforts, capabilities, experiences, and responsibilities so as to empower them to do what they deem necessary to be done to achieve the stated objectives; without being second guessed, undermined, and/or overridden by (micro) management from above.

- **INTEGRITY** – to be truthful & accurate in what data / information is presented.

- **HONESTY** – to be truthful & sincere in what is said.

- **DIGNITY** – to be truthful & humble in what is done and how it is accomplished.

> **EXAMPLE CASE** ... *Business ethics versus your personal morals.*

Consider the following real-world scenarios.

...

You are (verbally) requested by senior management, *"to just"*, invoice the customer for a portion of the project work that is yet to be done. Though, your boss has reassured you that it will be done first thing next month. But, it needs to be invoiced now, so as to improve the organization's financial results for this quarter's / End Of Financial Year's reporting.

You could raise your concerns, hesitate to do what is requested, or you could humbly decline to do this request. However, would your "high morals" stance be marking you as *"not a team player"*, and hence could you possibly be limiting your future career prospects at that organization (under that management's reign)? ... And, what would happen if you decided to escalate this issue, to those bosses above them?

...

There are two unrelated but simultaneous projects for the same customer. Both of these projects have been "sold" to the customer as being 1000 hours of effort each. However, during the implementation, one project is a lot easier than expected and only required 750 hours of effort, whereas the other project was harder than expected and required 1250 hours to complete. Your senior management tell you, as the Project Manager, *"to just"*, shift hours in the timesheet system from the over-hours project to the under-hours project, and then invoice the customer as though each project was in factuality a 1000 hour job. ... As there is *"no overall harm done"*, as the customer got what was promised, and the performing organization *"broke even"* on both projects.

...

There is a fixed price project for the customer, and at the same time, that customer also has an ongoing contract with the performing organization to provide a certain number of maintenance & support hours per year for their suite of products. However, during the project's implementation, it goes way over the hours of effort and will be a loss maker for the performing organization. Your senior management tell you, as the Project Manager, "*to just*", shift hours in the timesheet system from the over-hours project to the maintenance contract's job code. ... Given that, *"the customer is not using up all of those maintenance hours, and the project's product output would just use up those maintenance hours when it's released, so might as well just use those maintenance hours now during the product's development"*.

...

So, **HOW FAR IS YOUR SENSE OF REAL-WORLD BUSINESS ETHICS PREPARED TO BEND?** Before you decide on what is right & wrong, reconsider those previous examples and the following examples, and set these during a major financial crisis for your employer.

...

There are two unrelated but simultaneous projects for two unrelated customers. One project is going really well and is not consuming all of the planned hours of effort, but the other project has real problems and is burning through its hours of effort. Your senior management tell you, as the Project Manager, "*to just*", shift hours in the timesheet system from the over-hours project to the under-hours project, and then invoice each of these independent customers as per the timesheet factualities.

There are two unrelated but simultaneous projects for two unrelated customers, but there is a common body of work to be developed. Your senior management tell you, as the Project Manager, "*to just*", book both customers for the full amount of the common work. In essence, invoice double dipping.

...

So, **HOW BENDABLE IS YOUR ETHICS**, "*just to*" retain your job during such an economic crisis?   ... "*Oh, and those customers are your national government.*"

> "IT IS THROUGH ONE'S OWN BATTLES, THAT ONE DISCOVERS WHAT DOES AND DOES NOT WORK FOR THEM. ... GIVEN THE SITUATIONS AND THOSE PREVAILING CIRCUMSTANCES."

That sounds like a Sun Tzu "Art of War" style of retrospectively stating the in hindsight obvious, having experienced similar situations & circumstances.

NOTE: Each of the following example case "field test" scenario projects focuses on different real-world aspects of project management (both the good & the bad) that you are likely to encounter at some point in your PM career. ... And, sometimes, these situations will be experienced again & again, like "déjà vu"; yet, as one project finishes and the next project begins, the same underlying "high impact" mistakes will be made (and these repeated mistakes will not necessarily be made solely by the Project Manager, but by those even higher up the chain-of-command).

Therefore, the following chapters detail the occurrence of these situations, and how to cope with these various situations; subsequently, some sections of the following chapters may feel a bit like a scientific formulated text on the processes & procedures to be utilized as mitigation & correction techniques.

5. PM you ... Field Test 1

5.1. Overview

This "field test" examines the following topics with respect to real-world project management:

- Building a good first impression of the Project Manager in the eyes of the performing organization's senior management, Project Implementation Team, and general staff.

- Obtaining an understanding of the project; what is entailed, and what is the "lay of the land".

- The project value to the business greatly exceeding the investment in the implementation.

- Adapting a recognized project management methodology and SDLC techniques to the needs of the organization, and to the size & scale of the project to be undertaken.

- Adapting & improvising based on the project's available time duration and budget.

- Getting the project underway based on a laid out plan to move forwards.

- Costing the project for success, not pricing it to a certain death.

- Steps to rescue and recover an ailing project when it gets into trouble.

- How a project can get into trouble, and that differences in understanding of SDLC can lead a project to "failure".

- Coping with "failure", self-doubt, work related stress, and making those hard decisions.

- "Failure" is a possibility of fact.

5.2. Introduction

So, what now? ... Given that, at your disposal is all of your project management knowledge and training, thus far. ... Plus, the copious amounts of your past technical experience that you can always call upon in a pinch.   ... *"Though, will your technical past be a true predictor of your future success as a project manager?"*

Well, I guess there is nothing else to do, but put your personalized project management techniques (i.e. "PM you") to the trials & tribulations of a real-world SDLC project.

5.2.1. The Scenario

```
   Yes, ... the interview was a success, and today you start
your new job at this smallish company that wants to build
their own bespoke product to replace their (externally
sourced) aging primary business platform.
```

Though, the question still remains, hidden in the back of your self-evaluating mind, as to whether your past experiences (both managerial and technical) will be sufficient to lead your assigned project to a "deemed success", ... of being delivered in accordance with:

- **Performance Measures** – within the agreed [Time] frame and agreed [Cost] budget.

- **Customer Satisfaction** – with the agreed [Scope] coverage and agreed level of [Quality].

- **Effective & Efficient Utilization** – of those assigned [People] and allocated [Resources].

- **R.I.S.C. Management** – given the project's inherent <Risks>, and given the current situation & prevailing circumstances confronting the project (and the performing organization).

5.2.2. Building That First Impression

Building A Good First Impression, of PM you

As per [Section 4.2.2], Day 1 - Week 1, building that good first impression of PM you:

- ☑ **Dress appropriately for the occasion** – dress ready to achieve the business's objectives of delivering the project successfully while conforming to social norms.

- ☑ **Introductions all round** – politely & calmly introduce yourself to those persons you meet in the performing organization; note-padding each person's name (at least their first name or surname as per cultural norms), where they sit, and what their role is. Especially, those persons connected to the project(s) that you will be responsible for.

- ☑ **Meet The Team** – strive to build up the Project Implementation Team members' initial confidence that you know what you are doing; hence, present a demeanour of **confidence in** one's own abilities, but not arrogantly self-assured nor overconfident.

- ☑ **Mud map the project** – kick off with either the project sponsor, senior management, your direct supervisor, and/or the senior implementer(s) to determine what are;
    - the **basic details** of the project, i.e. the customer's needs & wants,
    - the **basic architectural** details of the project's product / system / application that is to be implemented, i.e. block diagram the major parts, and
    - the **basic** details of the **technologies and processes & procedures** underlying the project's product / system / application that is to be implemented.

- ☑ **Small talk alignment** – start **building** those **friendly working relationships** within the performing organization, especially with fellow project managers, and your immediate contacts such as your supervisor and the members of the Project Team.

- ☑ **Discover the true lay of the land** – as what was implied during the job interview may not have been the **true representation of the project's current situation & prevailing circumstances** confronting the project (and the performing organization).

For this scenario's project; during that first week of getting to know your way around the performing organization and building up those working relationships that will be essential for the success of the project, you discovered:

(1) The external source of the incumbent system has overtime reduced the depth of the rollout of new features & functionality via the release of periodic product upgrades; hence, this business application has fallen behind in the ever evolving competitive industry that your employer is engaged. … *"Let alone, that the product looks & feels like it was built during the last decade, being based on aging technology that is now experiencing compatibility problems with today's operating systems and web-browsers, so that the platform is one band-aid fix upon another fix"*.

(2) Those features & functionality change requests that your employer has submitted (and in some cases, paid for the development of), *"recently these haven't seen the light-of-day"* as *"it would appear that they are more concerned with the needs of their own domestic market and have forgotten all about us international customers."* … And, *"the product lacks the presentation & usability characteristics of modern day web applications"*.

(3) Helpdesk & technical support of the incumbent system by the external source has degraded in service turnaround times, now taking several days to weeks delay to respond to those difficult questions. … *"Let alone, the significant time zone differences between them and us"*.

(4) And, speculation is that, *"within a year or two, they are going to terminate that product line to concentrate on their other business. Thus, leaving us out in the cold"*.

Hence, the conversational conclusion is that your employer has no other viable options, other than to go it alone and develop their own application; because, the alternative products offerings are either just as aged as the incumbent product or the associated licensing & service agreements would be too costly for your employer (or more specifically the trickle-down cost to your employer's customers would be too excessive for the local market to accept). … And, that there is only 12 months within which to deliver the first publically usable release of this replacement system.

In addition to the project's technical situation, you also discovered that, other than industry specific Subject Matter Expertise to call on, there is currently no work-experienced SDLC capabilities within your employer's organization with which to start building the Project Implementation Team. Hence you, as the project's manager, are going to have to recruit the project development team from scratch (with the assistance & oversight of your employer's senior management).

You also discovered that, while "*formal project management qualifications and Agile project delivery experience*" were advertised as essential to your job's description, there is not a significant amount of SDLC project management experience within the organization. Hence you, as the "*Technical Project Manager*", will have to define the SDLC methodology that will be used by those future additions to this (pending) performing organization.

And, while there has been capital reserves set aside for the project, you have the niggling feeling that there is most probably not a deep-pocketed management reserve with which to call on, if the project happens to get into trouble.

Business Value versus Face Value

While your project may carry a relatively low dollar cost of implementation (i.e. "face value"), it could carry a high "business value" to the performing organization as a "keystone" or "linchpin" project.

That is, **while the project's cost of implementation maybe relatively low, the customer organization and/or the performing organization could have the business's future riding upon the successful outcome of that particular project.**

> *For example; this I.T. systems upgrade project may only have a price tag of $100k - $200k, yet it results in an essential part of the continued & effective operations of the organization's multi-million dollar business. If this "low budget" project was to perform badly, then the business could find that its operations are gravely affected (and its profitability declines noticeably).*
>
> *For example; this relatively "low budget" project is the first steppingstone to a larger program of work. Where the customer organization is, in essence, using this project as a litmus-test of the performing organization's capabilities & competencies. If this "low budget" project was to perform badly and/or fail, then the performing organization could be eliminated from consideration for the larger contract(s), and subsequently find itself scrounging for other work.*
>
> *For example; this "low budget" compliance project has to be rolled out by a specific date, else the performing organization would lose it certification / license, and subsequently not be able to legally operate in its primary market.*
>
> Which brings us to the next rule of "Adaptive & Proactive" SDLC Project Management.

RULE 40: A project's worth is not always proportional to its upfront cost of implementation.

5.3. And So, It Begins

5.3.1. Answers To The Opening Questions

Back in [Section 3.3] it was stated that; *"the purpose of the Initiating Phase is to define a potential new project based on the 'needs & wants' of the customer, to evaluate the proposed project's viability, and to then obtain formal signed off approval & authorization to commence the project."*

Additionally, there was listed a sequence of questions that needed to be diligently answered, so as to establish the project's realistic chances of success.

(1) What do **THEY** (the customer) **WANT**?

(2) What will **WE** (the implementer) **GIVE** (to the customer)?

What will **WE** (the implementer) **NOT** be **GIVING** (to the customer)?

(3) Approximately **HOW LONG** do **WE** (the implementer) **EXPECT** it to **TAKE**?

What will it **COST US** (the implementer) **TO GIVE** it (to the customer)?

(4) What will **WE** (the implementer) **GET IN RETURN** (from the customer)?

(5) What is **AT STAKE FOR US** (the implementer)?

The answers to these above questions should provide a good enough understanding of the project's primary stakeholders "wants", "needs", high-level "requirements", "desired outcomes", and the **"perceived benefits" of conducting the project.** These answers would subsequently be **recorded in** some formalized document to officially accept or reject the undertaking of the project; i.e. the **"Project Charter"**.

```
But alas, for this scenario's project, the Project Charter
is considered as a given, because the project "just has to be
done".
```

A **project kick-off meeting** is held, for your benefit as the first member of the Project Implementation Team to come on-board. … And, this meeting turns into a couple of weeks of coming together and going through the "Customer Requirements" document. The Customer Requirements document that is a hundred plus pages of bullet-point notes which are a collection of requirements intermixed with ideas (and self-questioning thoughts) shuffled together into quasi-sequential sections, with the insertion of clarification notes and the addition of extra requirements that were thought of during these "explanation" meetings.

Your past technical experience kicks in, and on the meeting room's large whiteboard you start sketching overview block diagrams of how the various parts of the system seem to fit together (from the perspective of your very limited topic relevant subject matter expertise). The end result being an artwork that is more Picasso-esque than a system engineering drawing; however, your "customer" the project sponsor and the in-house SMEs are impressed with seeing their system represented in an abstract-physical form.

By the conclusion of this meeting you realize that a few factors will influence how this project should be undertaken:

- ☠ Customer injected "**scope creep**" of adding extra requirements (as the need arises) is potentially going to be an ongoing issue during the project's life.

- ☠ **What is being defined** is the [Scope] **equivalent of a "limousine" when the available [Time]** would **only** be **sufficient to produce a "daily driver"**. Hence, work towards getting something onto the road in the shortest time, so as to have something functional to use.

Given the [Time] criticality of delivering a useable system as soon as possible, while faced with continued [Scope] clarification and the possible inclusion of additional features & functionality, then a **traditional Waterfall system development life cycle of defining the majority (if not all) of the requirements upfront, … prior to the sequential planning then implementation then testing of the deliverables,** … IMHO, is probably not going to work out very well for this particular project.

Hence, you figure that a **"rolling wave" Iterative system development life cycle of breaking the project up into major releases of specific features & functionality would be better suited to the situation at hand.** However, … …

☠ You have the growing suspicion that the Customer's Representative is more of a **traditionalist Waterfaller;** and as such, will want to define each & every requirement in detail before moving onto the Planning Phase. You also surmise that the Customer's Representative **could have acceptance issues with limiting the requirements definition to those high priority items that need to be known right now, and pushing back on the detailed definition of those lower priority features.** Thus, you have a niggling feeling that you are in for a bit of **"negotiated compromising"** on the [Scope] of the first release (and the subsequent releases).

> **Rolling Wave Planning** is where those activities to be performed in the immediate future are defined & planned out now in detail, and those activities to occur in the not so immediate future are only broadly defined & planned for (and these will be dealt with in more detail later on during the implementation of those earlier parts).

5.3.2. Adapt & Improvise on the Project's Documentation

For this scenario's project, you realize that producing all of those documents as per a PMBOK® based SDLC structure, shown previously in [Figure 23], would consume a considerable amount of the project's available [Time] and allocated [Cost] budget. Hence, you decide to minimize the documentation tree by hybridizing & amalgamating some documents to have summation information transfer purposes; based on, your past experiences of "*what really needs to be known*".   See [Figure 83].

"The sign of a leader is knowing when (and being prepared) to revise & adapt an existing plan | framework to better suit a given situation; when taking into consideration the prevailing circumstances of limited time, money, and people resources that are currently at hand."

The Project Charter

"*The purpose of the Initiating Phase is to define a potential new project based on the 'needs & wants' of the customer, to evaluate the proposed project's viability, and to then obtain formal signed off approval & authorization to commence the project*", [Figure 26].

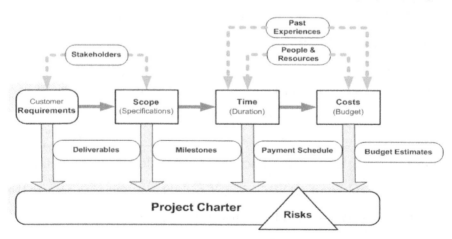

Figure 26: The Initiating Phase as a process to create the Project Charter.

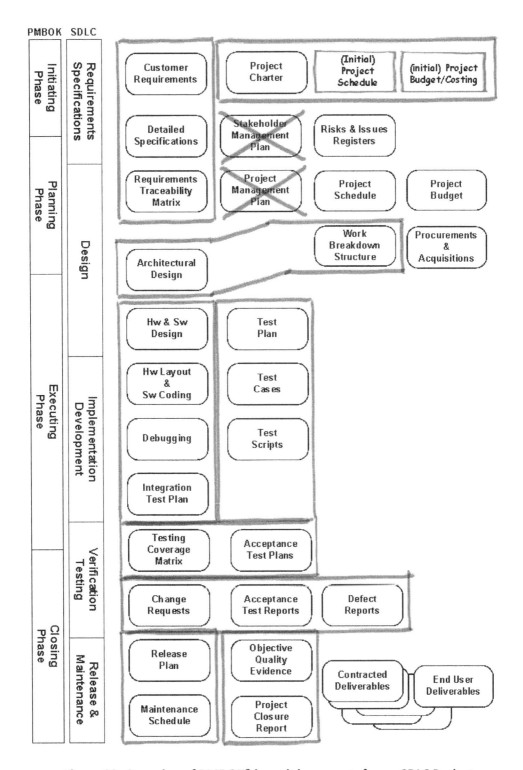

Figure 83: Grouping of PMBOK® based documents for an SDLC Project.

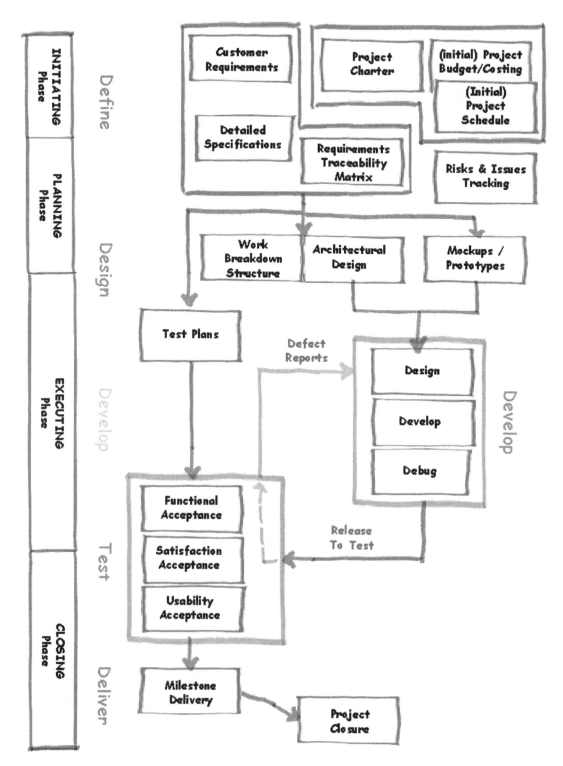

Figure 84: Grouping of documents to cope with the project's situational constraints.

So, ... what to put in this **Project Charter**:

(1) **Formally presented document** ... with an obviously titled cover page (with the performing organization's company branding and identifying the customer organization), table of contents, section headers, and consistent font style & size and text spacing to ease readability. Not just a hurriedly thrown together blank-format document of bullet-point notes; rather, the appearance of a professional document.

(2) **Identifying name** for the project ... a name that human-people can easily roll-off the tongue in conversations & discussions. Not just a serialized Project Number nor convoluted acronym which denotes the constituent parts of the outputted product.

(3) **Table Of Contents** ... to retain a sense of order and to aid in information discovery.

(4) **Introduction**

- **Background Information** ... an "executive summary" of the underlying 'needs & wants' as to why this project is proposed to exist. Briefly describing what is this proposed project's purpose, and what are the 'perceived benefits' from undertaking this project.

- **Project Goals** ... a summary of the 'desired outcomes' of conducting this project.

- **Measurable & Tangible Objectives** ... a bullet-point summation of the project's high-priority high-level [Scope] of deliverables and those hard-pressed constraints on the project's undertaking, such as; restricted [Cost] budget, unmovable delivery dates [Time], regulatory [Quality] standards compliance, limitations on the assignment of [People] and/or allocation of [Resources].

(5) *Optionally* ... **Extra Introductory Information** that would be beneficial to (be plagiarized by) senior management when putting together a submission to access externally sourced funding, such as via venture capitalists and/or government development grants. That is, the information that would be required to sell the proposed project to those potential financial investors, in a 5 minute presentation.

(6) **Project Constraints** … those project parameters, see the theory in [Section 1.2], that are *"set in stone"* (i.e. FIXED) from the outset of the commencement of the proposed project:

- **[Scope]** … those high-level features & functionality that are categorically essential for the project to deliver; else, the delivered system / application / service will not be "fit for use" for "the purpose that was intended" by the end-user customer.

    For example; a mobile phone absolutely must be able to make & receive bidirectional telephone calls. In comparison, all other functionality is relatively meaningless without this mandatory capability.

- **[Time]** … where there is some unmovable *"drop-dead"* End Date when this project absolutely must provide its deliverables by, or ELSE !!!

- **[Cost]** … where there is some form of financial limitation on this project; either the total funds available, the maximum "burn-rate" spend per month, and/or an assigned "bucket of money" that must be spent for the quarter / half-year / financial year (where the project cannot over-spend nor greatly under-spend else risk losing an equivalent portion of the next period's "bucket of money").

- **[Quality]** … where there is some [Quality] standards / certification that must be abided by for regulatory, statutory, and market expectations for the project deliverables to be deemed acceptable.

- **[People]** … the availability /quantity / skill-level of the necessary & available [People], both internal & external to the performing organization, will greatly influence the project's [Time] & [Scope] that can realistically be delivered.

- **[Resources]** … the availability / quantity / grade of the necessary & available [Resources], both internal & external to the performing organization, will greatly influence the project's [Time] & [Scope] that can realistically be delivered.

(7) Additionally, the **Priorities** of these project constraints may be specified.

(8) **Project Assumptions** … those things that are believed to be true about the project; such as, the availability of key skilled personnel, allocations of equipment & materials, pre-existing technical knowledge, stability of the underlying technologies, meaningful & valid data, the future economy, and environmental conditions… etc.

(9) **Project Risks** … those high-level risks involved with (and with not) undertaking this project; including those **inherent risks**, **management & control risks**, and **execution vs. operational risks**. See [Section 2.6.1]. Plus the project's **potential resultant impact outside of the performing organization and/or outside of the customer organization**; i.e. to the general public?

(10) **Project Scope Boundary** … those things that will be considered as being within the project's domain, and possibly listing those things that will be "out-of-scope".

- **High Level Requirements** … a generalized list of those high-level features & functionality that will be handed over to the Customer's Representatives during and at the end of the project, and possibly at various release points during the project's life. The receipt & acceptance of such deliverables will form the basis of the Acceptance Criteria for the project.

DO NOT venture into the specifics of the details (no matter how deep), as that is the purpose of the Detailed Specifications. … Just "a single sentence" description sufficient to establish the foundation of the "idea" of each high-level requirement.

(11) **Intellectual Property Rights** … who exactly **owns** what of those **intangible assets** of the "**knowledge learnt**" during the undertaking of the project and as a result of the process of creating the deliverables.

For example; inventions, patents, copyrights, trademarks, designs, source code, circuit layouts, algorithms … the know-how.

Clearly state what are the (IP) Intellectual Property Rights of each party involved with the project; including the customer organization, the performing organization, and any third parties (such as sub-contractors) who will be engaged to participate in the project.

- **Background IP** – is that intellectual property which existed prior to the contracted relationship commencing, or is independent of the contract.
- **Foreground IP** – is that intellectual property which resulted from or is generated pursuant to the contracted relationship.

(12) **Project Phases**

- **Delivery Milestones** … a list of the major releases; though, what exactly will be delivered in each specific release should be still open to negotiations (based on the situation & prevailing circumstances closer to the time of delivery). And, noting that a "rolling wave" methodology is being proposed for this scenario's project.
- **Summation Schedule** … a **Rough Overview Schedule** of "a time scale" sufficient to provide a logical guesstimation of the duration and the "level of effort" involved with conducting the project, so as to evaluate the duration viability of the undertaking. However, this **is NOT a definitive**, down to the last day & hour **valuation of the project's [Time]**.
- **ROM Budget** … a **Rough Order of Magnitude (ROM)** *"in the ball-park"* intelligent guesstimation (within at least a 20 – 25% range) of the [Costs] for the [People], [Resources], and [Time] involved. However, this **is NOT a definitive**, down to the last dollar & cents **valuation of the project's [Cost]**; but rather, a logical calculation for the purposes of evaluating the financial viability of conducting the project.

(13) **Payment Milestones** … a breakdown of the proposed partial / progress payments as per the major delivery milestones / releases.

For example; 5% at contract sign-off, 5% at completion of the Detailed Specifications & Project Management Plan, 5% at completion of the architectural designs, 35% at completion of the Site Acceptance Tests, 20% at completion of the User Acceptance Tests, 10% at the project closure sign-off, and the remaining 20% at the conclusion of the warranty period. ... But wait, that example is for an external customer waterfall project arrangement, whereas this scenario's project is for an internal customer using "rolling waves", thus the payment milestones would be based using windows of calendar time (such as monthly, quarterly, half yearly, yearly).

(14) **Special Conditions** such as

- o **Performance Bond** ... that needs to be paid by the performing organization as guarantee of their performance.
- o **Liquidated Damages** ... involved if the performing organization does not deliver on [Time], and what are the Terms & Conditions related to these liquidated damages.

(15) **Project Authorization Signature Block** ... a list of those persons who will approve the undertaking of this project. Also, include each person's job title, and which organization or business group they represent.

- o **Project Sponsor** ... the person who is paying for this project's undertaking.
- o **Cost Account Authorization** ... budgetary [Cost] approval for the commitment of [Resources], [People], and [Time].
- o **Points Of Contact** ... contact details, and their responsibilities on this project (i.e. Responsibilities Assignment Matrix).

For the scenario's project, ... the Project Charter, ROM summation schedule, ROM project budget, are all accepted by the Customer's Representative (i.e. your boss).

5.3.3. Adapt & Improvise on the Project's Duration & Budget

When limited [Cost] budget and/or limited [Time] duration are going to be delimiting factors for the project (i.e. the prominent project constraints) then the [Scope] of the project should be decomposed into re-organisable work packages (i.e. modules of functionality) with an estimated [Time] duration and a price-tag [Cost] associated with each work package. The representatives of the customer organization and the performing organization would then come together and figure-out (i.e. "negotiate") the most cost-effective combination & arrangement of work packages that will result in the optimal composition of features & functionality; i.e. maximizing the "Return On Investment". This selected [Scope] would then constitute the "Scope Boundary" for the first milestone release, which would subsequently be delivered as though it were a mini project.

As this first milestone release nears completion (delivery) then the Customer's Representatives would figure out exactly how much money they have remaining in their budget. Then as previously, the customer representative in consultation with the performing organization would determine the next combination of the remaining work packages that will provide the most cost-effective composition of desired features & functionality to be implemented in the next release.

And, so-on & so-forth for each subsequent release until there is either; not enough of the budget [Cost] remaining, not enough [Time] remaining for another release, or all of the desired [Scope] has been implemented to the satisfaction of the Customer's Representatives. By which time, the customer may have been able to pull together the necessary budget to fund some of the additional development (possibly loosening their paymaster's purse strings sufficiently to fund all of the outstanding work). However, firstly they need to see (and play with) tangible proof of the project's potential for success.

This is similar to a sprint backlog as used for an agile project; though, this would be for converting their waterfall mindset to an iterative implementation, including agile "sprints" of one month durations. See [Figure 85].

5.4. Getting Underway

5.4.1. Overview Schedule as an Attack Plan

In the project's schedule, you have planned for multiple agile implementation sprints (each of one month's duration), where each sprint will be targeted at to-be-defined Use Cases, then at the start of the next month the output of that last sprint will be evaluated by the Subject Matter Experts (i.e. "Recognizing" any problems & issues), and then subsequently "Reassessing & Revising" prior to being "Reapplied" in the next sprint.

Due to the significant proportion of the project's features & functionality which will be dependent on the Graphical User Interface (GUI), you propose that a web designer be brought onto the project as soon as possible, so as to produce wireframe mockups that will help to solidify the understanding & interpretation of the Customer Requirements (and form part of the Detailed Specifications).

Also, due to the previously mentioned quarterly financial limitations, you suggest forgoing the hiring of a senior system engineer / system architect for a while (and to perform this role yourself), until the Customer Requirements have stabilized. Subsequently, after the new year, the plan is for the senior systems engineer / database designer to join the project, then followed by the other implementers.

"Oh, the Project Manager doing non PM work, this is risky, in my opinion."

 DO NOT forget to include holiday breaks, and also to **align milestone releases with specified "Drop Dead" delivery dates**, such as the End Of Financial Year.

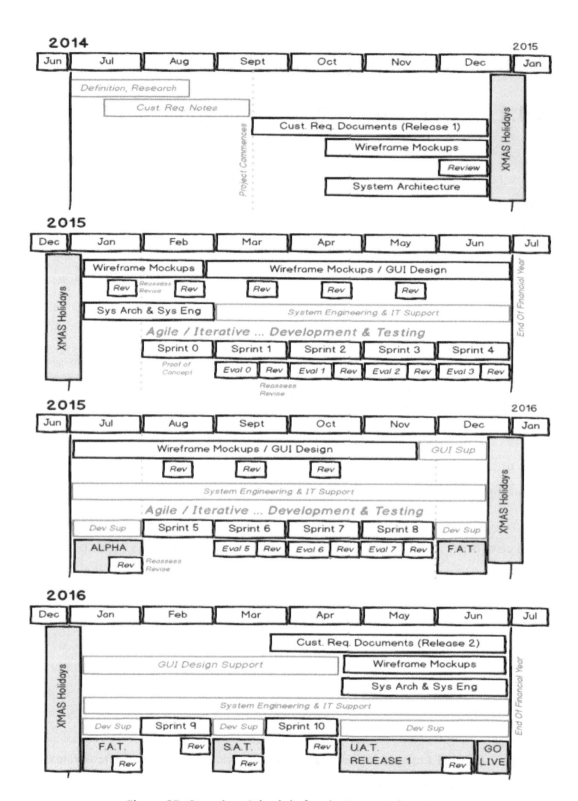

Figure 85: Overview Schedule for the Scenario's Project.

For the scenario's project, the long-term plan is for the members of the Project Implementation Team to remain with the organization for the coming years, so as to develop & deliver the proceeding major releases of the product; i.e. a year and half for the general availability of Release 1.0, then a year later Release 2.0, with a point release every quarter year after Release 1.0. has "gone live" to the paying public.

5.4.2. Costing The Project To Success

In the scenario's project budget, you calculate the Per Person Costs [Figure 86], the Yearly Staffing Costs [Figure 87], and the Monthly Spend & Total Yearly Cost [Figure 88].

PER PERSON COSTS

SALARIES (WAGES)	Yearly Gross Salary (On Project)	Monthly Gross Cost	Total Cost for Week	Total Cost Per Hour	Laptop / Workstation	Software	Total Equipment
PM, BA, Sys Arch	$ 150,000.00	$ 12,500.00	$ 2,884.62	$ 72.12	$ 1,250.00	$ 2,000.00	$ 3,250.00
GUI & Web Designer	$ 120,000.00	$ 10,000.00	$ 2,307.69	$ 57.69	$ 2,500.00	$ 2,500.00	$ 5,000.00
Snr Sys Eng / BA / Coder	$ 140,000.00	$ 11,666.67	$ 2,692.31	$ 67.31	$ 1,250.00	$ 750.00	$ 2,000.00
Sys Eng / DB Eng / Coder 1	$ 130,000.00	$ 10,833.33	$ 2,500.00	$ 62.50	$ 1,250.00	$ 500.00	$ 1,750.00
Sys Eng / DB Eng / Coder 2	$ 130,000.00	$ 10,833.33	$ 2,500.00	$ 62.50	$ 1,250.00	$ 500.00	$ 1,750.00
Junior Sys Eng / Coder	$ 100,000.00	$ 8,333.33	$ 1,923.08	$ 48.08	$ 1,250.00	$ 500.00	$ 1,750.00
(Part-time) Subject Matter Expert 1	$ 50,000.00	$ 4,166.67	$ 961.54	$ 24.04	$ -	$ -	$ -
(Part-time) Subject Matter Expert 2	$ 25,000.00	$ 2,083.33	$ 480.77	$ 12.02	$ -	$ -	$ -

MONTHLY OFFICE COSTS	Per Month
Office Rental	$ 5,000.00
Electricity	$ 500.00
Gas	$ 250.00
Water	$ 250.00
Telecommunications & IT Support	$ 550.00
Total Office Overhead Cost	$ 6,550.00
Total Staff In Floor Space	12
Monthly Office Cost Per Staff	$ 545.83
Weekly Office Cost Per Staff	$ 125.96
Hourly Office Cost Per Staff	$ 3.15

Figure 86: The scenario's (in-house) project Per Person Costs worksheet.

NOTE: that the Yearly Gross Salary per person (for that portion of the time that each person is scheduled to work on the project) should also include superannuation.

NOTE: Also include in the Per Person Costs, the distributed Overhead Cost of the workplace rental, electricity, gas, water, telecommunication & IT support. ... And, remember to include those one-off software & hardware establishment costs.

For this budget calculation, you also include a worksheet for the Yearly Staffing Costs, based on a 40 hours work week.

Yearly Staffing Costs

LABOUR RATES PER HOUR		
Role	Category	Cost Per Hour
PM, BA, Logistics	CAT1	$ 72.12
Graphics & Web Designer	CAT2	$ 57.69
Snr Sys Eng / BA / Coder	CAT3	$ 67.31
Sys Eng / DB Eng / Coder	CAT4	$ 62.50
Junior Sys Eng / Coder	CAT5	$ 48.08
(Part-time) Subject Matter Expert 1	CAT6	$ 24.04
(Part-time) Subject Matter Expert 2	CAT7	$ 12.02

OFFICE COST PER HOUR		
Role	Category	Value
PM, BA, Logistics	CAT1	$ 3.15
Graphics & Web Designer	CAT2	$ 3.15
Snr Sys Eng / BA / Coder	CAT3	$ 3.15
Sys Eng / DB Eng / Coder	CAT4	$ 3.15
Junior Sys Eng / Coder	CAT5	$ 3.15
(Part-time) Subject Matter Expert 1	CAT6	$ 3.15
(Part-time) Subject Matter Expert 2	CAT7	$ 3.15

Unit	Value
Hr/wk	40
Week / Yr	52
Mth/Yr	12
Week / Mth	4.33

PROJECT LABOUR RATES					
Role	Category	Cost / Hr	Cost / Wk	Cost / Mth	Cost / Yr
PM, BA, Logistics	CAT1	$75.26	$ 3,010.58	$ 13,045.83	$ 156,550.00
Graphics & Web Designer	CAT2	$60.84	$ 2,433.65	$ 10,545.83	$ 126,550.00
Snr Sys Eng / BA / Coder	CAT3	$70.46	$ 2,818.27	$ 12,212.50	$ 146,550.00
Sys Eng / DB Eng / Coder 1	CAT4	$65.65	$ 2,625.96	$ 11,379.17	$ 136,550.00
Sys Eng / DB Eng / Coder 2	CAT4	$65.65	$ 2,625.96	$ 11,379.17	$ 136,550.00
Junior Sys Eng / Coder	CAT5	$51.23	$ 2,049.04	$ 8,879.17	$ 106,550.00
(Part-time) Subject Matter Expert 1	CAT6	$27.19	$ 1,087.50	$ 4,712.50	$ 56,550.00
(Part-time) Subject Matter Expert 2	CAT7	$15.17	$ 606.73	$ 2,629.17	$ 31,550.00
			Per Yearly Staff Cost (12 MONTHS)		$ 897,400.00

Figure 87: The scenario's (in-house) project Yearly Staff Costing worksheet.

"Take note; in some organizations, using generic role rates may not be acceptable and hence individual cost based estimates will be required."

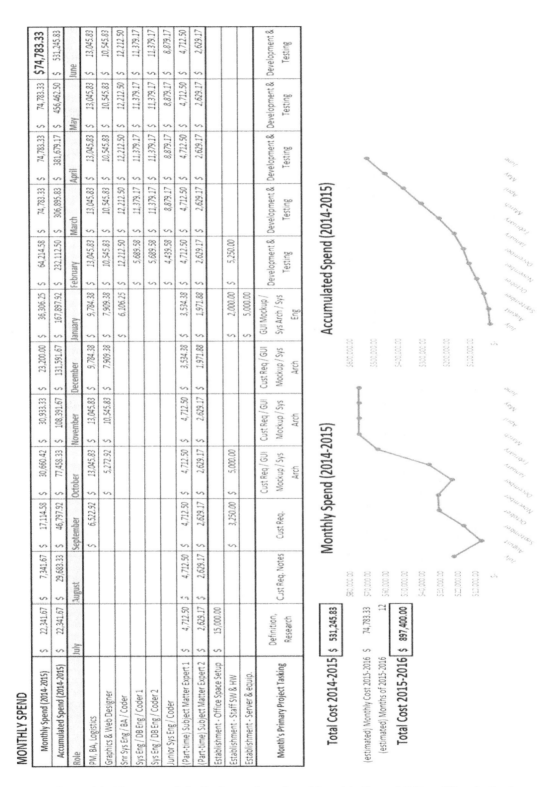

Figure 88: The scenario's (in-house) project Monthly Spend & Total Yearly Cost.

The example budget calculations in [Figure 86], [Figure 87], and [Figure 88] are for a project that is "in-house" | internal to the performing organization, where the Project Steering Committee are only interested in the "bare bones" costs involved with undertaking the project. However, **with a project** that is **for an external customer**, then these costings would **also include** "**Safety Margins**" and "**Profitability Margins**".

Safety Margins are those additional contingencies for cost outlays that potentially (most probably) will be incurred during the life of the project, such as:

- **Escalated Costs** ... are changes in costs | prices due to economic fluctuations over the duration of the project (over each financial year). *For example; exchange rates, inflation / Consumer Price Index.*

- **Project Risk** ... is a **percentage of cost buffering on the "Level Of Effort"** involved with conducting the project; i.e. the project may require more 'actual work to be done' than was planned for, so have some money set aside to pay for that. *For example, an additional 15% to the total cost of the implementers, and 10% for PM – BA – SME – SA.*

- **Cost Contingency** ... is an amount set aside to cover potential future expenses, such as providing **warranty support** and the repair of **latent defects**; where we are not sure about the actual amounts involved, but based on past experiences then there is an expectation that some costs will probably be incurred, i.e. a "known unknown".

    "Though, some unscrupulous senior managers, after the project's development has finished, may declare this contingency as revenue come profit; and unfortunately, sometime in the immediate future, when the customer puts in a warranty claim or requires a defect to be rectified then it will be discovered that there is no contingency fund remaining to cover this obligated maintenance period cost."

Depending on the performing organization's processes & procedures and based on the contractual agreement with the customer (that all costs are to be itemized), then these **Safety Margins would be contained within the Labour Rates Per Hour and the per unit Sell Price** for materials & 3rd party services that are incorporated into the project.

That is, the external customer is given a **"Time & Materials"** price, **"Fixed Price"**, or **"Cost Reimbursement"** based on these Safety Margins being embedded into the calculation.

- o **Time & Materials** – the amount to be paid is based on the charge-out rate for each person involved plus the sell-price of materials & 3rd party services used.

- o **Fixed Price** – the amount to be paid is not going to change from that stated in the quotation, irrespective of whether the actual costs incurred are lesser or greater. A modification being **Fixed Price with Economic Price Adjustments**.

- o **Cost Reimbursement** – the amount paid is the costs incurred plus some additional percentage or fixed margin. Could also add **Performance Incentives**.

Prior to presenting the final costings to the Customer's Representative, a "responsible" Project Steering Committee should also **add to the calculated Project Budget**, a **"Management Reserve"** and a **"Profit Margin"** so as to obtain an **Overall Budget**.

- **Management Reserve** ... a **"hidden" contingency** held back by senior management **"*just in case*"** the project encounters an emergency situation that exceeds those other contingencies; such as, **a "surprise" issue which falls within the [Scope] boundary** of the project and hence does not justify a Baseline Change. See [Section 3.6.5].

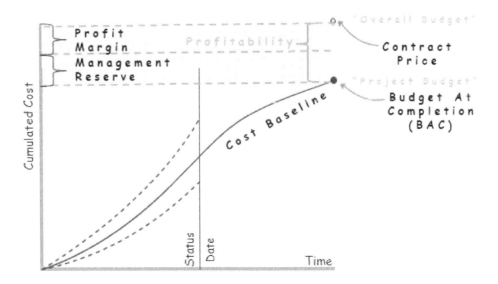

Figure 89: Relationship of Management Reserve & Profit Margin to the Project's S-Curve.

319

- **Profit Margin** ... (Net Profit) is the difference between what the performing organization is prepared for the project to [Cost], versus the Contract Price | Sale Price that they formally agreed to with the Customer's Representative; i.e. **the measure of the project's profitability, once all of the project [Costs] "expenses" and the received "revenues" have been accounted for.**

    While, technically, this is not the correct definition for Profit Margin, see the correct equation below, this makeshift definition is sufficient for the graphical explanation in [Figure 89].

    $$Profit\ Margin = \frac{Revenue - Expense}{Revenue} = \frac{Net\ Profit}{Revenue}$$

If during the project's Executing Phase, the Actual Costs of the implementation were to exceed the [Cost Baseline], see [Figure 90], and such excessive Actual Costs were to continue and not converge to conform with the [Cost Baseline] by the completion of the project, then this will eat into the profitability of the project. Potentially, this excess could **consume both the Management Reserve and the Profit Margin**, so that **the project becomes a loss-maker** for the performing organization; **possibly resulting in a negative cash-flow crisis** that **could consequently affect the business's liquidity and solvency.**

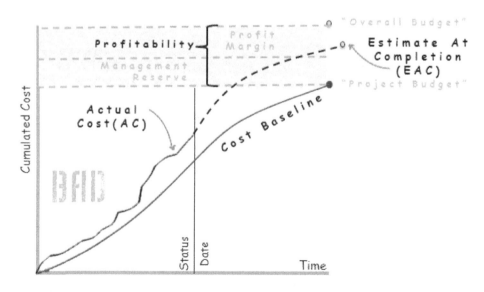

Figure 90: Relationship of Profitability to the Project's S-Curve.

The **project's budget / cost calculations and the associated Profit Margins** (and Management Reserve) is information that the performing organization's senior management definitely **DO NOT want to have known by the Customer's Representatives**. … Let alone, to have this information fall into a competitor's hands, nor be viewed by nosey employees & contractors.

5.4.3. Pricing The Project To A Certain Death

It is **during the project's Initiating Phase that the performing organization's Project Steering Committee** (i.e. senior management) **need to adjust the Profit Margin (and the Management Reserve) so as to "negotiate" an acceptable Contract Price with the Customer's Representatives**; and by so doing, hopefully win the business. Conversely, those Safety Margins (of Escalated Costs, Project Risk, and Cost Contingency) should not be the primary | sole means by which senior management adjust the negotiated Contract Price. … *"While leaving the paperwork percentage profit margin, untouched."*

"Some senior managers, when needing to win new business so as to achieve set targets for the financial year, will cut into both the Safety Margins and the Level Of Effort hours (as was estimated by the implementers). Thereby, reducing the cost of the work to be done, and coincidently the time allocated to do that work. Though, they will be very hesitant to reduce the project's scope that is being negotiated with the customer.

Subsequently, the project's Overall Budget (i.e. Contract Price) will be reduced to a level that will most probably be accepted by the customer, and thereby win this essential new business.

However, all that this 'cut to the bone' strategy has done is shift the burden for the project's profitability onto the Project Manager and the members of the Project Implementation Team, as unrecorded overtime hours that these people will just have to work, if this project is to have any chance of being completed on time and within the stipulated budget."

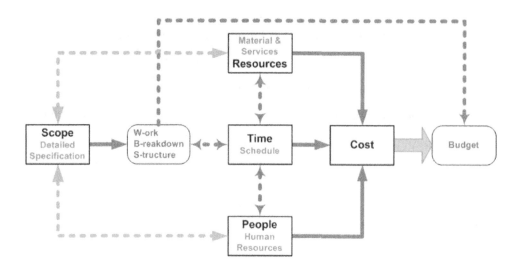

Figure 32: The Planning Process Model to determine the project's Cost and Budget.

The problem with this 'cut to the bone' strategy, is that while the "on paper" Level Of Effort [Time] has been reduced so as to decrease the [Cost] and subsequently to "*optimize*" the Contract Price | Budget, see [Figure 32], … by not reducing the project's [Scope] to compensate for these reductions in [Time] & [Cost], then the [People] implementers will still have the same amount of 'actual work to be done' as they had previously stated (when they were asked to provide estimates for task durations).

Once the project's Executing Phase is underway, these [People] implementers will be recording (in the timesheet system) the actual hours that they have worked on the scheduled [Scope] of tasks. This timesheet data and the Percentages Complete for that scheduled work will be used in the Monitoring & Control of [Time], see [Section 3.6.4], and the Monitoring & Control of [Costs], see [Section 3.6.5]. … … … … … … And, "*lo & behold*", the Earned Value Performance Measures of the Cost Performance Index (CPI) and the Schedule Performance Index (SPI) will both be trending "*badly*". See [Figure 91].

Consequently, senior management will be in a panic, as though, "*the sky is falling*".

Yet, in hindsight, if the originally provided estimates for the Level Of Effort hours were used for the [Time] & [Cost] calculations, then the project's performance thus far would have been "*perfectly fine*". See [Figure 92].

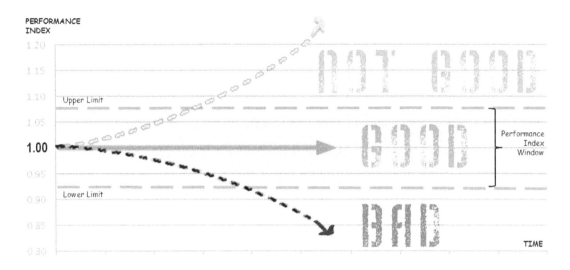

Figure 91: Performance Index trending badly away from the ideal of ONE.

From the plot below in [Figure 92], if the "original" Budget At Completion [Cost Baseline] had been used, then the Actual Cost would have progressed "*perfectly fine*" for the majority of the project's life to date. However, given that the "adjusted" [Cost Baseline] is being used, then the Actual Cost has been "*in trouble*" for all of the project's life to date.

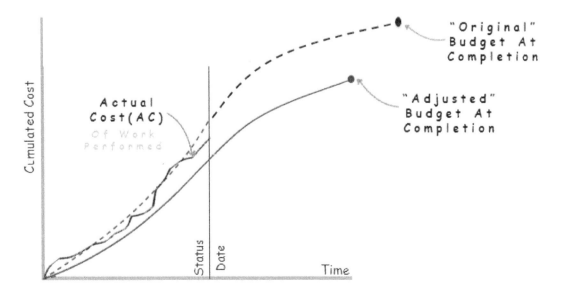

Figure 92: Actual Cost performance compared to the "original" (realistic) budget versus an "adjusted" (unrealistic) budget.

Earned Value – Measures

NOTE: The Actual Cost (AC) relative to the project's S-Curve and the corresponding Cost Performance Index (CPI) is, … meaningless, without also taking into consideration the Earned Value (EV) and the corresponding Schedule Performance Index (SPI), up to the same project status date. See [Figure 52].

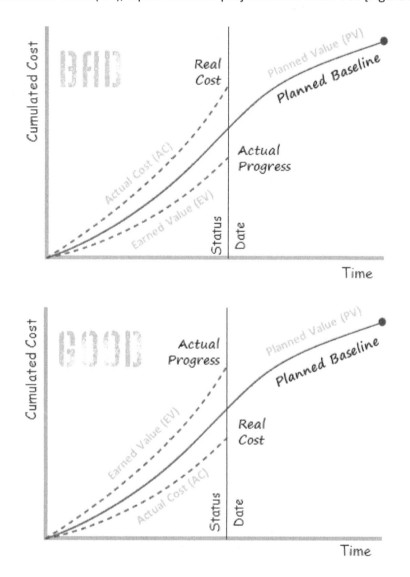

Figure 52: Actual Cost (AC), Earned Value (EV), and Planned Value (PV) versus time mapped as Earned Value Management S-Curves.

> **COFFEE BREAK DISCUSSION** ... *Moving dates on an inevitable confrontation.*
>
> At some point in your project management career, you will be given a project with a specific set of functionality, and you will be asked to put together a schedule and costings. In consultation with your potential implementers and in-house Subject Matter Experts, you will come up with estimates for the Level Of Effort and the task durations involved with undertaking this project. You will diligently prepare a "realistic" project schedule, checking it over with your potential implementers so as to verify that the schedule is feasible as far as they are concerned, and then you will calculate the associated project costings. After careful planning as to the [Time], [Cost], [People], and [Resources] required, you then hand this information over to your senior management for their review and for them to include a Management Reserve and Profit Margin.
>
> Then sometime later, you will either be handed back those estimates with reworked numbers, or the resultant numbers will be passed onto the Customer's Representatives as part of a contractual agreement. Subsequently, when you next see the resultant numbers, you are shocked to discover that there has been noticeable cuts made to the Level Of Effort, so as to produce sizeable reductions in the [Cost] and [Time]. On further investigation, you are astonished to find that the project's [Scope] has not be reduced accordingly to compensate for the reduction in the Level Of Effort. ... At this point, many a project manager will question their senior management about these reductions; and, occasionally there will be some negotiated adjustments made to increase some of these numbers. Though, often these numbers *"must stand as is"*, and so you (and your Project Implementation Team) will *"just have to deal with it"*.
>
> While most project managers will yield to their senior management's right to make such cuts, and just accept their fate of having to make do the best that they can, it is a rare breed of project manager who is prepared to stand their ground and argue the point. However, will such a stance brand that project manager as *"not a team player"*, being *"difficult to deal with"*, *"problematic"*, and/or *"unmanageable"*.

With the project having been signed off with these 'cut to the bone' [Time] & [Cost] baselines in place, the Project Manager and the Project Implementation Team will just have to work (struggle on) within these imposed constraints. Subsequently, to keep the project on track (SPI, CPI, and milestone date wise), the Project Implementation Team (and the Project Manager) have to work unclaimed overtime hours so as to compensate for those cuts to the Level Of Effort and durations; recalling that, the project's [Scope] has not been reduced, and hence the project's "actual Level Of Effort" required has also not reduced from those original estimates that were provided by the implementers.

"When Level Of Effort hours and durations are reduced in a budget spreadsheet, this in no way means that the actual amount of work to be done is also reduced; as these cuts will still have to be redeemed."

The longer that the project goes on, then the more and more that the project's actual performance slips away from those baselines, until the reality of facts sinks in to all involved with the project, that these [Time] & [Cost] baselines cannot be achieved without some major changes, specifically to the [Scope] of the deliverables. At which point, the Project Manager (and the Project Implementation Team) bear a disproportionate amount of the blame for the project's failings. ... Even though, they and others know that, "*if only those original estimates had been used and not those butt-plucked figures, then*". ... But alas, who ends up being held accountable ??

*"Consider a 6 months (26 weeks) project with a 5 people team, at a charge out rate of $100 per hour per person doing a 40 hour week. ... 26 x 5 x 100 x 40 = 5200 hours @ $520K project cost. Then senior management agree to a 20% price reduction for the same quantity of scope; but, the Board Of Directors stipulate that charge out rates can't go below $100 per person per hour. Hence, who makes up for the (5200 * 20%) 1040 hours shortfall ?? The implementers will, 1040hr / 5 people / 26 weeks = 8 hrs extra work per team member per week. ... And, then (key) implementers start to resign due to their having to work unacceptable levels of unpaid overtime for extended periods."*

5.5. All Ahead, Full ... *"Iceberg !!"*

With the project now motoring along in the Executing Phase, the project's monitoring & control KPIs (of CPI & SPI) may start to indicate that the project is getting into some trouble; subsequently, it will be necessary for the Project Manager and the Project Implementation Team to take steps to rescue this project from a pending disaster.

Though sadly, those senior managers who *"adjusted"* the project's [Time] & [Cost] so as to *"optimize"* the project's negotiated Contract Price, see [Section 5.4.3], they will often forget that, *"if the project had used those original estimates"* for the Level Of Effort, *"then the project would not be in this trouble"*, ... according to the implementers, that is.

5.5.1. Project Rescue ... Getting Out Of Trouble

An implementer's solution to problems is often one of, *"put your head down, and work harder and longer hours"*; whereas, the Project Manager needs to figure out how the Project Team can *"work smarter"* so as to rescue this project. See [Section 2.5.1].

Project Rescue

1.. **RECOGNIZE** ... Before anything can be done to resolve the project's trouble, it must be **realized that a problem exists and that the problem is occurring** (or is about to occur). That is, the appropriate involved persons must acknowledge & accept that the actual performance / results does not compare well with the expected / planned outcomes.

2.. **REASSESS** ... Before being able to do something "appropriate" to resolve the problem one must; **evaluate** exactly **what happened**, determine **the cause** of the differences between what was expected to have happened and what has actually happened.

3.. **REVISE** ... Now that there is an understanding of what went on and what went wrong then in sequential order; **decide whether it is necessary to do something about it**, decide whether something has to be done **straight away or can it wait** for a more opportune moment, **decide what exactly has to be done**, decide **what is** the newly **expected** performance, **revise the expected outcomes**, and **let people know** these.

4.. **REAPPLY** … **implement** that which was **decided to be done**.

Schedule Duration Compression

Alas, it is decided that the project's duration needs to be reduced so as to bring the current delivery dates back into line with the agreed [Time Baseline] milestones, [Section 3.6.4]. There are a few different techniques that can be used for reducing the delivery duration so as to align with the targeted delivery milestones.

1) **Crashing "C.P.R."** – is compressing the schedule's duration **by shortening the time allocated for critical path** tasks that directly affect the resultant delivery date, via:

    o **paying additional (C)ash** to the implementers to increase the priority of those tasks that have to be completed sooner, and allowing for additional overtime.

    o **adding more (P)eople** to work on those tasks that have to be delivered immediately.

    o **adding more (R)esources or extending the allocated usage** of those resources that are vital for the completion of those tasks.

NOTE: The **Crashing technique only works for tasks on the critical path**; i.e. those tasks (with zero "float") that if not completed by the task's individual due-date will directly result in the project / release being delivered later.

 The **Crashing technique is ineffective when applied to non-critical path tasks**. In fact, this technique if applied inappropriately will be detrimental to the project because it will unduly increase [Cost] and its execution can have negative effects on the Project Implementation Team's morale.

2) **Fast Tracking** – is compressing the schedule's duration **by performing some tasks in parallel and glossing over those tasks that are not evident in the deliverables**. However, this could potentially reduce [Quality].

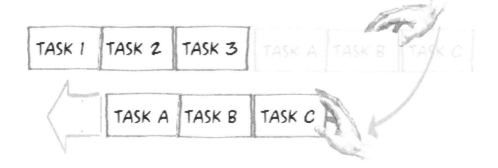

NOTE: The Fast Tracking technique requires that the shifted tasks are **not sequentially dependent**, ... and, it works best for tasks associated with the critical path; but it could be used en masse for all tasks in the release.

NOTE: Those "non-evident" parts such as documentation, source code comments, and peer reviews are often the first victims of fast tracking. Followed by a reduction in unit testing, and eventually a reduction in system testing; thus, the project's inherent [Quality] declines.

 The **Fast Tracking technique can result in more problems than it was intended to solve**, because of the potential that rework will be required on the work that was done, due to:

- the **consequential reduction in [Quality]**, and
- the possibility that the **deliverables may not have been built with future re-use in mind**, instead being tailored "hard coded" specifically for the targeted milestone. ... And, not for the needs of future milestones.

Fast Tracking is inherently risky to the project's long-term success.

3) **Scope Minimization** – *"De-scoping"* is compressing the schedule's duration by officially (and/or unofficially) **not including certain features & functionality in the time-sensitive release**; i.e. scope re-baselining. Instead these features & functionality shall either; be included in a later release, or be removed completely from the project's deliverables.

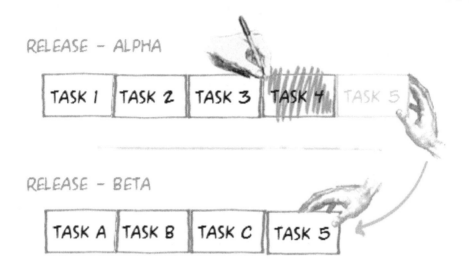

NOTE: The Scope Minimization technique is most effective with an iterative / agile project life cycle. ... *"Though, this technique is a catalyst for conflict with waterfall-centric stakeholders."*

 The **Scope Minimization technique could potentially result in the schedule duration problem being shifted towards the end of the project** as more and more tasks are reassigned to later releases.

In addition, partial payments at specific milestone dates could be reduced or delayed, compensation and/or penalty clauses may be invoked for the minimalist deliverables with its missing features & functionality.

Thus, what's now in & what's now out could potentially cause future debates and disagreement; so **better update the Detail Specifications accordingly to correspond to the agreed changes and seek re-approval**.

5.5.2. Project Recovery ... Redefining The Plan

The sad reality of facts is that, sometimes the project cannot be rescued from its negative predicament (possibly due to the current situation & prevailing circumstances, in combination with the inflexibility induced by those imposed project constraint baselines); consequently, some form of project recovery will have to be undertaken.

Project Recovery

Project Recovery consists of the following steps:

1.. **RECOGNIZE ... Detect when the project is having difficulty maintaining balance** between the project variables and those project constraints. See [Section 2.7]. And, involve the senior implementers in the evaluation of this predicament.

2.. **ACCEPT ... Do not ignore the fact** that there is **a problem** that should be **dealt with now**, and that it should not be put off until tomorrow or to some later date (or in the worst case to *"a head in the sand"* never).   ... *"aka, management by avoidance, of dancing around the issue."*

3.. **ACKNOWLEDGE ... Lay out the true predicament to the relevant primary stakeholders** (i.e. the Project Working Group and the Project Steering Committee).

 As the Project Manager, **DO NOT decide to take it upon yourself to inform the Customer's Representative of the project's current negative predicament without having firstly conferred with the members of the Project Steering Committee** (and the Project Working Group), as they will need to **decide if & when it would be *"strategically optimal"* to acknowledge the current situation** to those involved external parties.

"To be a truly successful project manager then one needs to be PROACTIVE by having prepared in advance competent answers to the following questions. ... what have you already done to try to rescue this situation, what do you recommend that we do now to get out of this mess, and what is your Plan-B and Plan-C ??"

4.. **DON'T PANIC** … **Back off** the level of implementation effort that is currently being applied, *"think the whole situation through"*, and redeploy efforts to effectively & efficiently *"work the problem"* to regain stabilized control.

5.. **DON'T RUSH** … **Do not hurriedly throw whatever is available** (i.e. Cash, People, Resources, Time) **at the problem** in an attempt to overwhelm and mitigate the failing project as this **most probably won't be effectively & efficiently targeted at the root cause(s)** of the problem(s), and **won't necessarily produce the desired outcome**(s).

 STOP for a moment and give serious consideration as to whether mounting this project recovery could end up dragging down other projects and potentially sinking the performing organization.

"Bail-outs, fire-fighting and playing catch-up to rescue a failing project can have unforeseen knock-on effects due to the impact of redirecting much needed Cash, People, Resources, and/or Time away from other projects, … and, away from Business As Usual activities."

6.. **GET A GRIP** … **Identify the cause(s)** of the project's failing, and subsequently **make a clear statement about the identified cause(s)** of these failings.

- ☐ Was it due to **misunderstandings or misinterpretations of the customer**'s needs, wants, concerns, expectations, and perspectives?
- ☐ Was it due to **incomplete requirements, misconstrued requirements**, and/or **ambiguous specifications**?
- ☐ Was it due to **underestimating** the scale of the project?
- ☐ Was it due to **unforeseen technical difficulties**?
- ☐ Was it due to the **unavailability of** the necessary **[People] and/or [Resources]** when these were planned / expected **to participate**?
- ☐ Was it due to **poor performance of the Project Implementation Team**?
- ☐ Was it due to **poor management by the Project Manager**?

- [ ] Was it due to **poor management by the Project Working Group** (possibly caused by a lack of coordination & cooperation)?
- [ ] Was it due to **poor management by the Project Steering Committee** (possibly caused by a lack of direction, too many changes in direction, over direction)?
- [ ] Was it due to **circumstances beyond the project's control**?

With these "due to" answers in mind, ensure that solutions to these are considered when coming up with the Rescue Plan (and also, for the Recovery Plan).

7.. **A RESCUE PLAN** ... What things can be done right now without having to involve the Customer's Representative, so as to come up with **a reasonable, responsible, and sensible sounding Recovery Plan**.

8.. **BASELINE CHANGES** ... To be able to come up with a **viable Recovery Plan** will necessitate that one or more of the **existing project baselines** (i.e. project constraints) will have to be **changed and possibly changed dramatically**. Traditionally, this means that the baselines of [Scope], [Time], and/or [Cost] will have to be reassessed, and subsequently restructured "re-baselined"; else [Quality], [People], and/or [Resources] will bear a disproportionate amount of the burden of the enacted Recovery Plan, yet still <Risk> project failure.

It **must** be **accept**ed that those **existing baselines are now unrealistic** to achieve; because, **to insist on the conformance with** all of **those old project constraints** (baselines) **is a continuation of those ingredients that already got the project into trouble.**   ... *"which is a recipe for disaster."*

Are there any self-imposed restrictions on the performing organization, such as:

- [ ] **What [Time] remains available** to complete the project?
- [ ] **What [Cost] budget remains available** to complete the project?
- [ ] **What [People] remain available** to complete the project?
- [ ] **What [Resources] remain available** to complete the project?

Are there any restrictions mandated by the customer organization, such as:

- ☐ **What [Scope] is "*a must have*"**, and what is the **priority order** of that [Scope] which remains to be done?

- ☐ **What [Time] must** the project deliverables **definitely be provided by** or ELSE?

- ☐ **What [Cost] budget must be stayed under** to remain financially viable?

- ☐ **What [Quality]** acceptance criteria **must be satisfied in order** for the deliverables and associated artefacts **to be accepted**?

- ☐ **What [People] will still be available** to undertake the remainder of the project? Additionally, **what skilled [People] will be required** to implement that [Scope] which is selected to be done?

- ☐ **What [Resources] will still be available** for the remainder of the project? Additionally, **what quantities & grade of [Resources] will be required** to implement that [Scope] which is selected to be done?

9.. **VIABILITY** ... Based on the information collected so far, answer these questions;

- ☐ **Can this project still be success**fully completed?

- ☐ Can this project **still be considered as a worthwhile endeavour**?

- ☐ **Does** undertaking **this** project **still make sense**?

Given the answers to those questions, then should the **recommendations** presented **to the Project Steering Committee to pass judgement on**, be;

(a) **proceed with the project as** it currently **is**,

(b) **revise** the project's **baselines**,

(c) **recover only a portion** of the project,

(d) **restart** the project as a **"do over"**, OR

(e) **terminate** the project henceforth?

10. **AGREE THE TERMS** … The Project Steering Committee (in consultation with the Project Working Group) has to decide what is the best strategic options for the Performing Organization; then, to **sit down with the Customer's Representatives to work out and agree on the details of the** "Terms Of Action" (i.e. what conceptually is to be done), **or to come to an amicable resolution on how to terminate the project** (or that portion of the project that is no longer viable).

11. **A REALISTIC PLAN** … If it is agreed by both the Project Steering Committee and the customer representative(s) that the project should be recovered then, internal to the performing organization answer the following questions:

    ☐ **What [Scope]** can **realistically** be **included**?

    ☐ **What** is the **priority for** implementing this **selected [Scope]**?

    ☐ **What** portions of the **existing implementation can be reused**?

    ☐ **What portions of the** project's **old plan**ning can **be reused**?

    Redo the project's **schedule to determine realistic [Times]** and to obtain **revised milestone dates**. … **Revise this schedule** based on the re-**allocation of available [People] & [Resources]**. … **Determine realistic [Costs]** due to these revisions.

Beware of the risk **of recutting estimates too low to minimise the [Costs], and thereby risk not achieving deliverables a second or subsequent time.**

DO NOT forgo (necessary) **documentation and/or testing** due to the need to speed up the project, as this will **only result in future failings of the project** (especially during acceptance testing and the receipt of the deliverables).

12. **NEGOTIATE** … The internally produced & approved Project Recovery Plan has to be presented as a **proposed plan** to the customer organization's representatives, then **negotiated ("compromise") with both sides** on what they "**can and can't do without**".

13.. **REVISE THE PLAN** … Based on the outcome of those negotiations, then the Project Recovery Plan will need to be **reworked, internally reviewed & approved**, and then **re-presented** to the Customer's Representative **for another round of negotiations**.

Be aware that the Customer's Representative(s) **may try to retain more of those pre-existing project baselines than was agreed to** during those discussions of the "Terms Of Action".

14.. **REACH AN AGREEMENT** … With both sides now agreed that the negotiations are completed, then the Project Recovery Plan (aka "**Baseline Change Request**") will need to be **signed off to signify the approval & authorization of the revised baselines**. And, hence supersede (invalidate) any pre-existing baselines prior to this new agreement.

The project could get itself **into** another **trouble**d situation **if some** of the **stakeholders are still working to the** previous **(old) baselines** while others are working to the new & approved set of baselines.

Hence, **the (new) approved baselines for the project must be made known (and enforced) to all** of the project's stakeholders.

15.. **SO WHAT NOW** … With the signed off revised plans in hand, there should be a **mini kick-off meeting or stand-up meeting** held **with** the **Project Implementation Team and the Project Working Group** (and if need be with the Project Steering Committee) **to inform them of**; the project's **change of direction**, what are the **new baselines**, and **how these changes will affect them**.

16.. **GET TO IT** … The Project Implementation Team should now recommence the Executing Phase in earnest, with respect to the "new & approved" baselines.

Watch out for project stakeholders who try to get "**scrapped**" **Scope components reintroduced after the Project Recovery Plan (and Baseline Change Request) has been signed off & authorized.**

17.. **BE VIGILANT** ... **Continuously monitor the project for** the **reoccurrence** of those causes that derailed the project previously, as these problems may arise again.

Which brings us to an addendum to "Adaptive & Proactive" SDLC Project Management.

**RULE 6^C: Just because you fixed up after a
BAD thing doesn't mean that
BAD things can't happen anymore.**

5.5.3. The Scenario ... Failure To Gain Altitude

For this scenario's project, the Project Charter has been approved (though with the [Cost] budget and [People] allocation noticeably reduced from that originally planned), the supporting computing infrastructure has been put in place, the planning & design application software has been acquired, and (via your personal networking contacts) the GUI designer has been hired to start mocking up wireframes of the (stabilized) Customer Requirements. … And, you have even sounded out some previous work colleagues about their potential future availability to work on this project.

The Customer's Representative (i.e. your boss) still insists on holding almost-daily meetings to define all of the front-end requirements in detail, even though some of this functionality is for later releases. Thus covering in depth the requirements for that part of the system which they (as the SMEs) understand very well, being how the user interface should look & feel and how it should operate from the end-user's perspective. … … However, they show little interest in the definition of the system which would underlay that user interface functionality.

With the continuation of these daily gatherings to refine the **Customer Requirements** (with the addition of even more notes), you draw up a breakdown structure of the various constituent modules of the intended system and map out how these bits inter-relate (i.e. an overview starting point for a **Work Breakdown Structure - WBS**, and software architecture), and you sort the growing mountain of requirements into a spreadsheet of worksheets that correspond to each of those modules (i.e. forming the beginnings of a **Requirements Traceability Matrix - RTM**).

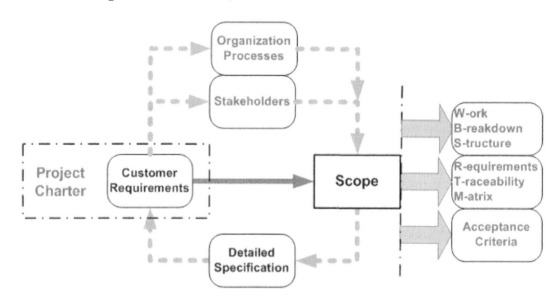

Figure 27: Process to refine the project's [Scope] and the Detailed Specifications.

Though at present, you are functioning more as a business analyst & part-time systems architect (rather than as a project manager) as you bullet point the requirements, sort these points into the various worksheets which correspond to the numerous work packages / modules, and draft an overview module breakdown diagram (which is starting to look like a high-level Software Object Tree or primitive database schema), plus some initial system architecture designs.

These are long days, but satisfactory progress seems to be getting made; as the plan to bring on that senior systems engineer at the start of the new year (followed by the rest of the implementers in the following months) looks likely to happen.

The first of the GUI mockups start being presented to the SME-Customer (i.e. your boss), with the subsequent updates following the review meetings, where more "clarification" notes are added to the growing list of requirements. Though most importantly, your customer appears to be happy with the work presented thus far, as they repeatedly state how this product betters all of its competitors with its feature set.

Then you are sidelined for a week while you put together the documentation for this project to apply for a government development grant. … And, all the while, the requirements "clarification" meetings continue on a regular basis. Subsequently, you haven't got the available work-hours to completely fulfil your daily combined duties as business analyst - systems architect | engineer - project manager. Thus, you are left in a state of perpetual catch up; but, that's okay, you will do what you have always done (especially when you were an implementer) and work harder & longer hours. … Your spouse and child at home will (have to) understand this, as they have always done in the past.

However, … … as more and more of the intended system is defined, you (and your GUI designer) start to perceive the project as the equivalent of an iceberg, with a lot more beneath the surface than was initially perceived.

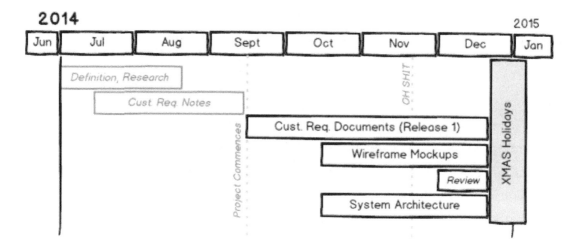

Figure 93: Overview Schedule for the Scenario's Project ... "Oh Shit" moment.

Then, ... a few weeks out from the first all-in progress & presentation review meeting, you are shocked to discover that there has been a disconnection in perspective of what are the deliverables for that target date. As per the overview schedule, you had planned to present only the wireframe mockups for Release 1; however, your customer (i.e. your boss) is expecting wireframe mockups for the entire system (including those future releases), even though some sections are yet to be defined by the SMEs.

You explain your differences in expectations, but are only given a few days to get back to them on whether it can all be done by then. ... And in passing, a Sword Of Damocles, comment is made that, "*if all of these mockups can't be delivered by then, well, we'll have to consider cancelling this project, or outsourcing it overseas*".

5.6. Coping With ... "Failure"

5.6.1. Self-Doubt and Questioning Why

```
   For the scenario's project, you as the Project Manager
have just been "kicked in the guts" as it has been revealed
that your project's progress has not met with senior
management's (and the customer's) expectations.  Even though,
you and your team member(s) have been working long hours
producing the deliverables; while having to deal with the
continuous "clarification" of the requirements.
```

Other than feeling knocked backwards by such news, any (non-narcissistic) project manager will start doubting them self and questioning their own abilities; especially if they have followed a prescribed project management methodology accordingly, and they have done what (they believed that) they were supposed to do.

Question: ☐ **Did you really know what** this primary stakeholder **was expecting**?

☐ **Did you really keep** this primary stakeholder **informed of progress**?

☐ **Did you really provide** this primary stakeholder with **what they needed to know**, by **when** they needed to know, **in a form** that they **could use**?

☐ **Did you make sure things were done** by **when** you told this primary stakeholder that these things would be done?

Or, did you just assume that they knew this stuff; given how tightly the customer has been involved with the project (on an almost daily basis), as though they were a member of the Project Implementation Team.

While questioning yourself and doubting your abilities (and potentially that of your team members), there is something else to realize. ... Analogously, the first delivery milestone is the equivalent of the first game in a sporting championship year.

Unless the project has only one or very few progress marker milestones, then there should still be sufficient opportunities to turn things around; provided that, these selected milestones are close enough together to provide real benefit as performance feedback points.

Alternatively, the project constraint baselines could be so tightly restrained that there is absolutely no leeway if & when things go the slightest bit wrong. Hence, is it not better that the project fail now during the early phases of its life; because, with so tight baselines it was inevitable that the project would eventually run out of manoeuvring room. Given that, with any project of a length greater than a few months, then the project situation & prevailing circumstances of the performing organization (and customer organization) will change, ... and, sometimes these changes will be of a significant order.

Though, if as (captain) coach of this team, you're going to "get the boot" after only the first game of the season, then in all honesty, things were set from the start to eventually fail.

5.6.2. Differences In Understanding Of SDLC

Alas for this scenario's project, you realize that your customer does not understand (or is not accepting of) the concept of implementing the project via the planned rolling waves of agile's iterative releases; but rather, they see the project from a waterfaller's perspective of having to define all of the requirements (in detail) at the beginning.

Which brings us to the next rule of "Adaptive & Proactive" SDLC Project Management.

RULE 41: Differences in understanding of the underlying principles of the SDLC methodology being used will derail a project.

THEORY ... *Project Layering and Slicing.*

For the scenario's project, one of the notable causes of the problems that are being encountered is due to the difference in perspective of how the project is being broken up for implementation; where the primary stakeholder (i.e. Customer's Representative) conceptually perceives the project as being implemented via "layering", see [Figure 94].

For example, consider the ISO Layers Model, for the mobile phone case:

- *Layer 7:*   *Application Layer (i.e. user interface),*
- *Layers 5 & 6:*   *Session Layer and Presentation Layer*
- *Layers 3 & 4:*   *Network Layer and Transport Layer*
- *Layers 1 & 2:*   *Physical Layer and Link Layer*

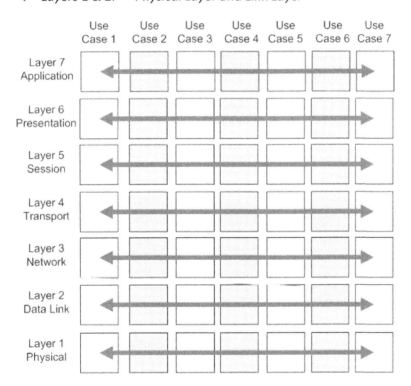

Figure 94: Project broken up via "Layering".

With a "layering" approach, **specialists may concentrate on specific layers** across all Use Cases; and, **each layer would be developed simultaneously and** much of each layer has to be **completed before the product can be evaluated** / delivered.

In the scenario project's case, due to the SMEs in-depth knowledge of how the User Interface (i.e. Application Layer) should function then their focus is entirely on this layer.

With a "slicing" approach, see [Figure 95], then the Project Implementation Team can **collaboratively work on selected Use Cases** across the various layers, so as **to produce a usable deliverable** (*e.g. a "proof-of-concept"*) **that can be** customer **evaluated** / delivered **earlier**; instead of having to wait for most of each layer and the Use Cases to be finished. Subsequently, **necessary scope changes can be made at an early stage before a significant proportion of the development is underway**.

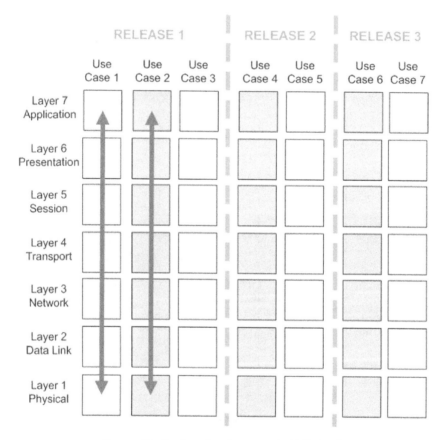

Figure 95: Project broken up via "Slicing".

In the scenario projects case, The Plan was to utilize a "slicing" approach with sequential rolling wave releases.

Rolling Wave versus Waterfall

With the project having a fluidic [Scope] baseline, but with the expectation that the project will be done using a waterfall methodology, then how can a [Time] baseline be expected to be fixed when the [Scope] is variable ??

If however, the project was being implemented using agile's rolling waves, then selected portions of [Scope] could be fixed to establish realistic [Time] baselines for each release's delivery milestones.

Though, if an agile methodology is to be used then this does not mean that you can take a waterfall structured project of upfront defining all of the requirements to the nth degree, ~~prior to~~ while simultaneously undertaking the implementation, then expecting to lob in additional (primary) requirements and redefining existing requirements, with *"aj-jile"* circumventing the need for Baseline Change Request Management.

Baseline Change Request Management

Baseline Change Request management is the review & authorization of the inclusion of changes (predominately [Scope] changes) **that will impact on the agreed [Time] & [Cost], and/or the agreed assignment of [People] & [Resources], and/or the agreed [Quality] & [Scope] of the deliverables.** Given that such changes will directly affect the contractually agreed [Baselines], then such **Baseline Change Requests (BCR) need to be reviewed and approved by the Project Steering Committee**, as well as with input from the Project Working Group and from the Project Implementation Team.

```
But alas, here is a problem with this scenario's project
in that the Project Steering Committee is also the project's
customer.  Hence, a "just do it" attitude can be mandated on
down; thereby, curtailing the process of proper Baseline
Change Request Management, and subsequently, selectively
fluidizing those project defining baselines.
```

5.6.3. Work Related Stress

```
The insomniac ticking, on the wall, indicates that it's
ten past four. It is as dark outside, as it was a few
hours ago, when you awoke in a pool of your own sweat.
```

Was that a "panic attack" (aka an anxiety attack), potentially caused by chronic work related stress of being faced with the hard decisions of saying that, *"it can't be done"* when knowing that to say so could potentially be the end-of-the-road for you and possibly the members of your project team; versus saying, *"it can be done"* when under those current [Baseline] constraints the project will most probably end as a massive failure.

"Personally, it is better to say that, 'it will be difficult to achieve, but these are the things I will do to try to make it work."

Yet, no matter how the Project Manager tries to balance those triple constraints of [Scope] vs [Time] vs [Cost], it just doesn't work out; given that the duration is too short, the budget is too small, for the requirements that keep growing. See [Section 2.7].

Different project managers (as well as implementers who are faced with impossible delivery workloads) may react negatively to work related stress in various ways:

```
Looking for a solution at the bottom of a whisky glass,
while cigarette butts smoulder in a coffee cup ashtray.

Secretly sobbing in the car park, not knowing what to do,
nor whom to turn to.

Throwing the job in, even though there is nowhere else
lined-up to go.

And sadly, but true, committing suicide, because it all
must have got too much for them to bear any longer.
```

Now, these reactions may seem extreme; but, shockingly these are a collection from the textbook author's associates past experiences (since, the GFC – Global Financial Crisis).

Though, these extreme reactions are not caused solely due to work related stress; but also, in part due to one's own home life situation and personal circumstances.

"Why is accepting (and stating) the facts that it can't be done as was planned, considered as a 'failure' ?? The actual failure would be in lying that it can be done, when you honestly know that it can't be done that way. It is better to explain the difficulties, explain the proposed approach, document the issues (including names & dates of the decision makers), and accept stretched targets. … … If anyone calls 'failure' when the project slips, then refer them back to the amended targets and the documented issues. … … The real failures are those people who believe that they can make impossible things happen by simply saying, "just do it !!"

Why Do People React To "Failure" In Different Ways

People will react to what they perceive as "failure" in different ways, based on their:

- **Self-Expectation** … Deep down each person has a (private semi-realistic) idea of what their performance should have been able to achieve; and, if the outcome does not match the expectation then this can result in feelings of self-doubt, inadequacy, and/or depression.

- **Return On Investment** … With all endeavours where time & effort is required to be spent (such as continual overtime work to meet stringent deadlines), you will have to give up partaking in something else or at least reduce your interests in other things for the interim. When the outcome from that endeavour does not balance with what has been surrendered, then it is only natural to question whether it was all worth it, and even to become angry with the party(s) who put them in that situation.

- **What else is riding on the outcome** … If the successful outcome of the endeavour is a determining factor as to whether a person obtains a much needed payment, and/or an inflation-matching pay rise, or just having a job, then difficulties could arrive if this is not forthcoming. Hence, failure could have noticeable financial impacts on that person, and subsequently negative consequences to their standard of living.

- **Work versus Home Life Balance** … Everyone has to balance his or her work life versus home life, and weigh up which is more important to him or her (at that particular point in their live). It may seem essential to the performing organization that the project has to be completed by a specific deadline, but is it personally as important to them when compared to what outside of work could be lost or be execrably damaged by doing so.

    In addition, everyone does not live and work in isolation; what is happening in their personal home life can and will eventually affect their work life, and vice versa. If the situation happens to be bad at home then this can affect their emotional state at work. Similarly, a bad work life can exert pressure on their home life, which can eventually result in an unpleasant death-spiral of misery at home and at work, where something will finally have to give (potentially, depressingly so).

    Which brings us to the next rule of "Adaptive & Proactive" SDLC Project Management.

    RULE 42: **Everyone has two parts to their life, being work and home. … As long as one of these is okay, then eventually the other will come good, and you'll end up with a happy worker. But, if both work and home are not okay, then you will have a constantly unhappy worker.**

    RULE 42^A: **A constantly unhappy worker will contagiously make those around them unhappy too.**

- **I did all and more than what was expected & required of me** … How do they feel & react when they believe that their consistent performance should have been sufficient to achieve the objectives, yet factors beyond their control wrecked their chances.

    "And, now you're putting the blame for the failure on, me !!"

- **Cultural, ethnic, and community background** ... Today's workplace is a multicultural environment with peoples from diverse cultural, ethnic, and community backgrounds; and, each of these groups will have a slightly different perspective on what differentiates "A Winner" from "A Loser", what it means "to win" and "to lose", and where the consideration (or opinion) that one is a "failure" is a "loss of face".

- **Their personality traits** ... Everyone reacts differently "to failure", more specifically towards criticism for having "failed", and to the pressures of finding a resolution that is satisfactory to the person(s) who are stacking on the "negative criticism".

At the most basic level, all humans are programmed with a "fight or flight" instinct; but what happens when (physically) fighting is not a social | political option, and when there seems to be nowhere else to run? Alas, the stress builds up until something has to give; consequently, a conscious or subconscious "panic attack" can occur. ... And, making resolute decisions during such an anxiety attack or straight afterwards (more often than not) does not result in the best long-term decisions.

Strategies For Dealing With Work Related Stress

The following are commonly suggested ways of handling work related stress:

- ☑ Have a **balanced diet** and not burying your sorrows in unhealthy "comfort" junk food.
- ☑ Doing **regular exercise**, sweating out some of that stress, breathe in some fresh air.
- ☑ Getting a good night's **sleep**, and that means, packing it in early and going to bed.
- ☑ **Talking** the situation over with someone who will listen.
- ☑ **Social Participation** of doing something (non-work related) with other people, and not isolating yourself away (to battle on with the problem alone).
- ☑ **Disconnecting from work** and routinely **doing something you like to do** (irrespective of how age inappropriate or juvenile it may seem to others). ... *"All work and no play makes Jack a, stressed out sack of bones"*.

- ☑ **Putting together a short prioritized to-do-list of things that must be done**, where lower priority to-do-list items can wait for tomorrow's list. ... Taking it one day at a time; or rather, one to-do-list item at a time.

- ☑ **Doing only one thing at a time**, because multi-tasking of trying to get multiple things done at the same time ends up with nothing being done satisfactory; but, don't do each thing to perfection, because **good-enough is probably sufficient to move on** to the next thing to be done.

- ☑ **Delegate, Delay, Delete** ... that is, don't do everything by yourself, allow / ask others to help. If it doesn't need to be done right now (with little to no negative consequences) then leave it to another time. If it can wait to another time then does it really need to be done at all?

- ☑ **Look for solutions to the current situation at hand**, and not mentally stockpiling more issues & problems that could eventuate.

And, as the great philosopher, Ferris Bueller once said, *"Life moves pretty fast; if you don't stop and look around once in a while, you could miss it."*[3]

Project Management Dealing With Work Related Stress

From a project management perspective, strategies for handling stress would include:

- ☑ **Discussing the project's difficulties over with your immediate supervisor** (i.e. the Program Manager); trying to get them to understand the current situation & prevailing circumstances which make the project currently inviable (from your perspective), and strive to have them assist you in requesting (possibly arguing for) the project baselines to be adjusted accordingly. ... Unfortunately, sometimes there is no intermediary between the Project Manager and the Customer's Representative (as per the scenario project's case), and hence such a discussion may fall on death ears.

---

[3] Ferris Bueller, "Ferris Bueller's Day Off", Paramount Pictures, 1986.

> *"Though, being new to an organization, you probably won't realise what kind of boss you will be working for until you are heavily involved with the project. By which time you may find that they are volatile, autocratic, narcissistic, and/or have a psychopathic personality where they have a 'just do it' attitude to making impossible things happen, without any consideration to compromising."*

- ☑ If support from management above is evidently not forthcoming, then **seeking advice / input from your peer project managers** (who are at the same performing organization) could result in the consideration of alternative solutions; such as, proposing the reallocation of [People] and [Resources] for a short period of [Time].

    > *"Though, don't go robbing Peter to pay Paul, as this will fail them all."*

- ☑ Alternatively, **seeking the advice of workmates** / friends in similar industries and workplaces, even consider **calling upon former mentors** who may suggest alternative solutions that you may not have considered (or given real consideration to).

    > *"Though, their response will be analogous to your younger days when you sought advice from a friend about what to do about a boyfriend / girlfriend relationship that wasn't working out. That being, their impartial response often corresponded with what you deep down knew you had to do."*

- ☑ **Talking it over with your spouse / partner**, because any hastily made decision about work could have a direct (negative) effect on them (your family).

    Which brings us to an addendum to "Adaptive & Proactive" SDLC Project Management.

    ## RULE 42[B]: When in trouble, talk to someone about it; do not bottle it up inside, as there could be disastrous consequences when the pressure eventually bursts.

However, after all of this talking it over and contemplating (the what, when, where, and so forth), the resolution will come down to variations on only a few options.

1) **Battle on through** with things exactly as these currently are.

2) **Throw it in, and walk away**. ... Or, at least devise a personal exit strategy.

3) **Accept the facts** that it can't be done as was originally planned, and **propose alternative** modifications to The Plan, i.e. how to re-baseline the project. ... And, then work off of senior management's response.

Whatever you do, you need to do something; as to continue on with such chronic work related stress of being constantly on edge is not good for one's health, nor for those in close proximity, nor for one's work & personal relationships.

Making The Hard Decisions

For all of that stress that has built up, it is simply a matter of fact, that eventually a decision will have to be made; whether that decision be made voluntarily or forced.

Something will have to be decided; hence, the Project Manager could choose to:

1) Say that *"it can be done"* as was planned, and strive to bulldoze on through. However, the members of the Project Implementation Team will in due course validate that decision; because, if the implementers are faced with impossible delivery workloads, then they may individually decide to "throw it in and walk away", or give up trying.

2) The Project Manager could decide to "cut & run"; but, what will this do to that project manager's reputation? ... And, here is an ethical conundrum; decide to stay and work on a project which you have little to no faith that it can be successfully completed as was planned, or to abandon your fellow team mates to a fate which you have probably just pushed to the edge with your departure.

> *"Though, walking away is often the best response to illogical commands of 'just do it' for something which is not possible to achieve."*

3) The Project Manager could admit to the reality that the project can't be done as was originally planned, and hence state that modifications need to be made to The Plan, i.e. re-baselining the project. Then, let senior management decide how to respond to the project's actual situation. ... Senior management could either; initiate Project Recovery, as per [Section 5.5.2], or they could demand that *"it just be done, OR ELSE"*.

> *"I'm in the same boat. I gave them a realistic estimate to complete the project and was told that this was unacceptable. So they tried to enforce staff ramp down and the project slipped. Then the client complained, so we tried to rehire the staff, as the project slipped even more, and the costs continued to mount up. We are now 3 months later than the original estimated dates and budget. But, the guys who sold the deal and signed off on the contract are still employed."*

Such threats against the Project Team is a double-ended knife in that; by getting rid of the Project Manager and/or members of the Project Team, then this will result in greatly reducing the project's current progress & stability, and disrupting the motivation & morale of those who remain on the Project Team. Alternatively, the members of the Project Team could privately decide to ride-it-out, and look externally for other employment opportunities (and then leave on their own terms).

> *"Yep, I've witnessed a project get screwed because another long-term senior engineer resigned. Subsequently, senior management were in a panic, trying to encourage that particular engineer to stay (at least to the end of the project) while privately shoring up those other remaining engineers. But, anyone who was paying attention would have realized that this situation had been in the making for some time, given the implementers dissatisfaction with the heavy burden of their workload due to them having to make up for all of those on-paper hours that were cut from the project when the Contracted Price was signed off."*

5.7. "Failure" a Possibility of Fact

Well, the scenario's project has not worked out exactly as was planned. After the Project Manager officially stated that all of those requested deliverables could not be produced by the stipulated date, the senior management (aka the customer | your boss) made the decision to continue on with the project as a waterfall arrangement, with all of the GUI mockups to be produced before moving forwards with any of the implementation. Hence;

"Thank you, for all of your efforts on the project; but, I've decided that it would be better if we got in a couple more GUI designers, in your place." Henceforth, your contract as project manager has been discontinued.

Alternatively, you tender your resignation; because, you have already found new employment elsewhere.

Alternatively, "come hell or high water" you decide to push on with this project, even though it has been reorganized from (what you believe is the best solution of a) rolling wave to a waterfall arrangement.

…

Yes, that last project was not the success that you (and they) had planned it to be. … And, here is **an undeniable reality of fact**, … **not every project will be a success**.

Not every project that you will manage (let alone work on) will be deemed "a success"; projects will fail due to various reasons, given the current situation & prevailing circumstances, as well as the flexibility & inflexibility of the project constraints [Baselines].

Just as, a sporting team is not going to win the championship year after year, let alone win every game in a single season.

5.7.1. "Failure" Is Opinionated

Every once and a while, you as a project manager will have a losing project; i.e. a project that was considered "a failure", because it did not satisfy the project's primary stakeholders (and especially the project sponsors) expectation and/or stipulation on the project's [Baselines].

That being, the project, as per the theory in [Section 1.2], did not satisfy some or all of the following:

(1) **SCOPE** ... did the project produce the expected results and does the resultant **deliverables contain the agreed features & functionality?**

(2) **TIME** ... was the project's output delivered to the customer (i.e. the persons requesting the product, service, or goods) when it was agreed to be delivered? That is, **did the project deliver to its milestone dates?**

(3) **COST** ... did the cost of the project not exceed the budget that was allocated for undertaking the project? That is, **was the project a profitable endeavour?**

(4) **QUALITY** ... did the resultant **deliverables meet the agreed Acceptance Criteria?**

Additionally,

(5) **PEOPLE** ... did the project **effectively & efficiently utilize those persons** that were **assigned** to do work on the project?

(6) **RESOURCES** ... did the project **effectively & efficiently utilize those material & services resources** that were **allocated** to be used by the project?

Recalling that, **the determination of whether a project is deemed a Success | Failure is based entirely on the expectations & perspectives of the people involved with the project**; i.e. "FAILURE IS OPINIONATED".

5.7.2. Project Autopsy

In this scenario project's case, the project did not succeed due to a combination of the following factors:

- ☒ The **[Scope]** of the requirements were **continually evolving | in flux**.

- ☒ The **[Cost] budget** was **insufficient for the scale of the project** that was to be undertaken.

- ☒ The limited [Cost] budget restrained the **allocation of [People] being insufficient** for the amount of work to be done.

- ☒ The mandated delivery **[Time] milestone date being too short a duration** given the [Scope] of work to be delivered by the limited number of [People] doing the work on the project.

Additionally,

- ☒ **Differences in understanding** of how best to implement the project given those above considerations, hence a rolling wave structure versus a waterfall arrangement.

- ☒ **Misunderstandings and misinterpretations** between the project sponsor (aka project steering committee – project SME – the boss), the Project Manager (aka business analyst – system architect), and the Project Implementation Team.

> Which brings us to the next rule of "Adaptive & Proactive" SDLC Project Management.
>
> **RULE 43:** **Irrespective of the best laid plans, failure is always a possibility.**
>
> **RULE 43^A:** **Project management is about minimizing the possibilities for failure; however, project management cannot always guarantee success.**

COFFEE BREAK DISCUSSION ... *Even a successful project can end in failure.*

During ~~my~~ our careers we have witness projects that were destine to eventually fail due to the misalignment of the project variables and project constraints; such as, the agreed [Scope] exceeding the capacity of the allotted [Time] duration, and/or the allocated [Cost] budget, and/or the assignment of [People] & [Resources], and/or the mandated [Quality] standards, ... and, other such combinations of project variables and constraints.

However sometimes, a perfectly fine project that was clearly on the road to success, also failed due to situations and prevailing circumstances that were beyond the control of the Project Manager and the Project Implementation Team (and for that matter, not the Project Working Group, and not even the Project Steering Committee).

Project failures have been due to such things as:

☒ All or the majority of the members of the Project Team had to be hurriedly reassigned to work on a strategically higher priority project; irrespective of the fact that their own project was oh-so close to release or a successful finish.

☒ An economic crisis hit the parent organization (and thereby the performing organization), consequently a hasty downsizing in the organization's operating costs had to occur, and subsequently "the baby was thrown out with the bathwater" as all or the majority of the members of the Project Team had to be made redundant.

☒ The parent organization purchased or was purchased by another organization, and during the subsequent analysis of the overlap between these businesses it was found that the project is very similar to an existing project / product in the combined organization, and hence the project was terminated due to being surplus.

☒ The Sales Department | Business Development Group misread the market, and subsequently what is being developed is not what the market is actually wanting.

☒ Senior management were involved with some underhanded business practices and on discovery by either the parent organization (or worse the customer organization) the contract was hastily terminated, and hence the project ended unceremoniously.

6. PM you ... Field Test 2

6.1. Overview

This field test examines the following topics with respect to real-world project management:

- Just because the last project was a Success | Failure does not necessarily mean that the next project will turnout the same way.
- Start each project logically & methodically from a "clean slate" beginning.
- For a 'Fixed Price' contract project, a fair amount of the Planning Phase may need to be conducted during the Initiating Phase.
- Determining the [Scope], what will and will not be given to the customer.
- Determining the [Time], how long it is expected to take to conduct the proposed project.
    - Project schedule layout, contents & components, feasible estimations.
    - Spreading the partial payment milestones for a "positive cash flow".
    - Integrating the project schedule into the Program Roadmap.
- Determining the [Cost] and what is the expected budget to conduct the proposed project.
    - Calculating the overall budget based on the schedule's determined Level Of Effort, the inclusion of burdens & margins (i.e. safety, management, profit).
- Importance of multiple acceptance testing over the duration of the implementation and delivery lifecycle.

6.2. Introduction

The Success | Failure of one project does not make or break the validity of your personalized project management techniques (i.e. "PM you"); as it needs to evolve based on the trials & tribulations of those real-world SDLC projects that you encounter.

6.2.1. The Scenario

```
So, onto the next project, the 'Fixed Price' development
of a new system from concept to production rollout.
```

Though, the question still remains, hidden in the back of your self-evaluating mind, as to whether your recent setback on your previous project will be a forbearer of bad tidings for this newly assigned project.   *... "Given that, your past technical expertise was no real predictor of your successes and failures on that project."*

> Which brings us to the next rule of "Adaptive & Proactive" SDLC Project Management.

> **RULE 44:** Just because your last project was a success (or a failure) does not necessarily mean that your next project will also be a success (or a failure).

> **RULE 44^A:** Each project has its own unique set of project variables & constraints; as well as, differences in the situation & prevailing circumstances.

"The reality is, no matter how skilled the Project Manager and the Project Team, sometimes bad things do happen. The cause of the project's failure may have been within their grasp, or it may have been outside of their

control. The important thing now is to have some idea of what went wrong, why it went wrong; and hopefully, learn from this, so as to minimalize the potential occurrence of those bad things in the future."

As demonstrated in Field Test 1 [Chapter 5], a project can fail for many and varied reasons, see [Section 5.7]; however, for now, the important thing is to "**learn the lessons from the past**", so that the Success | Failure of the next project is not left simply to chance.

Which brings us to the next rule of "Adaptive & Proactive" SDLC Project Management.

RULE 45: Learning from your mistakes is smart; but, learning from the mistakes of others is wise.

6.2.2. Beginning From A Clean Slate

As per the previous project, start the new project with what apparently worked:

1.. **Building that first impression** of the project, the performing organization, the Project Implementation Team, and you as the Project Manager, [Section 5.2.2].

2.. **Answer those opening questions** that will define the new project based on the 'needs & wants' of the customer, evaluate the viability of the proposed project, and then obtain formalized approval & authorization to commence the project. That is, create **the Project Charter** and get it signed off, so as to begin in earnest, see [Section 5.3.2].

3.. **Based on the [Scope] Boundary** of work uncovered thus far, **estimate a Rough Order of Magnitude for** the project's **[Time] duration and** the **"Level Of Effort" involved**, see [Section 5.4.1] and the example case in [Section 3.3.9]. ... Calculate the **Overall Budget [Cost]** including the **Management Reserve** and **Profit Margin**, see [Section 5.4.2]; though, try to ensure that the project is not priced to death, see [Section 5.4.3].

361

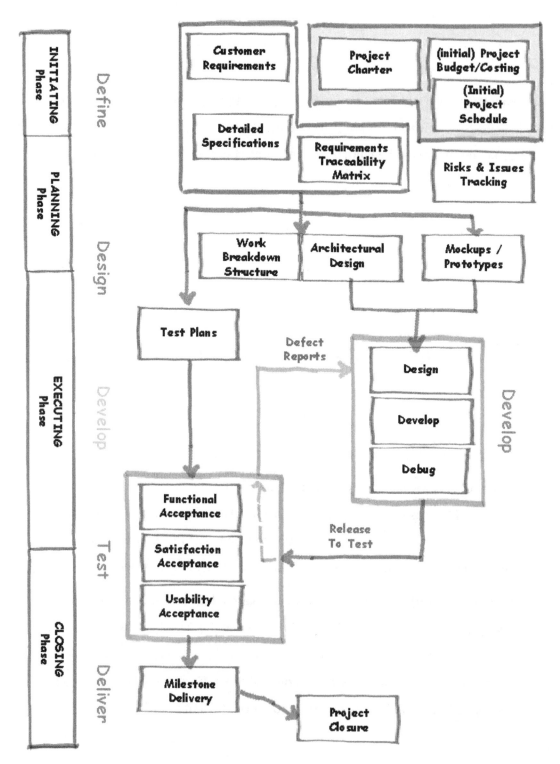

Figure 84: Grouping of documents to cope with the project's situational constraints.

6.3. And So, It Begins

6.3.1. What Do They Want?

During the project's **Initiating Phase**, the **customer's requirements** (including the associated notes and reference materials) should be **accumulated** for the representatives of the customer & performing organizations to come together **to review, explain,** and **discuss**. Subsequently, there should be an **understanding** as to the **customer's "perceived benefits", "desired outcomes", and "expected business changes"** from conducting this project; as ideas should be forming with respect to the primary Use Cases that will need to be implemented in order to satisfy the **customer's "needs & wants"**. See [Section 3.3.3].

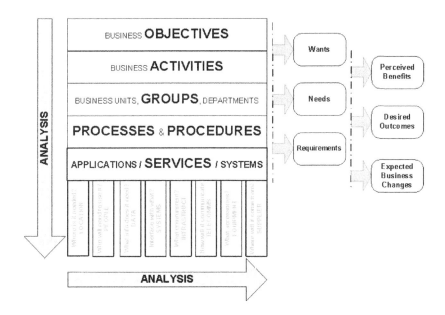

Figure 25: Top Down Analysis to determine "Wants", "Needs", and "Requirements".

And, hopefully by now, the **representatives from both the customer & performing organizations** (are starting to) **have a "common understanding" and "mental alignment"** with respect to; the **interpretations of the requirements**, what are the **desired deliverables**, the **intended milestones**, and the **scale of the project** to be undertaken. Which is all then summarized in the **Project Charter**.

6.3.2. What'll We Give & Not Give To Them?

With the **Customer Requirements** now having been **agreed to**, for the purpose of intent, being not binding on either the performing & customer organizations (**a "wish list" per se**); then the **performing organization** would put together a **high-level list of "what will & won't be delivered"**. Depending on the situation and industrial circumstances, this response may be included as part of the Project Charter (with the level of detail as per a glossy sales & marketing brochure), or it may be stipulated that it be expanded upon as a feature-by-feature list with appropriate levels of explanation in a **Detailed Specifications** (aka Functional Specifications – FS, Detailed Business Requirements Specifications – DBRS).

> In some situations, this Detailed Specifications document may be incorporated into the response to a Request For Tender (RFT), a Request For Proposal (RFP), or a Request For Quotation (RFQ).

Whatever form and wherever this document resides, **the level of technical detail within this specifications should be sufficient to provide the Customer's Representatives with enough confidence that the performing organization understands what is being asked of them, that they demonstrate a level of knowledge sufficient for the proposed endeavor, and that they have a good idea as to what needs to be done**. ... Without committing the performing organization to what could be (in hindsight) a mistaken technical specification.

Subsequently, the creation of this (Detailed) Specifications would **involve the Business Analyst (BA)**, the **Systems Architect (SA)**, and relevant **Subject Matter Experts (SME)** working together with the **Project Manager (PM)**, ... and, **senior implementers**, so that **realistic & technically feasible solutions are considered** and summarized; instead of the technically unachievable being proposed (by Sales & Marketing | Business Development).

While defining the **(Detailed) Specifications**, the Project Manager could **cross match** the **Customer Requirements** with the features & functionality point entries in the **Requirement Traceability Matrix (RTM)** spreadsheet, so as to ensure the coverage of "what's in, what's out, and what needs to be decided upon" is recorded for clarification.

At this point, depending on the project's payment method to be utilized (i.e. 'Fixed Price', 'Time & Materials', or 'Cost Reimbursement'), then different levels of due diligence consideration would be given to determining the proposed project's [Time] & [Cost].

However, there firstly needs to be some understanding as to **how does this project** (and its deliverables) **fits into the "big picture"** (overall scheme) **of what the customer & performing organizations are trying to accomplish**; i.e. where the project fits into the Program of Work. Noting that, both **organization**s, at any given time, **only have certain quantities of capital [Cost], [Time], [People] & [Resources] at their disposal**; thus, would the organization **be strategically better served to pay to use other organizations' capabilities & capacities rather than utilizing their own**, or, **forgoing this project entirely**?

> **Program** (of Work) ... is where **multiple projects** are **coordinated** to contribute towards **achieving** a **common business objective**.
>
> **Portfolio** ... is where a **collection of projects & programs** are coordinated to achieve the organization's "**strategic goals**" (i.e. greater business objectives).

Project Charter (Extract)

```
This project is the first stage of a multi-staged program
of work, the stages being; (1) deployment of the baseline
functionality, (2) customization of the provided solution
with advanced features, (3) future enhancements that have
yet to be defined as these will most probably be defined
once Stage 1 is completed and Stage 2 is underway.

Where each stage will not commence until the predecessor
stage has been signed off as successfully delivered.

The project for Stage 1 has been broken up into the
following major activities:

I.     Analysis and design of the system integration.
```

II. Application development on an in-house test platform.

III. Functional Acceptance Test (FAT) using the in-house test platform.

IV. Deployment into the customer's "Ready For Deployment" network, and Satisfaction Acceptance Test (SAT).

V. Deployment into the customer's "Ready To Go Live" network, and Usability Acceptance Test (UAT).

VI. Public Go Live activity which is to occur on Monday **31-10-2015**.

VII. Stage 2 and Stage 3 are out-of-scope of this Project Charter's Statement Of Work.

VIII. Helpdesk support training will not be included as part of the Stage 1 deliverables.

IX. The associated Administration & Maintenance Manuals and Help Desk Support Guides will not be included as part of the Stage 1 deliverables.

X. The performing organization is only responsible for providing the FAT testing facility; all other testing facilities will be provided, supported, and maintained by the customer organization.

The Stage 1 project milestones and associated partial payment milestones have been agreed to align with the culmination of those major project activities listed.

Given that this scenario's project is going to be a **'Fixed Price' contract with a** stipulated **End Date** (when the first release of the resultant product deliverable must go live), see [Section 3.4.4], **then this project's Initiating Phase will need to incorporate a relatively detailed amount of planning that would normally be done during the Planning Phase.**

In Depth Planning Conducted During The Initiating Phase

In [Section 2.3] on the project life cycle and [Chapter 3] on the PMBOK® theory, there was presented a relatively straight forward relationship of the project's Initiating Phase flowing into the project's Planning Phase. However, for some projects, specifically those involving a **'Fixed Price' contract**, then **much of the detailed planning** (i.e. the Detailed Specifications, and the Architectural Design) **may need to be undertaken during the Initiating Phase so as to produce a detailed Project Schedule, and thereby calculate an accurate Overall Budget | Contract Price**. See [Section 5.4.2] & [Section 5.4.3].

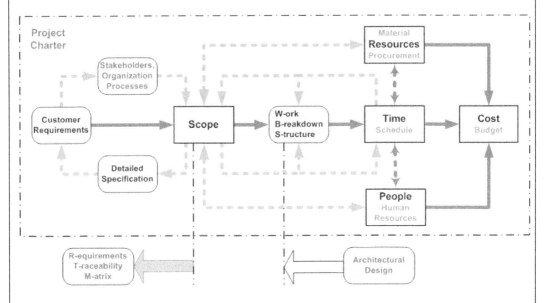

Figure 96: Project Initiating & Planning Phases merged to determine [Time] & [Cost].

For a **'Fixed Priced' contract then due diligence [Time] & [Costs] estimates are necessary**; however, these can only be determined **by having a good understanding** of the project's **[Scope] and with the realistic allocation of the [People] & [Resources] that would be required for the implementation & delivery to occur**. See [Figure 96].

Whereas, for 'Time & Materials' or 'Cost Reimbursement' based projects then the quoted price only needs to be within at least 20-25% range of that which would be produced by a relatively detailed analysis; i.e. a Rough Order of Magnitude (ROM).

6.3.3. How Long Is It Expected To Take?

1st Pass Schedule

During this Initiating Phase, the BA – SA – SME – Senior Implementers (and PM) should **transform those initial ideas & concepts** of the project's composition **into modularized Architectural Design sketches of** how this thing should probably go together; thereby, determining the **constituent parts** as denoted in the **Work Breakdown Structure (WBS)**. ... And hopefully, aligning the project management's [Time] Schedule and subsequent [Cost] Budget with the implementers' perspective of how the [Scope] can be delivered.

Thus, with the aid of the BA – SA – SME – Implementers, progressively elaborating the 1st Pass schedule to determine the major work activities to be done. See [Section 3.4.4].

"Woow Man, that is such a way-out there radical idea to get the implementers involved with helping to determine the estimates for those technical & production activities; instead of having non-technical people guesstimating (or some would say, 'randomly selecting') the Time, People, and Resources involved with such areas of technical expertise."

2nd Pass Schedule

With the 1st Pass Schedule having taken form (though containing no actual start dates nor end dates, no named allocations of people & resources, and no real task duration estimations), just the listing of all of the major activities and the identification of the primary Use Cases that will need to be done. Then, with the aid of the BA – SA – SME – Implementers, a 2nd Pass Schedule (with the sequencing & parallelizing of tasks) should be produced, focusing on the relatively accurate estimation of the Level Of Effort involved with each of those identified major activities and primary Use Cases.

Suggest, using a technique such as **'Weighted Average'** where the senior implementers | SME provide for each task an estimate of; **(t_o) optimistic** best case, **(t_m) most likely** normal case, and **(t_p) pessimistic** worst case.

$$\text{Weighted Average} = \frac{(t_o + 4t_m + t_p)}{6}$$

While this 'Weighted Average' calculation does provide more realistic estimates than the simple putting forward of single values, this technique is a bit more pessimistic with its numbers. Have a Wiki at, **PERT – Project (Program) Evaluation Review Technique**.

 I would recommend that, the Project Manager **obtain as accurate as possible estimates for task duration and Level Of Efforts [Time], because these estimates will directly affect the project's [Cost] budget calculation.**

When obtaining estimates for task durations | Level Of Effort, then:

- ☒ **DO NOT accept too long a duration | effort estimate for a single task**, because when this long duration | effort task is broken down into its constituent parts then the resultant child-tasks often increase or decrease the parent-task's length to something closer to what really will be required for its implementation.

    "Most people are prone to giving overly optimistic estimates; while, others may give as big an estimates as they think will be accepted (as a form of their own private safety buffer), and others will provide estimates that they think the asker wants to hear."

- ☒ **DO NOT seek too small a duration | effort estimate for a single task**, because too fine a granularity will result in micro-managing the [People] and an overly complex & bulky schedule. Instead, **go to the level of tasks that are apparent steps towards a deliverable** and not down to the minor steps along the way.

- ☑ **Set a maximum limit on the estimated duration | effort for any individual task** (e.g. 80 hours), **anything quoted as longer than this duration has to be broken down into shorter duration sub-tasks.** Though, if the project is of a relatively short duration (e.g. one month) then the maximum estimation limit per task maybe set as a couple of days; hence, any task longer than that has to be subdivided into smaller tasks.

When scheduling task duration | effort, then:

- ☑ **Set realistic workday durations per person**, because not every project team member is going to be able to do a full 7 to 8+ hours worked per day on his or her assigned tasks. While they may do a full day's work, there will be meetings to attend, workmates to assist, other competing projects requiring attention, administrative activities, and short notice leaves of absence (i.e. sick leave, single days annual leave). And, **keep track of the individuals "resource utilization", as assigning people 15 hour days is not conducive to their work-home life-balance nor the viable success of the project**. Hence, try the "**Resource Levelling**" tool on (a copy of) the 2nd Pass Schedule.

- ☑ **Ensure that** both **Project Management (PM – Snr Mgmt.) and Technical Authorities (BA – SA – SME) are allocated time** in the project's schedule; i.e. appear as distinct tasks that are to be charged to the project's Cost Account Code. When scheduling these 'Project Management' and 'Technical Authority' supporting tasks, then calculate these hours as a percentage (5-15% based on the project's complexity / riskiness) of the Implementation total hours, and spread these over the project's life. … And, these **"support hours" are independent of any dedicated tasks in the project's schedule that are assigned to specific 'Project Management' and 'Technical Authority' persons** (such as for reviews, meetings, reporting); because, it is a certainty that these two parties will spend some unexpected time supporting the project's activities.

- ☑ Ensure that all of the **holiday breaks** (including public holidays) **are marked in the project's schedule calendar**; also, note when particular Project **Team members** would be **unavailable** due to their taking **annual leave**, or being **assigned to other projects / Business As Usual activities**, … and, their full-time or part-time work status.

- ☑ **If certain features & functionality** (or the order of implementation) **won't be clearly understood until closer to the time of the actual implementation** (as per an **agile** based project), **then allocate blocks of work effort hours** (i.e. **Sprints**) during which these 'To Be Defined' tasks will be undertaken by the Project Implementation Team.

Later on, during the execution of each Sprint the implementers will self-organize which tasks will be included in the particular sprint. See [Figure 31] reproduced below.

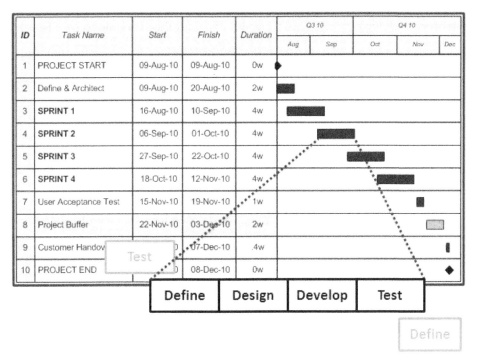

Figure 31: A waterfall schedule with the inclusion of an agile implementation.

- ☑ **Add contingency buffering to the 'Critical Path'** (and to those feeder paths where there is reasonable likelihood of some delays occurring, such as with the delivery of hardware componentry); but, **do not buffer every possible path through the project**.

- ☑ **Choose milestones** (tasks assigned to zero days duration) **that correspond to events of notable interest to the project's primary stakeholders**; *e.g. the Start and/or the End of each major sequence of activities (such as the FAT, SAT, UAT testing), when funds need to be paid (such as international travel tickets), Progress Review meetings, progress payments for the work completed thus far, and the project's Start Date & End Date.*

Which brings us to the next rule of "Adaptive & Proactive" SDLC Project Management.

RULE 46: The estimation of task duration, is when size really does matter.

Schedule Feasibility

With the **'draft' 2nd Pass Schedule** now **completed**, then **talk it through with the senior implementers** to determine whether they also consider the schedule to be a **feasible** plan; thereby, aiming to **obtain "buy-in"** from the representatives of the (potential) Project Implementation Team.

"A project manager who does not consult with the implementers, and continually assigns task durations that are not feasible to successfully accomplish the given work, will soon lose the respect of the Project Team, and will garner a reputation as someone who is … devoid of intelligence."

Having made feasibility adjustments to the 2nd Pass Schedule, then **run this schedule past the Program Manager** so as to ensure that **no components** have been **left out** (*e.g. production of the mandated Objective Quality Evidence, Project Auditing, Progress Update Meetings & Reports, Post Completion Reviews … etc*). That is, stuff that the implementers (and the Project Manager) may have forgotten about.

Also, **ensure that the project's sequence of activities & milestone dates are compatible with those other projects** that are currently in the pipeline, and those projects that will be in the pipeline at the same time as this project's implementation (i.e. on the program "roadmap"); thereby, **aligning this project with the forthcoming Program Of Work**.

Additionally, there may be **strategic business considerations** such as; **financial windows** (*e.g. the End Of Financial Year*) where it is highly advantageous that the project has **payment milestones** that occur either **prior to or after certain dates**, that only certain amounts of money be spend in certain periods of time in which case only a certain range of **Level Of Effort hours can be exerted in that particular period** (*e.g. per quarter year*), and **progress payments** (thus far) **exceed the project's costs incurred** (to date).

Once it is understood how this project relates to other projects' milestones, the availability of [People] & [Resources], and to other strategic business considerations, then the schedule should be adjusted accordingly. … And, an updated revision produced.

Schedule Layout & Contents

For the scenario's project, the schedule is as shown in [Figure 97].

ID		Task Name	Work	Duration	Start	Finish	Predecessors	Resource Names	Cost
1		PROJECT: EXCASE 2	4,937 hrs	218.94 days	Mon 16/02/15	Tue 15/12/15			$909,542.5(
2		Task Milestones	0 hrs	218.94 days	Mon 16/02/15	Tue 15/12/15			$0.0(
3		Input - START DATE	0 hrs	0 days	Mon 16/02/1	Mon 16/02/1			$0.0(
4		Part 1 - Analysis & Design - Start	0 hrs	0 days	Mon 16/02/1	Mon 16/02/1	73SS		$0.0(
5		Part 1 - Analysis & Design - Finish	0 hrs	0 days	Fri 03/04/1!	Fri 03/04/1	73FF		$0.0(
6		Part 2 - Implementation - Start	0 hrs	0 days	Fri 20/03/1!	Fri 20/03/1	104SS		$0.0(
7		Part 2 - Implementation - Finish	0 hrs	0 days	Mon 03/08/1	Mon 03/08/1	104FF		$0.0(
8		Part 3 - Factory / Functionality Acceptance Test - Start	0 hrs	0 days	Mon 03/08/1	Mon 03/08/1	196SS		$0.0(
9		Part 3 - Factory / Functionality Acceptance Test - Finish	0 hrs	0 days	Mon 17/08/1	Mon 17/08/1	196FF		$0.0(
10		Part 4 - Site / Satisfaction Acceptance Test - Start	0 hrs	0 days	Mon 17/08/1	Mon 17/08/1	226SS		$0.0(
11		Part 4 - Site / Satisfaction Acceptance Test - Travel Date	0 hrs	0 days	Fri 11/09/1!	Fri 11/09/1	240FF+7 days		$0.0(
12		Part 4 - Site / Satisfaction Acceptance Test - Finish	0 hrs	0 days	Fri 18/09/1!	Fri 18/09/1	226FF		$0.0(
13		Part 5 - User / Usability Acceptance Test - Start	0 hrs	0 days	Fri 18/09/1!	Fri 18/09/1	299SS		$0.0(
14		Part 5 - User / Usability Acceptance Test - Travel Date	0 hrs	0 days	Tue 29/09/1	Tue 29/09/1	226FF+7 days		$0.0(
15		Part 5 - User / Usability Acceptance Test - Finish	0 hrs	0 days	Sat 10/10/1	Sat 10/10/1	299FF		$0.0(
16		Input - GO LIVE DATE	0 hrs	0 days	Sat 31/10/1	Sat 31/10/1			$0.0(
17		Part 6 - Go Live - Start	0 hrs	0 days	Fri 23/10/1!	Fri 23/10/1	347SS		$0.0(
18		Part 6 - Go Live - Finish	0 hrs	0 days	Mon 16/11/1	Mon 16/11/1	347FF		$0.0(
19		Part 7 - Warrantee - Start	0 hrs	0 days	Sat 31/10/1	Sat 31/10/1	397SS		$0.0(
20		Part 7 - Warrantee - Finish	0 hrs	0 days	Thu 10/12/1	Thu 10/12/1	397FF		$0.0(
21		Part 8 - Project Closure	0 hrs	0 days	Tue 15/12/1	Tue 15/12/1	406FF		$0.0(
22		Progress Payment Milestones	0 hrs	184.81 days	Fri 03/04/1!	Tue 15/12/15			$0.0(
23		PART 1 - Analysis & Design	0 hrs	0 days	Fri 03/04/1!	Fri 03/04/1	103		$0.0(
24		PART 2 - Implementation - Sprint 1 Progress Update	0 hrs	0 days	Fri 08/05/1!	Fri 08/05/1	133		$0.0(
25		PART 2 - Implementation - Sprint 2 Progress Update	0 hrs	0 days	Fri 05/06/1	Fri 05/06/1	140		$0.0(
26		PART 2 - Implementation - Sprint 3 Progress Update	0 hrs	0 days	Fri 03/07/1	Fri 03/07/1	147		$0.0(
27		PART 3 - FAT - (Functional) Factory Acceptance Test	0 hrs	0 days	Mon 17/08/1	Mon 17/08/1	225		$0.0(
28		PART 4 - SAT - (Satisfaction) Site Acceptance Test	0 hrs	0 days	Fri 18/09/1!	Fri 18/09/1	298		$0.0(
29		PART 5 - UAT - (Usability) User Acceptance Test	0 hrs	0 days	Sat 10/10/1	Sat 10/10/1	346		$0.0(
30		PART 6 - Go Live	0 hrs	0 days	Fri 13/11/1	Fri 13/11/1	386		$0.0(
31		PART 8 - Project Closure	0 hrs	0 days	Tue 15/12/1	Tue 15/12/1	418		$0.0(
32		Major Deliverables	0 hrs	12 days	Mon 17/08/15	Wed 02/09/1!			$0.0(
33		Acceptance Test Procedures	0 hrs	0 days	Mon 17/08/1	Mon 17/08/1	224		$0.0(
34		System Hardware	0 hrs	0 days	Wed 02/09/1	Wed 02/09/1	248		$0.0(
35		System Media	0 hrs	0 days	Wed 19/08/1!	Wed 19/08/1	231		$0.0(
36		Licenses & Software Keys	0 hrs	0 days	Thu 20/08/1	Thu 20/08/1	235		$0.0(
37		End User Documentation	0 hrs	0 days	Mon 24/08/1	Mon 24/08/1	239		$0.0(
38		Progress Deliverables	0 hrs	193.44 days	Mon 23/03/15	Tue 15/12/15			$0.0(
39		Part 1 - System Integration Analysis - Report	0 hrs	0 days	Mon 23/03/1	Mon 23/03/1	91SS		$0.0(
40		Part 1 - System Integration Analysis - Review Minutes	0 hrs	0 days	Mon 30/03/1	Mon 30/03/1	97		$0.0(
41		Part 2 - Sprint 1 Progress Update	0 hrs	0 days	Fri 08/05/1!	Fri 08/05/1	132		$0.0(
42		Part 2 - Sprint 2 Progress Update	0 hrs	0 days	Fri 05/06/1!	Fri 05/06/1	139		$0.0(
43		Part 2 - Sprint 3 Progress Update	0 hrs	0 days	Fri 03/07/1!	Fri 03/07/1	146		$0.0(
44		Part 2 - Progress Review / Test Readiness Review - Minutes	0 hrs	0 days	Wed 29/07/1!	Wed 29/07/1	189		$0.0(
45		Part 2 - Acceptance Test Procedures	0 hrs	0 days	Mon 03/08/1	Mon 03/08/1	194		$0.0(
46		Part 3 - Factory / Functionality Acceptance Test - Review Minutes	0 hrs	0 days	Fri 07/08/1!	Fri 07/08/1	212		$0.0(
47		Part 3 - Factory / Functionality Acceptance Test - Report	0 hrs	0 days	Thu 13/08/1	Thu 13/08/1	219		$0.0(
48		Part 4 - Site / Satisfaction Acceptance Test - Review Minutes	0 hrs	0 days	Fri 11/09/1!	Fri 11/09/1	285		$0.0(
49		Part 4 - Site / Satisfaction Acceptance Test - Report	0 hrs	0 days	Fri 18/09/1!	Fri 18/09/1	297		$0.0(
50		Part 5 - User / Usability Acceptance Test - Review Minutes	0 hrs	0 days	Mon 05/10/1	Mon 05/10/1	333		$0.0(
51		Part 5 - User / Usability Acceptance Test - Report	0 hrs	0 days	Sat 10/10/1	Sat 10/10/1	345		$0.0(
52		Part 6 - Go Live Completion - Review Minutes	0 hrs	0 days	Tue 03/11/1	Tue 03/11/1	383		$0.0(
53		Part 6 - Go Live Completion - Report	0 hrs	0 days	Fri 13/11/1	Fri 13/11/1	395		$0.0(
54		Part 8 - Project Close Out Report	0 hrs	0 days	Tue 15/12/1	Tue 15/12/1	417		$0.0(
55									
56		Stage 1 - Activity Execution	4,537 hrs	218.94 days	Mon 16/02/15	Tue 15/12/15			$829,542.5(
57		PART 0 - Project Kick off	17 hrs	0.63 days	Mon 16/02/15	Mon 16/02/1!	3		$2,700.0(
73		PART 1 - Analysis & Design	216 hrs	33.5 days	Mon 16/02/15	Fri 03/04/1!	57		$46,460.0(
104		PART 2 - Implementation (Agile Sprints)	3,305 hrs	95.75 days	Fri 20/03/1!	Mon 03/08/1!			$548,325.0(
196		PART 3 - FAT - (Functional) Factory Acceptance Test	120 hrs	10.31 days	Mon 03/08/15	Mon 17/08/1!			$18,557.5(
226		PART 4 - SAT - (Satisfaction) Site Acceptance Test	276.5 hrs	24.44 days	Mon 17/08/15	Fri 18/09/1!			$73,962.5(
299		PART 5 - UAT - (Usability) User Acceptance Test	144.5 hrs	15.94 days	Fri 18/09/1!	Sat 10/10/1			$41,452.5(
347		PART 6 - Go Live	174.5 hrs	17.94 days	Fri 23/10/1!	Mon 16/11/1!			$51,377.5(
397		PART 7 - Warrantee Support	240 hrs	30 days	Sat 31/10/1!	Thu 10/12/15	377		$38,890.0(
406		PART 8 - Project Closure	43.5 hrs	2.94 days	Fri 11/12/1!	Tue 15/12/15	397		$7,907.5(
419		Stage 1 - Management Support (5% Total Hours)	200 hrs	25 days	Mon 16/02/1	Fri 20/03/1	56SS	Management Suppo	$40,000.0
420		Stage 1 - Technical Authority Support (5% Total Hours)	200 hrs	25 days	Mon 16/02/1	Fri 20/03/1	56SS	Technical Authority	$40,000.0

Figure 97: Scenario's project schedule with milestone dates & costs.

373

"Gee, I really have to zoom in here to examine the details. … … … … … WHAT!! You're using MS-Project to calculate costs!! This is one situation where you should definitely use a spreadsheet, because a spreadsheet is better at the intricacies of the cost calculations; whereas, a schedule is better suited as a tool for tracking work versus time, the determination of the percentage complete, and the assignment of people & resources to tasks.

In the spreadsheet, I would have a worksheet (or worksheets) for the major packages of work that were determined during the progressive elaboration of the Work Breakdown Structure (WBS) and the schedule. Once the worksheets have represented the project schedule, then I'd accumulate this cost information into a summation worksheet at the front of the spreadsheet to produce a total Contract Price."

Agreed, … however, this schedule is being used for the Initiating & Planning Phases estimation purposes; and, the inclusion of [Cost] in this example does demonstrate how the [People] & [Resources] assigned to tasks accumulate to drive the project's [Cost].

Also, this schedule's Costing is not the accurate calculation of the project's Contract Price; rather, it provides an "in the ballpark" estimate of what the project is potentially going to cost, excluding Safety Margin (i.e. Cost Escalations, Cost Contingencies, and Risk Percentages), Management Reserve, and Profit Margin. See [Section 5.4.2].

> For the scenario's project, from the resultant schedule, the Level Of Effort is accumulated to 4940 hours, calendar duration of 10 months, and an Internal Total Cost of $910K which when adding a 25-50% mark-up (depending on the industry and the complexity / riskiness of the endeavour) for Safety Margin + Management Reserve + Profit Margin is a $1.14M to $1.37M Contract Price … initial guesstimation.
>
> *"So, let's see what a budget calculation spreadsheet comes up with."*

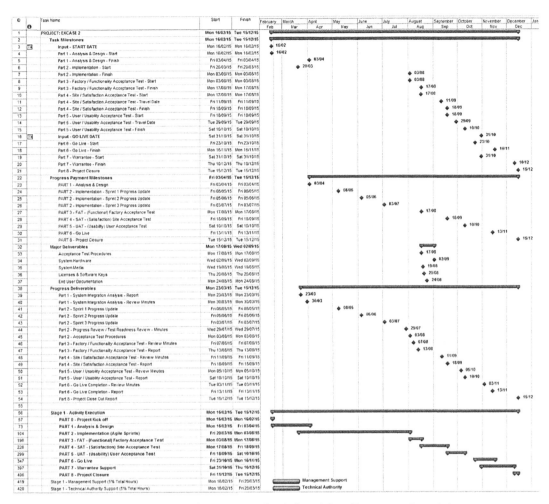

Figure 98: Scenario's project schedule with milestones & task bars.

As an aside, ... how the project's schedule is laid out is up to the schedule's creator (i.e. the Project Manager), as influenced by how the performing organization's senior management (including the Program Manager) expect to see project schedules presented.

Though personally, I like to have at the beginning | top of the schedule; a **clear identifier of the project's name**, then indent accordingly, a **list of the Major Milestones** (including any defined Start Date & End Date constraints), a **list of the Progress Payment Milestones**, a **list of the Major Deliverables**, a **list of the** (proof of) **Progress Deliverables**, then a **rolled-up summation of** the project's **tasking activities**. See [Figure 98].

This kind of **schedule layout provides a sense of "visibility in control"**, as the project's primary stakeholders don't have to go trawling through the entire schedule looking for information, because what is of major concern to them is detailed right there at the beginning of the schedule. ... And, when the "Activity Execution" tasks are rolled-up to summarization form, then there is a nicely presented flow of Define – Design – Develop – Test – Deploy (+ Warranty Support), which goes a long way to satisfying the concerns of those "waterfaller" primary stakeholders.

ID		Task Name	Start	Finish
1		PROJECT: EXCASE 2	Mon 16/02/15	Tue 15/12/15
57		PART 0 - Project Kick off	Mon 16/02/15	Mon 16/02/15
73		PART 1 - Analysis & Design	Mon 16/02/15	Fri 03/04/15
104		PART 2 - Implementation (Agile Sprints)	Fri 20/03/15	Mon 03/08/15
196		PART 3 - FAT - (Functional) Factory Acceptance Test	Mon 03/08/15	Mon 17/08/15
226		PART 4 - SAT - (Satisfaction) Site Acceptance Test	Mon 17/08/15	Fri 18/09/15
299		PART 5 - UAT - (Usability) User Acceptance Test	Fri 18/09/15	Sat 10/10/15
347		PART 6 - Go Live	Fri 23/10/15	Mon 16/11/15
397		PART 7 - Warrantee Support	Sat 31/10/15	Thu 10/12/15
406		PART 8 - Project Closure	Fri 11/12/15	Tue 15/12/15
419		Stage 1 - Management Support (5% Total Hours)	Mon 16/02/15	Fri 20/03/15
420		Stage 1 - Technical Authority Support (5% Total Hours)	Mon 16/02/15	Fri 20/03/15

For the scenario's project, the individual labour rates to be used (as per the performing organization's standard rates) are:

ID		Resource Name	Std. Rate
1		BA - SA	$200.00/h
2		Sys Lead Eng	$195.00/hr
3		Doco Support	$125.00/h
4		Proj Mgmt	$185.00/h
5		Sw Lead Eng	$175.00/h
6		Sw Eng 1	$150.00/h
7		Sw Eng 2	$140.00/h
8		Sw Eng 3	$130.00/h
9		Sw Eng 4	$120.00/h
10		Sw Eng 5	$110.00/h
11		Logistics Mgr	$155.00/hr
12		Test Eng	$125.00/h
13		Test Lead Eng	$150.00/h
14		Technical Authority	$200.00/h
15		Management Support	$200.00/h

Hence, this project will require; 1x Project Manager, 1x Business Analyst | System Architect, 1x Lead Systems Engineer, 6x Software Engineers, 2x Testers, 1x Document Writer, and 1x Logistics Coordinator. See [Figure 99].

ID	Task Name	Work	Duration	Start	Finish	Predecessors	Resource Names	Cost
57	PART 0 - Project Kick off	17 hrs	0.63 days	Mon 16/02/15	Mon 16/02/15	3		$2,700.00
58	Meeting Preparation / Presentation	4 hrs	0.5 days	Mon 16/02/1	Mon 16/02/1		Proj Mgmt	$740.00
59	Meeting Participation	13 hrs	0.13 days	Mon 16/02/15	Mon 16/02/15	58		$1,960.00
60	Meeting Participation	1 hr	0.13 days	Mon 16/02/1	Mon 16/02/1		Proj Mgmt	$185.00
61	Meeting Participation	1 hr	0.13 days	Mon 16/02/1	Mon 16/02/1	60SS	Sys Lead Eng	$195.00
62	Meeting Participation	1 hr	0.13 days	Mon 16/02/1	Mon 16/02/1	60SS	BA - SA	$200.00
63	Meeting Participation	1 hr	0.13 days	Mon 16/02/1	Mon 16/02/1	60SS	Sw Lead Eng	$175.00
64	Meeting Participation	1 hr	0.13 days	Mon 16/02/1	Mon 16/02/1	60SS	Sw Eng 1	$150.00
65	Meeting Participation	1 hr	0.13 days	Mon 16/02/1	Mon 16/02/1	60SS	Sw Eng 2	$140.00
66	Meeting Participation	1 hr	0.13 days	Mon 16/02/1	Mon 16/02/1	60SS	Sw Eng 3	$130.00
67	Meeting Participation	1 hr	0.13 days	Mon 16/02/1	Mon 16/02/1	60SS	Sw Eng 4	$120.00
68	Meeting Participation	1 hr	0.13 days	Mon 16/02/1	Mon 16/02/1	60SS	Sw Eng 5	$110.00
69	Meeting Participation	1 hr	0.13 days	Mon 16/02/1	Mon 16/02/1	60SS	Test Lead Eng	$150.00
70	Meeting Participation	1 hr	0.13 days	Mon 16/02/1	Mon 16/02/1	60SS	Test Eng	$125.00
71	Meeting Participation	1 hr	0.13 days	Mon 16/02/1	Mon 16/02/1	60SS	Docu Support	$125.00
72	Meeting Participation	1 hr	0.13 days	Mon 16/02/1	Mon 16/02/1	60SS	Logistics Mgr	$155.00

Figure 99: Scenario's project schedule – Part 0 – Project Kick Off.

"WOOH!! This portion of the schedule has individual tasking activities down to hourly resolution; this is way too detailed a breakdown for a planning schedule, as this could result in the micro-management of the project. And, just how detailed should you go for every group of tasks?"

Agreed, ... however, depending on what is stipulated by the performing organization's senior management (including the Program Manager) and what the schedule is to be used for, then the schedule's creator (i.e. the Project Manager) may be required to include such a detailed breakdown. In this case, to form the foundations by which to build an accurate calculation of the project's costing.

Also, recall that a project is deemed "a success" or "a failure" based entirely on the expectations & perspectives of the project's primary stakeholders; hence, if "the bosses" require that the schedule contains such detailed breakdowns for certain activities, then so be it.

Which brings us to the next rule of "Adaptive & Proactive" SDLC Project Management.

RULE 47: It is best to give the project's primary stakeholders what they want & need; rather than, quibbling over the level of detail.

Schedule of Project Parts

ID		Task Name	Work	Duration	Start	Finish	Predecessors	Resource Names	Cost
73		PART 1 - Analysis & Design	216 hrs	33.5 days	Mon 16/02/15	Fri 03/04/15	57		$46,460.00
74		Part 1 - System Integration Analysis & Design	214 hrs	33.25 days	Mon 16/02/15	Thu 02/04/15			$46,090.00
75		Travel To Site	16 hrs	2 days	Mon 16/02/15	Wed 18/02/15			$7,800.00
76		Travel Coordination	8 hrs	1 day	Mon 16/02/1	Tue 17/02/1		Logistics Mgr	$1,240.00
77		TRAVEL DATE	0 hrs	0 days	Tue 17/02/1	Tue 17/02/1	76		$0.00
78		Travel To Site	8 hrs	1 day	Tue 17/02/1	Wed 18/02/1	77	Sys Lead Eng	$1,560.00
79		Travel to & from site Cost	0 hrs	0 days	Wed 18/02/15	Wed 18/02/15	78		$2,000.00
80		Accommodation & Allowance Cost	0 hrs	0 days	Tue 17/02/1	Tue 17/02/1	78SS		$3,000.00
81		Analysis of Current Front Office System	24 hrs	3 days	Wed 18/02/1	Mon 23/02/1	75	Sys Lead Eng	$4,680.00
82		Analysis of Current Back Office System	56 hrs	7 days	Mon 23/02/1	Wed 04/03/1	81	Sys Lead Eng	$10,920.00
83		Travel From Site	8 hrs	1 day	Wed 04/03/1	Thu 05/03/1	82	Sys Lead Eng	$1,560.00
84		Write Report	60 hrs	10 days	Thu 05/03/15	Thu 19/03/15	83		$11,800.00
85		Write Report	40 hrs	5 days	Thu 05/03/1	Thu 12/03/1		Sys Lead Eng	$7,800.00
86		Support	20 hrs	2.5 days	Thu 05/03/1	Tue 10/03/1	85SS	BA - SA	$4,000.00
87		BUFFER - Time	0 hrs	5 days	Thu 12/03/1	Thu 19/03/1	85		$0.00
88		Internal Review Report	8 hrs	1 day	Thu 19/03/1	Fri 20/03/1	84	BA - SA	$1,600.00
89		Document Presentation Preparation	4 hrs	0.5 days	Fri 20/03/1	Mon 23/03/1	88	Doco Support	$500.00
90		Review - Doco	1 hr	0.13 days	Mon 23/03/1	Mon 23/03/1	89	Proj Mgmt	$185.00
91		DELIVERABLE - Send Analysis & Design Report to Cust Rep to Ev	0 hrs	5 days	Mon 23/03/1	Mon 30/03/1	90		$0.00
92		Review Meeting - Analysis & Design Report	12 hrs	0.5 days	Mon 30/03/15	Mon 30/03/15	91		$2,320.00
93		Review Participation	4 hrs	0.5 days	Mon 30/03/1	Mon 30/03/1		Sys Lead Eng	$780.00
94		Review Participation	4 hrs	0.5 days	Mon 30/03/1	Mon 30/03/1	93SS	BA - SA	$800.00
95		Review Participation	4 hrs	0.5 days	Mon 30/03/1	Mon 30/03/1	93SS	Proj Mgmt	$740.00
96		Minute Review Meeting	2 hrs	0.25 days	Mon 30/03/1	Mon 30/03/1	92	Proj Mgmt	$370.00
97		DELIVERABLE - Send Minutes to Cust Rep	0 hrs	0 days	Mon 30/03/1	Mon 30/03/1	96		$0.00
98		Update Report	16 hrs	2 days	Tue 31/03/1	Thu 02/04/1	96	Sys Lead Eng	$3,120.00
99		Internal Review Report	4 hrs	0.5 days	Thu 02/04/1	Thu 02/04/1	98	BA - SA	$800.00
100		Document Presentation Preparation	2 hrs	0.25 days	Thu 02/04/1	Thu 02/04/1	99	Doco Support	$250.00
101		Review - Doco	1 hr	0.13 days	Thu 02/04/1	Thu 02/04/1	100	Proj Mgmt	$185.00
102		Send Updated Analysis & Design Report to Cust Rep	0 hrs	0 days	Thu 02/04/1	Thu 02/04/1	101		$0.00
103		MILESTONE - Closeout Report & Invoicing	2 hrs	0.25 days	Thu 02/04/1	Fri 03/04/1	74	Proj Mgmt	$370.00

Figure 100: Scenario's project schedule – Part 1 – Analysis & Design.

For this scenario's project schedule, see [Figure 100], the Lead Systems Engineer will need to travel to the customer's site to investigate how the to-be-delivered product should be integrated into the customer's existing systems & infrastructure. As per the project's contracted pre-arrangement, the customer is to be charged for travel time (to & from site), the cost of the travel method used, and for the project team member's (away from home) accommodation & travel allowance expenses.

Such costs of travel, accommodation, and expenses may or may not be included in the project's schedule, depending on how the schedule is being used to determine; the sequence of activities to be conducted, the [Time] duration & Level Of Effort involved, the allocation of [People] & [Resources], and the preliminary [Cost] estimations.

Though noting that, it is **common for the Project Manager to be responsible for the project's logistics** of such things as; initiating the booking of travel – transport –

accommodation, vendor & supplier relationships, procurement of software & hardware componentry, inventory controller, warehousing arranger, packaging obtainer, test facilities booker, import & export licenses organizer, … and, "last minute run around coordinator". That is, **the Project Manager's role is as "Project Success Facilitator"**, possibly functioning as a **"jack of all trades"** coverall goalkeeper for the project.

Hence, if the inclusion of such tasks as *"Travel to & from site Cost"* and *"Accommodation & Allowance Cost"* are sufficient to remind the Project Manager (and Senior Management) that these things need to be budgeted for and coordinated at specific times, then it is a good idea to add these to the schedule. … As **it is those forgotten activities that can greatly affect the project's chances of being successfully delivered on [Time] and within [Cost] budget**. … *"Shh-it, I forgot to budget for & book the airline tickets."*

Also note, in [Figure 100], the *"Milestone – Closeout Report and Invoicing"* (i.e. progress payment) for the work conducted up to this point. Whenever possible, **strive to have progress payment milestones that result in a "positive cash flow" for the performing organization**. Where, the **accumulated amount of customer payments (REVENUE) up to that point would exceed the performing organization's accumulated EXPENSE** of conducting the project up to that same point.

```
For this scenario's project schedule, see [Figure 99] and
[Figure 100], the Cost-To-Date for Part0 ($2700), plus
Part1 ($46460), plus a proportion of Management &
Technical Support ($12000) equals ($61160), which equates
to 6.7% of the Internal Total Cost.  Hence, set a progress
payment of 10-15% of the Contract Price, negotiating for
the higher percentage.
```

"But, what about all of those hours of work that were involved with the project's Initiating Phase to get the contract signed off?"

Ideally, arrange for a 'Time & Materials' mini-contract for the project's Initiating Phase up to contract sign-off, after which the 'Fixed Price' contract comes into effect. Though, the

379

project's Initiating Phase costs may be incorporated into the labour rates that are used for all projects, so that the costs of "LOST" projects is spread across those "WON" projects.

ID	Task Name	Work	Duration	Start	Finish	Predecessors	Resource Names	Cost
90	PART 2 - Implementation (Agile Sprints)	3,299 hrs	95.5 days	Fri 20/03/1!	Mon 03/08/1!			$547,215.0(
91	Part 2 - Sprint 1	960 hrs	20 days	Fri 03/04/1!	Fri 01/05/1!	59		$132,000.0(
92	Sw Eng Lead	160 hrs	20 days	Fri 03/04/1	Fri 01/05/1		Sw Lead Eng	$28,000.0(
93	Sw Eng 1	160 hrs	20 days	Fri 03/04/1	Fri 01/05/1	92SS	Sw Eng 1	$24,000.0(
94	Sw Eng 2	160 hrs	20 days	Fri 03/04/1	Fri 01/05/1	92SS	Sw Eng 2	$22,400.0(
95	Sw Eng 3	160 hrs	20 days	Fri 03/04/1	Fri 01/05/1	92SS	Sw Eng 3	$20,800.0(
96	Sw Eng 4	160 hrs	20 days	Fri 03/04/1	Fri 01/05/1	92SS	Sw Eng 4	$19,200.0(
97	Sw Eng 5	160 hrs	20 days	Fri 03/04/1	Fri 01/05/1	92SS	Sw Eng 5	$17,600.0(
98	Part 2 - Sprint 2	960 hrs	20 days	Fri 01/05/1!	Fri 29/05/1!	91		$132,000.0(
99	Sw Eng Lead	160 hrs	20 days	Fri 01/05/1	Fri 29/05/1		Sw Lead Eng	$28,000.0(
100	Sw Eng 1	160 hrs	20 days	Fri 01/05/1	Fri 29/05/1	99SS	Sw Eng 1	$24,000.0(
101	Sw Eng 2	160 hrs	20 days	Fri 01/05/1	Fri 29/05/1	99SS	Sw Eng 2	$22,400.0(
102	Sw Eng 3	160 hrs	20 days	Fri 01/05/1	Fri 29/05/1	99SS	Sw Eng 3	$20,800.0(
103	Sw Eng 4	160 hrs	20 days	Fri 01/05/1	Fri 29/05/1	99SS	Sw Eng 4	$19,200.0(
104	Sw Eng 5	160 hrs	20 days	Fri 01/05/1	Fri 29/05/1	99SS	Sw Eng 5	$17,600.0(
105	Part 2 - Sprint 3	960 hrs	20 days	Fri 29/05/1!	Fri 26/06/1!	98		$132,000.0(
106	Sw Eng Lead	160 hrs	20 days	Fri 29/05/1	Fri 26/06/1		Sw Lead Eng	$28,000.0(
107	Sw Eng 1	160 hrs	20 days	Fri 29/05/1	Fri 26/06/1	106SS	Sw Eng 1	$24,000.0(
108	Sw Eng 2	160 hrs	20 days	Fri 29/05/1	Fri 26/06/1	106SS	Sw Eng 2	$22,400.0(
109	Sw Eng 3	160 hrs	20 days	Fri 29/05/1	Fri 26/06/1	106SS	Sw Eng 3	$20,800.0(
110	Sw Eng 4	160 hrs	20 days	Fri 29/05/1	Fri 26/06/1	106SS	Sw Eng 4	$19,200.0(
111	Sw Eng 5	160 hrs	20 days	Fri 29/05/1	Fri 26/06/1	106SS	Sw Eng 5	$17,600.0(
112	Part 2 - Evaluation Testing	171 hrs	45.13 days	Fri 01/05/1!	Fri 03/07/1!			$23,835.0(
113	Part 2 - Evaluate - Sprint 1	57 hrs	5.13 days	Fri 01/05/1!	Fri 08/05/1!	91		$7,945.0(
114	Evaluate Test	40 hrs	5 days	Fri 01/05/1	Fri 08/05/1		Test Eng	$5,000.0(
115	Assistance	8 hrs	1 day	Fri 01/05/1	Mon 04/05/1	114SS	Test Lead Eng	$1,200.0(
116	Assistance	8 hrs	1 day	Fri 01/05/1	Mon 04/05/1	114SS	Sys Lead Eng	$1,560.0(
117	Progress Summary	1 hr	0.13 days	Fri 08/05/1	Fri 08/05/1	114	Proj Mgmt	$185.0(
118	DELIVERABLE - Progress Summary to Cust Rep	0 hrs	0 days	Fri 08/05/1	Fri 08/05/1	117		$0.0(
119	Part 2 - Evaluate - Sprint 2	57 hrs	5.13 days	Fri 29/05/1!	Fri 05/06/1!	98		$7,945.0(
120	Evaluate Test	40 hrs	5 days	Fri 29/05/1	Fri 05/06/1		Test Eng	$5,000.0(
121	Assistance	8 hrs	1 day	Fri 29/05/1	Mon 01/06/1	120SS	Test Lead Eng	$1,200.0(
122	Assistance	8 hrs	1 day	Fri 29/05/1	Mon 01/06/1	120SS	Sys Lead Eng	$1,560.0(
123	Progress Summary	1 hr	0.13 days	Fri 05/06/1	Fri 05/06/1	120	Proj Mgmt	$185.0(
124	DELIVERABLE - Progress Summary to Cust Rep	0 hrs	0 days	Fri 05/06/1	Fri 05/06/1	123		$0.0(
125	Part 2 - Evaluate - Sprint 3	57 hrs	5.13 days	Fri 26/06/1!	Fri 03/07/1!	105		$7,945.0(
126	Evaluate Test	40 hrs	5 days	Fri 26/06/1	Fri 03/07/1		Test Eng	$5,000.0(
127	Assistance	8 hrs	1 day	Fri 26/06/1	Mon 29/06/1	126SS	Test Lead Eng	$1,200.0(
128	Assistance	8 hrs	1 day	Fri 26/06/1	Mon 29/06/1	126SS	Sys Lead Eng	$1,560.0(
129	Progress Summary	1 hr	0.13 days	Fri 03/07/1	Fri 03/07/1	126	Proj Mgmt	$185.0(
130	DELIVERABLE - Progress Summary to Cust Rep	0 hrs	0 days	Fri 03/07/1	Fri 03/07/1	129		$0.0(

Figure 101: Scenario's project schedule – Part 2 – Implementation (Top Half).

For this scenario's project schedule, see [Figure 101], the implementation is to be undertaken as three agile sprints of one month duration each, to be conducted by the Software Engineering Team of six team members; then, followed immediately by evaluation (feedback) testing by an "impartial" Test Team.

That is, see [Figure 102], the Project Implementation Team are being allocated blocks of 'Level Of Effort' and 'time duration' during which they're to complete the development & evaluation testing of the features & functionality that they self-organized the Define – Design – Develop – Test.

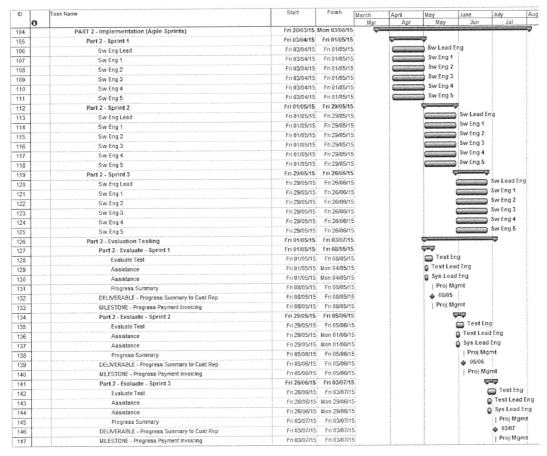

Figure 102: Scenario's project schedule – Part 2 – Implementation, Task Bars.

Hence, in the scenario's project, the agile Project Implementation Team is given a prioritized list of features & functionality that they are to deliver, i.e. a 'Product Backlog'. At a time closer to the actual implementation, The Project Implementation Team then hold their own 'Planning Meeting' per sprint, with the Project Manager (and/or the BA – SA) acting as the 'Product Owner', where the Project Implementation Team select a limited number of features & functionality to be put into the 'Sprint Backlog', which they then self-coordinate the implementation. See [Section 3.5.3].

Thus, this is an example of how an agile based mini-project can be undertaken as part of a larger waterfall project.

As an aside, in addition to the 'Progress Summary' updates being provided to the Customer's Representative at the conclusion of each sprint, ... **for the Monitoring & Control Phase, the Project Manager would track the entire project's progress via the use of Earned Value Performance Measures KPIs of CPI & SPI**, [Section 3.6.5]; whereas, the **Project Implementation Team would track the progress of each sprint via the use of the Task Board & the Burndown Chart**, [Section 3.5.4] and [Section 3.5.6].

ID	Task Name	Work	Duration	Start	Finish	Predecessors	Resource Names	Cost
148	Part 2 - System Integration	52 hrs	78.88 days	Fri 20/03/1!	Thu 09/07/15			$99,900.00
149	Hardware Acquistion & Delivery	16 hrs	42 days	Fri 20/03/1!	Tue 19/05/15			$12,880.00
150	Request Hardware	8 hrs	1 day	Fri 20/03/1	Mon 23/03/1	88	Sys Lead Eng	$1,560.00
151	Review Hardware Request	2 hrs	0.25 days	Mon 23/03/1	Mon 23/03/1	150	BA - SA	$390.00
152	Hardware Componentry Cost	0 hrs	0 days	Mon 23/03/1!	Mon 23/03/1!	151		$10,000.00
153	Order Hardware Componentry	2 hrs	0.25 days	Mon 23/03/1	Tue 24/03/1	151	Logistics Mgr	$310.00
154	BUFFER - Delivery Time	0 hrs	40 days	Tue 24/03/1	Tue 19/05/1	153		$0.00
155	Handling Hardware Delivery	4 hrs	0.5 days	Tue 19/05/1	Tue 19/05/1	154	Logistics Mgr	$620.00
156	System Integration	36 hrs	36.88 days	Tue 19/05/15	Thu 09/07/15	149		$87,020.00
157	End User Licensing	4 hrs	0.5 days	Tue 19/05/15	Wed 20/05/1!			$80,780.00
158	Web-Server – End User License - Cost	0 hrs	0 days	Tue 19/05/1!	Tue 19/05/1!			$15,000.00
159	Database-Server – End User License - Cost	0 hrs	0 days	Tue 19/05/1	Tue 19/05/1!			$65,000.00
160	Co-ordination	4 hrs	0.5 days	Tue 19/05/1	Wed 20/05/1	158	Sys Lead Eng	$780.00
161	Application Setup	32 hrs	4 days	Fri 03/07/1!	Thu 09/07/15	157,126		$6,240.00
162	Web-Server – Setup	8 hrs	1 day	Fri 03/07/1	Mon 06/07/1		Sys Lead Eng	$1,560.00
163	Database-Server – Setup	8 hrs	1 day	Mon 06/07/1	Tue 07/07/1	162	Sys Lead Eng	$1,560.00
164	Setup Verification & Dry Run Testing	8 hrs	1 day	Tue 07/07/1	Wed 08/07/1	163	Sys Lead Eng	$1,560.00
165	BUFFER - Time & Cost	8 hrs	1 day	Wed 08/07/1	Thu 09/07/1	164	Sys Lead Eng	$1,560.00
166	Part 2 - Dry Run Test	153 hrs	81.63 days	Thu 02/04/15	Mon 27/07/1!			$21,085.00
167	Acceptance Test Procedures (1st Pass)	40 hrs	5 days	Thu 02/04/1	Thu 09/04/1	74	Test Eng	$5,000.00
168	Internal Review Acceptance Test Procedures	8 hrs	1 day	Thu 09/04/1	Fri 10/04/1	167	Test Lead Eng	$1,200.00
169	Internal Review Acceptance Test Procedures	8 hrs	1 day	Fri 10/04/1	Mon 13/04/1	168	Sys Lead Eng	$1,560.00
170	Update Acceptance Test Procedures (2nd Pass)	16 hrs	2 days	Mon 13/04/1	Wed 15/04/1	169	Test Eng	$2,000.00
171	Dry Run Test	48 hrs	4 days	Thu 09/07/1!	Wed 15/07/1!	170,148		$6,760.00
172	Conduct Test	24 hrs	3 days	Thu 09/07/1	Tue 14/07/1		Test Eng	$3,000.00
173	Support	8 hrs	1 day	Thu 09/07/1	Fri 10/07/1	172SS	Test Lead Eng	$1,200.00
174	Support	8 hrs	1 day	Thu 09/07/1	Fri 10/07/1	172SS	Sys Lead Eng	$1,560.00
175	BUFFER - Time & Cost	8 hrs	1 day	Tue 14/07/1	Wed 15/07/1	172	Test Eng	$1,000.00
176	Write Test Report / Quality Evidence	33 hrs	3 days	Wed 15/07/1!	Mon 20/07/1!	171		$4,565.00
177	Write Report	24 hrs	3 days	Wed 15/07/1	Mon 20/07/1		Test Eng	$3,000.00
178	Support	4 hrs	0.5 days	Wed 15/07/1	Wed 15/07/1	177SS	Test Lead Eng	$600.00
179	Support	4 hrs	0.5 days	Wed 15/07/1	Wed 15/07/1	177SS	Sys Lead Eng	$780.00
180	Review Doco	1 h	0.13 days	Thu 16/07/1	Thu 16/07/1	179	Proj Mgmt	$185.00
181	Send Invitation to Cust Rep for Test Readiness Review	0 hrs	5 days	Mon 20/07/1	Mon 27/07/1	176		$0.00
182	Part 2 - Test Readiness Review	41 hrs	4.63 days	Mon 27/07/1!	Mon 03/08/1!	166		$6,025.00
183	Review Meeting - Test Readiness Review	16 hrs	1.5 days	Mon 27/07/1!	Tue 28/07/15			$2,620.00
184	Review Participation	4 hrs	0.5 days	Mon 27/07/1!	Mon 27/07/1		Test Eng	$500.00
185	Review Participation	4 hrs	0.5 days	Mon 27/07/1	Mon 27/07/1	184SS	Test Lead Eng	$600.00
186	Review Participation	4 hrs	0.5 days	Tue 28/07/1	Tue 28/07/1	184SS,185	Sys Lead Eng	$780.00
187	Review Participation	4 hrs	0.5 days	Tue 28/07/1	Tue 28/07/1	184SS,186	Proj Mgmt	$740.00
188	Minute Review Meeting	2 hrs	0.25 days	Wed 29/07/1	Wed 29/07/1	183	Proj Mgmt	$370.00
189	DELIVERABLE - Send Minutes to Cust Rep	0 hrs	0 days	Wed 29/07/1	Wed 29/07/1	188		$0.00
190	Update Acceptance Test Procedures	16 hrs	2 days	Wed 29/07/1	Fri 31/07/1	189	Test Eng	$2,000.00
191	Internal Review Report	4 hrs	0.5 days	Fri 31/07/1	Fri 31/07/1	190	Test Lead Eng	$600.00
192	Document Presentation Preparation	2 hrs	0.25 days	Fri 31/07/1	Fri 31/07/1	191	Doco Support	$250.00
193	Review Doco	1 h	0.13 days	Mon 03/08/1	Mon 03/08/1	192	Proj Mgmt	$185.00
194	DELIVERABLE - Send Acceptance Test Procedures to Cust Rep	0 hrs	0 days	Mon 03/08/1	Mon 03/08/1	193		$0.00
195	MILESTONE - Closeout Report & Invoicing	2 hrs	0.25 days	Mon 03/08/1	Mon 03/08/1	182	Proj Mgmt	$370.00

Figure 103: Scenario's project schedule – Part 2 – Implementation (Bottom Half).

For this scenario's project schedule, see [Figure 103], the implementation part also contains procurement of the hardware componentry and the end-user software licenses, which will all eventually be turned over to the Customer.

As per some project's contractual agreements, the purchase of such componentry may be invoiced separately to the customer, or incorporated into the next progress payment.

Notice the buffering tasks for the hardware delivery and systems integration.

```
For this schedule, also note the inclusion of a 'Dry Run
Test', a 'Test Readiness Review', and the sending off to
the customer of the 'Acceptance Test Procedures'.
```

It is ~~highly advantageous~~ **essential to conduct a private "dress rehearsal" | "dry run" step through of all of the test case procedures, prior to the customer's involvement in the Acceptance Testing**. Remembering that, perception is a significant contributor to the project being deemed "a success"; hence, *"practice makes perfect"*, and (more specifically) **practice uncovers those obvious mistakes & silly defects that would degrade the customer's perception of [Quality] deliverables**.

Additionally, by 'Dry Run' testing using the proposed 'Acceptance Test Procedures' that are to be sent to the Customer's Representative for review & acceptance, then this will help to verify & validate those test procedures; as well as, how the Objective Quality Evidence (OQE) is to be recorded. … And, if during the 'Dry Run' testing a sufficient amount of OQE were to be produced (and if necessary) made available to the Customer's Representatives for inspection prior to conducting the Acceptance Tests, then this could result in the Customer's Representatives having a lot more confident attitude towards accepting the deliverables, … which, in turn could mean that the Customer's Representatives are more forthcoming with signing-off on the progress payment milestones up to this point.

```
For this scenario's project schedule, see [Figure 101] and
[Figure 103], the Cost for Part2 ($548300) alone, equates
to 60.3% of the Internal Total Cost. However, Part2 spans
a duration of 4 months which could be a noticeable amount
of time for the performing organization to go without a
progress payment; hence, the inclusion of a progress
payment after each sprint's evaluation testing.
```

In circumstances where there is potentially a long time between the handover of customer deliverables as per the [Scope] of deliverables, then **recommend injecting progress markers into the project's schedule**; where the Customer's Representative may be invited to come witness demonstrations of the progress made thus far, or interim OQE could be provided as evidence of progress, or "preview" versions of the application | product could be given to the Customer's Representatives for them to "play with". ... And so doing, **attach "minor" progress payments to these markers**; thereby, **covering the performing organization's costs of conducting the project's implementation** up to this point.

```
For this scenario's project, suggest that there be
progress payments at the end of each sprint's evaluation
testing of (60% divided by 4 months equal) 15% of the
Contract Price.  Though, the customer may be hesitant on a
15% payout for sprints with no milestone deliverables, so
be prepared to negotiate on a 12.5% payout for each, with
22.5% on completion of the FAT; given that by the end of
Part3, 73% of the Internal Total Costs have been incurred.
```

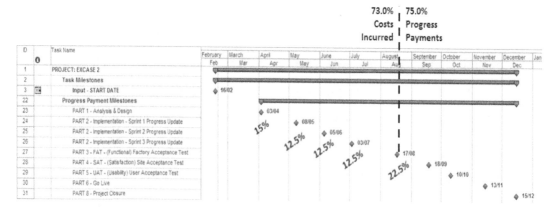

Figure 104: Scenario's project schedule – progress payments to conclusion of FAT.

Try to keep a "positive cash flow" with respect to Progress Payments (to date) **exceeding those Costs Incurred, try to "evenly space"** these payments, and try to maintain similar sized payments. ... Which can be *"easier said than done"* as the Customer's Representative will most probably be negotiating to delay the majority of the payments until as late as possible; however, the customer must understand that, if the performing organization goes bust (potentially due to the financial difficulties resulting from such late payments), then the customer organization risks not receiving their commissioned deliverables.

Which brings us to the next rule of "Adaptive & Proactive" SDLC Project Management.

RULE 48: *"You must never try to make all the money that's in a deal. Let the other fellow make some money too, because if you have a reputation for always making all the money, you won't have many deals."* — J Paul Getty

```
For this scenario's project, the Functional Acceptance
Test will be conducted at the performing organization's
premises using their own in-house equipment (and possibly
including the project's procurements) that were used for
the 'Dry Run Test'; however, this time the tests will be
conducted in front of the Customer's Representatives.
```

ID	Task Name	Work	Duration	Start	Finish	Predecessors	Resource Names	Cost
196	PART 3 - FAT - (Functional) Factory Acceptance Test	120 hrs	10.31 days	Mon 03/08/15	Mon 17/08/15			$18,557.50
197	Commencement Meeting - With Cust Rep Represenatives	2 hrs	0.06 days	Mon 03/08/15	Mon 03/08/15	104		$327.50
198	Review Participation	0.5 hrs	0.06 days	Mon 03/08/1	Mon 03/08/1		Test Eng	$62.50
199	Review Participation	0.5 hrs	0.06 days	Mon 03/08/1	Mon 03/08/1	198SS	Test Lead Eng	$75.00
200	Review Participation	0.5 hrs	0.06 days	Mon 03/08/1	Mon 03/08/1	198SS	Sys Lead Eng	$97.50
201	Review Participation	0.5 hrs	0.06 days	Mon 03/08/1	Mon 03/08/1	198SS	Proj Mgmt	$92.50
202	Conduct FAT	56 hrs	4 days	Mon 03/08/15	Fri 07/08/15	197		$9,240.00
203	Conduct Test	24 hrs	3 days	Mon 03/08/1	Thu 06/08/1		Test Eng	$3,000.00
204	Support	8 hrs	1 day	Thu 06/08/1	Fri 07/08/1	203SS	Sys Lead Eng	$1,560.00
205	Support	24 hrs	3 days	Mon 03/08/1	Thu 06/08/1	203SS	Sys Lead Eng	$4,680.00
206	Review Meeting - Test Completion Review	10 hrs	0.75 days	Fri 07/08/15	Mon 10/08/15	202	Sw Lead Eng	$1,680.00
207	Review Participation	2 hrs	0.25 days	Fri 07/08/1	Fri 07/08/1		Test Eng	$250.00
208	Review Participation	2 hrs	0.25 days	Fri 07/08/1	Fri 07/08/1	207SS	Test Lead Eng	$300.00
209	Review Participation	2 hrs	0.25 days	Fri 07/08/1	Fri 07/08/1	207SS,208	Sys Lead Eng	$390.00
210	Review Participation	2 hrs	0.25 days	Fri 07/08/1	Mon 10/08/1	207SS,209	Proj Mgmt	$370.00
211	Minute Review Meeting	2 hrs	0.25 days	Fri 07/08/1	Fri 07/08/1	207	Proj Mgmt	$370.00
212	DELIVERABLE - Send Minutes to Cust Rep	0 hrs	0 days	Fri 07/08/1	Fri 07/08/1	211		$0.00
213	Write Test Report / Quality Evidence	50 hrs	5.25 days	Mon 10/08/15	Mon 17/08/15	206,211		$6,940.00
214	Write Report	24 hrs	3 days	Mon 10/08/1	Thu 13/08/1		Test Eng	$3,000.00
215	Support	4 hrs	0.5 days	Mon 10/08/1	Mon 10/08/1	214SS	Sw Lead Eng	$700.00
216	Support	4 hrs	0.5 days	Mon 10/08/1	Mon 10/08/1	214SS	Sys Lead Eng	$780.00
217	Internal Review Report	4 hrs	0.5 days	Thu 13/08/1	Thu 13/08/1	214	Test Lead Eng	$600.00
218	Document Presentation Preparation	2 hrs	0.25 days	Thu 13/08/1	Thu 13/08/1	217	Doco Support	$250.00
219	DELIVERABLE - Send Factory Acceptance Test Report to Cust Re	0 hrs	0 days	Thu 13/08/1	Thu 13/08/1	218		$0.00
220	Update Acceptance Test Procedures	8 hrs	1 day	Thu 13/08/1	Fri 14/08/1	211,219	Test Eng	$1,000.00
221	Internal Review Report	2 hrs	0.25 days	Fri 14/08/1	Mon 17/08/1	220	Test Lead Eng	$300.00
222	Document Presentation Preparation	1 hr	0.13 days	Mon 17/08/1	Mon 17/08/1	221	Doco Support	$125.00
223	Review - Doco	1 hr	0.13 days	Mon 17/08/1	Mon 17/08/1	222	Proj Mgmt	$185.00
224	DELIVERABLE - Send Updated Acceptance Test Procedures to C	0 hrs	0 days	Mon 17/08/1	Mon 17/08/1	223		$0.00
225	MILESTONE - Closeout Report & Invoicing	2 hrs	0.25 days	Mon 17/08/1	Mon 17/08/1	224,212	Proj Mgmt	$370.00

Figure 105: Scenario's project schedule – Part 3 – Functional Acceptance Test (FAT).

"Umm, think you've made a mistake in the schedule, as Task 205 should have the 'Sw Lead Eng' doing support and not the repeat of the 'Sys Lead Eng'. ... Just a $20/hr = $480 over costing."

Hence why the schedule should be reviewed prior to being used for further purposes; especially, prior to being presented to the Customer's Representatives, as such mistakes (when discovered by the customer) can raise questions about the accuracy & validity of the rest of the schedule, and subsequently the correctness of the Contracted Price.

...

As stated previously in [Section 3.5.5], **the purpose of the Functional Acceptance Test (FAT) is the reproducible verification that every in-scope feature & functionality described in the approved Detailed Specifications has been successfully implemented in accordance with the agreed Acceptance Criteria** (as was prescribed in the reviewed & approved Acceptance Test Procedures). Subsequently, **prior to conducting the FAT**, it is **advisable to hold a 'Commencement Meeting'** with the Customer's Representatives, so that the performing organization can layout before them, both the approved Detailed Specifications and the approved (FAT) Acceptance Test Procedures, ... and, just in case any concerns arise, have the 'Dry Run Test' OQE ready to be presented as contrary evidence.

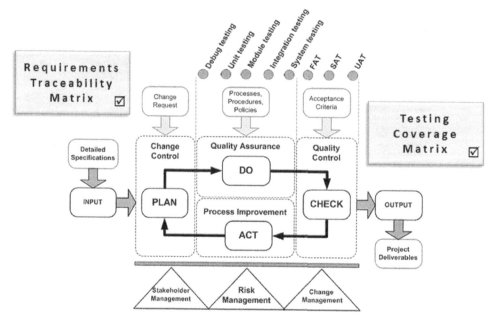

> In essence the Requirements Traceability Matrix (RTM) and the Testing Coverage Matrix (TCM) are the listings of the inputs and outputs to the Quality Assurance Processes and the Quality Control Procedures used to determine **the "correctness" of the project deliverables** that were agreed to. This "correctness" is subsequently **determined via the undertaking of a sequence of acceptance tests, and the records of such testing forms the project's Objective Quality Evidence (OQE)**. See [Section 3.7.3].
>
> During the FAT, in the associated entries of the TCM, mark the PASS : FAIL of each test; thereby, identifying what features & functionality do & don't operate in accordance with the agreed Acceptance Criteria. The TCM may also include references to any waivers if these were granted for specific failed tests; i.e. the test failed but this failure has been agreed to be acceptable under the current circumstances.

The important thing is to **keep the Customer's Representatives focused on verifying that the project's deliverables do comply with the functionality as defined in the approved Detailed Specifications; and NOT, their interpretation of its "fitness for use"**. Because, potentially **the FAT is the first opportunity that the Customer's Representatives have to really use the deliverables**; and, during such use **they may realize that what was defined is not what they now want**. Subsequently, **they may incorrectly report certain features & functionality as Defects instead of Change Requests**. See [Section 3.6.6] & [Section 3.6.7].

At the **finish of the FAT**, a **'Test Completion Review' meeting** should be held between the customer & performing organizations' representatives to **discuss & note any concerns with respect to the test results, and how the FAT was conducted**. By the end of this meeting, **the Project Manager's objective is to have the Customer's Representatives sign-off on the completion of the FAT**, as well as the resulting **FAT Report** (including OQE, deviations, waivers, and concessions), **and then the sign-off on the associated milestone progress payment for the conclusion of the FAT**. Conversely, the Performing Organization does not want this progress payment to be delayed due to differences of opinions over *"what was specified"* and *"what is actually wanted"*.

Deviations, Waivers, Concessions, Defects and Change Requests

During the Functional Acceptance Test (FAT), and the Satisfaction Acceptance Test (SAT), the following may be generated due to failed or incomplete tests:

- **Deviation** ... written permission for a departure from the approved processes, procedures, and/or specifications used to produce or test a deliverable.
- **Waiver** ... written permission for a non-conformance so that a deliverable may be accepted.
- **Concession** ... a "special case" approval signed off by the representatives of the performing organization (*e.g. the Project Manager and/or the Quality Authority*); this concession grants the release to the customer of a deliverable (with or without repair) that is known to have specific non-conformances to the approved baseline. Such a concession is granted because, while the non-conformance is technically not a match with the approved baseline, the non-conformance does have the characteristics of the approved baseline; *e.g. increasing the storage capacity of the installed hard-disk due to market unavailability of the specified older model.*
- **Defect** ... where some feature or functionality of a deliverable does not conform to the agreed Acceptance Criteria (nor to the approved Detailed Specifications), and hence this defect needs to be rectified.
- **Change Request** ... where some feature or functionality, as defined in the approved Detailed Specifications, will not result in the desired outcomes, or the agreed Acceptance Criteria is found to be inappropriate, subsequently the agreed [Scope], [Time] schedule, [Cost] budget, and/or [Quality] standards need to be changed.

The reason that the above are listed here is because, during the project's execution, dealing with these will **consume some amount of the 'Management Support' and 'Technical Authority Support' activities** that were included in the schedule. Just witness **how much time can/will be consumed in activities involved with Baseline Changes, Defect Reviews, and Engineering Changes**. See the workflows in [Figure 65] [Figure 66].

ID		Task Name	Work	Duration	Start	Finish	Predecessors	Resource Names	Cost
226		PART 4 - SAT - (Satisfaction) Site Acceptance Test	276.5 hrs	24.44 days	Mon 17/08/15	Fri 18/09/15			$73,962.50
227		Part 4 - Media Shipped To Site	32 hrs	4 days	Tue 18/08/15	Mon 24/08/15			$12,430.00
228		System Media	13 hrs	1.63 days	Tue 18/08/15	Wed 19/08/15			$4,735.00
229		Media Preparation	8 hrs	1 day	Tue 18/08/15	Wed 19/08/15	196	Sys Lead Eng	$1,560.00
230		Presentation Preparation & Packaging	3 hrs	0.38 days	Wed 19/08/15	Wed 19/08/15	229	Doco Support	$625.00
231		DELIVERABLE - Shipment Coordination / Export Licensing	2 hrs	0.25 days	Wed 19/08/15	Wed 19/08/15	230	Logistics Mgr ?	$2,550.00
232		Licenses & Software Keys	7 hrs	0.88 days	Wed 19/08/15	Thu 20/08/15	228		$3,675.00
233		Media Preparation	4 hrs	0.5 days	Wed 19/08/15	Thu 20/08/15		Sys Lead Eng	$780.00
234		Presentation Preparation & Packaging	2 hrs	0.25 days	Thu 20/08/15	Thu 20/08/15	233	Doco Support	$500.00
235		DELIVERABLE - Shipment Coordination	1 hr	0.13 days	Thu 20/08/15	Thu 20/08/15	234	Logistics Mgr ?	$2,395.00
236		End User Documentation	12 hrs	1.5 days	Thu 20/08/15	Mon 24/08/15	232		$4,020.00
237		Media Preparation	8 hrs	1 day	Thu 20/08/15	Fri 21/08/15		Doco Support	$1,000.00
238		Presentation Preparation & Packaging	3 hrs	0.38 days	Fri 21/08/15	Mon 24/08/15	237	Doco Support	$625.00
239		DELIVERABLE - Shipment Coordination	1 hr	0.13 days	Mon 24/08/15	Mon 24/08/15	238	Logistics Mgr ?	$2,395.00
240		Part 4 - Hardware Shipped To Site	34 hrs	12.25 days	Mon 17/08/15	Wed 02/09/15	206FS+5 days		$7,850.00
241		Packaging Materials	0 hrs	0 days	Mon 17/08/15	Mon 17/08/15			$250.00
242		Package for Shipping	18 hrs	2.25 days	Mon 17/08/15	Wed 19/08/15	241		$2,870.00
243		Package	8 hrs	1 day	Mon 17/08/15	Tue 18/08/15		Sys Lead Eng	$1,560.00
244		Assist	8 hrs	1 day	Mon 17/08/15	Tue 18/08/15	243SS	Test Eng	$1,000.00
245		Assist	2 hrs	2 hrs	Wed 19/08/15	Wed 19/08/15	243SS	Logistics Mgr	$310.00
246		Shipment Coordination / Export Licensing	8 hrs	1 day	Mon 17/08/15	Tue 18/08/15		Logistics Mgr	$1,240.00
247		Hardware Dispatch	8 hrs	1 day	Tue 18/08/15	Wed 19/08/15	246	Logistics Mgr	$1,240.00
248		DELIVERABLE - Shipment Of Hardware	0 hrs	10 days	Wed 19/08/15	Wed 02/09/15	247,242		$2,250.00
249		Part 4 - Travel To Site	32 hrs	12 days	Wed 02/09/15	Fri 18/09/15			$20,680.00
250		TRAVEL DATE	0 hrs	0 days	Fri 11/09/15	Fri 11/09/15	11		$0.00
251		Travel Coordination	8 hrs	1 day	Wed 02/09/15	Thu 03/09/15	250SS-7 days	Logistics Mgr	$1,240.00
252		Travel To Site (Sys Eng)	8 hrs	1 day	Fri 11/09/15	Fri 18/09/15	250SS	Sys Lead Eng	$1,560.00
253		Travel To Site (Sw Eng)	8 hrs	1 day	Fri 11/09/15	Tue 15/09/15	250SS	Sw Lead Eng	$1,400.00
254		Travel To Site (Mgmt)	8 hrs	1 day	Mon 14/09/15	Tue 15/09/15	250SS	Proj Mgmt	$1,480.00
255		Travel to & from site Cost (Sys Eng)	0 hrs	0 days	Fri 11/09/15	Fri 11/09/15	250SS		$2,000.00
256		Travel to & from site Cost (Sw Eng)	0 hrs	0 days	Fri 11/09/15	Fri 11/09/15	250SS		$2,000.00
257		Travel to & from site Cost (Mgmt)	0 hrs	0 days	Fri 11/09/15	Fri 11/09/15	250SS		$2,000.00
258		Accommodation & Allowance Cost (Sys Eng)	0 hrs	0 days	Fri 11/09/15	Fri 11/09/15	250SS		$3,000.00
259		Accommodation & Allowance Cost (Sw Eng)	0 hrs	0 days	Fri 11/09/15	Fri 11/09/15	250SS		$3,000.00
260		Accommodation & Allowance Cost (Mgmt)	0 hrs	0 days	Fri 11/09/15	Fri 11/09/15	250SS		$3,000.00
261		Part 4 - System Integration	52 hrs	3.5 days	Wed 02/09/15	Mon 07/09/15	248,227	?	$9,660.00
262		Hardware Setup	16 hrs	1 day	Wed 02/09/15	Thu 03/09/15	240		$2,960.00
263		Handling Hardware Delivery	8 hrs	1 day	Wed 02/09/15	Thu 03/09/15		Sys Lead Eng	$1,560.00
264		Assist	8 hrs	1 day	Wed 02/09/15	Thu 03/09/15	263FF	Sw Lead Eng	$1,400.00
265		Application Setup	36 hrs	2.5 days	Thu 03/09/15	Mon 07/09/15	262		$6,700.00
266		Web-Server – Setup	4 hrs	0.5 days	Thu 03/09/15	Thu 03/09/15		Sys Lead Eng	$780.00
267		Database-Server – Setup	4 hrs	0.5 days	Thu 03/09/15	Fri 04/09/15	266	Sys Lead Eng	$780.00
268		Setup Verification & Sanity Testing	4 hrs	0.5 days	Fri 04/09/15	Fri 04/09/15	267	Sys Lead Eng	$780.00
269		BUFFER - Time & Cost	8 hrs	1 day	Fri 04/09/15	Mon 07/09/15	268	Sys Lead Eng	$1,560.00
270		Assist	16 hrs	2 days	Thu 03/09/15	Mon 07/09/15	263FF	Sw Lead Eng	$2,800.00

Figure 106: Scenario's project schedule – Part 4 – Satisfaction Acceptance Test (SAT).

For the scenario's project, before the (site) Satisfaction Acceptance Test (SAT) can take place, all of the relevant procured & produced hardware & software have to be shipped to the customer's site. Once onsite this hardware & software will have to be installed & integrated into the customer's systems & infrastructure.

These onsite activities will take time, a lot more time than can normally be expected; therefore "*double it*" the duration (nay "*triple it*"), because this is the customer's facilities where things are done differently with various groups involved to get things achieved / rectified. Hence, buffer ("*by any other name*"); because, to fail with the in-situ installation & setup of the project's deliverables can be a rather problematic, costly, and a reputation damaging situation. Worse, is if the performing organization's representatives who were sent to the customer's site, if they have to leave to fix things up, then return later on to do the installation & setup properly; as *"who exactly is going to pay for this do over work"*, because it is possibly coming out of the performing organization's own pocket.

Thus, **the importance that must be placed on those pre-SAT preparation activities**, such as the creation of the **system media** (which contains the application software), preparing the hardware componentry, the **packaging** of such software & hardware, the **shipping** to the customer's site of all of the required materials, followed by the onsite **"professional" installation & setup** of the in-situ project deliverables. Because, there could be serious ramifications for the performing organization if it goes wrong; hence, for a highly critical customer's project, give serious consideration to conducting a local "dress rehearsal" | "dry run" setting up of the SAT (independent of the FAT facilities), so as to verify that everything is ready to go (including the tools that will be needed to do the job properly).

"Oh, how embarrassing and unprofessional it is to get to site and then realize that, you've forgotten the thingamabob."

 DO NOT allow accountant types to *"save a few bucks"* on these pre-SAT activities, by cutting back on the quality of the products & services used, and the Level Of Effort involved; **because, when onsite it will be regretted.**

"It is better to arrive early, over prepared, and have to wait; rather than taking it easy, under attired, and arrive late."

Which brings us to the next rule of "Adaptive & Proactive" SDLC Project Management.

RULE 49: Effective packaging, timely delivery, and efficient installation come to typify the performing organization's quality image, and hence influences the customer's perception of accept-ability & expect-ability of those project's deliverables.

COFFEE BREAK DISCUSSION … *Away from home travel, don't be an A.C.A.*

When planning & coordinating away (from home) travel for project team members, don't think entirely from 'A Cost Accountant' perspective; remember that, the person who is being sent away (from home) for some period of time on company business, they also have a home life that will be affected by such travel plans.

Question:

- How will their partner be burdened by having to handle their kids, alone?
- Will they miss their son's or daughter's first spoken words, first steps taken, school play, dance recital, championship game, a life unfolding?
- What about their outside-of-work social & community commitments, who will take care of these things?
- What are they giving up, so as to go on this trip on the businesses behalf?

Unfortunately, this is something that is often taken for granted when planning away (from home) travel for projects; because, the reality is that it ain't just tasks & numbers in a schedule come spreadsheet, it is actual people's lives that are affected.

And, when continually taken for granted, this away (from home) travel can be sufficient aggravation catalyst for that inflicted project team member to 'raise the middle finger' and say, "*I resign, I'm out of here, Jerks*"; leaving the project and the performing organization in an untimely predicament.

However, with just a bit of courteous forethought, discussion, and arranging with the intended traveler, this away (from home) trip can be made a bit more palatable.

- How about, for an international flight, sending them business class so that they can actually get some quality sleep, instead of being shoved back in economy class with no chance of real rest; yet expecting them to meet with the customer, all fresh.
- How about, sending them a day or two earlier so that they can have a day to rest

and partly get over that jet-lag; given that, they were sent economy class, thereby saving enough on business class to easily pay for those extra day's accommodation.

- ☐ How about, instead of mandating that they must fly out early on Monday morning, they get to choose when they leave, so that they could have the weekend exploring the destination city, or leave on Sunday afternoon having said a proper goodbye to the family. … And, split the difference on the additional nights' accommodation.

- ☐ How about, with an extended period away, that they have the option for their partner (and family) to come along, or to come visit for a few days. If they so choose; would the price difference between them voluntarily swapping business class for economy cover the cost of getting an extra seat for their partner's travel?

- ☐ How about, for only a few dollars more, putting them up at *"somewhere nice with civilized life"* instead of being *"stuck in a dog box with nothing to do"*.

- ☐ How about, giving them a daily travel allowance (corresponding to the destination city's cost of living), which enables them to *"afford more than just bread & water"*.

- ☐ How about, coordinating travel plans so that fellow staff (from different projects & groups) stay together in the same place and intermix after hours.

- ☐ How about, when senior management are in town they drop-in to say *"hi"* and take the away (from home) staff member out to lunch or to dinner; instead of, ignoring their existence.

Now, as a mere Project Manager, you possibly won't have much say over the performing organization's travel rules & arrangements; however, it would not hurt to put a case forward to senior management (on the project team members behalf).

Sometimes "penny pinching" on travel, accommodation, and travel allowance can result in a more costly problem than those few dollars saved; as the affected project team member could very well decide to throw it all in and walk away, or their performance is sub-par because they are too tired or they just don't want to be there anymore.

ID	ⓘ	Task Name	Work	Duration	Start	Finish	Predecessors	Resource Names	Cost
271		Part 4 - Onsite Test	65.5 hrs	3.56 days	Mon 07/09/15	Fri 11/09/15	261		$12,117.50
272		Commencement Meeting - With Cust Rep Representatives	1.5 hrs	0.06 days	Mon 07/09/15	Mon 07/09/15			$277.50
273		Review Participation	0.5 hrs	0.06 days	Mon 07/09/15	Mon 07/09/15		Sys Lead Eng	$97.50
274		Review Participation	0.5 hrs	0.06 days	Mon 07/09/15	Mon 07/09/15	273SS	Sw Lead Eng	$87.50
275		Review Participation	0.5 hrs	0.06 days	Mon 07/09/15	Mon 07/09/15	273SS	Proj Mgmt	$92.50
276		Conduct SAT	56 hrs	3 days	Tue 08/09/15	Thu 10/09/15	272		$10,360.00
277		Conduct Test	24 hrs	3 days	Tue 08/09/15	Thu 10/09/15		Sys Lead Eng	$4,680.00
278		Assist	24 hrs	3 days	Tue 08/09/15	Thu 10/09/15	277SS	Sw Lead Eng	$4,200.00
279		Support & Record Taking	8 hrs	1 day	Tue 08/09/15	Tue 08/09/15	277SS	Proj Mgmt	$1,480.00
280		Review Meeting - Test Completion Review	6 hrs	0.25 days	Fri 11/09/15	Fri 11/09/15	276		$1,110.00
281		Review Participation	2 hrs	0.25 days	Fri 11/09/15	Fri 11/09/15		Sys Lead Eng	$390.00
282		Review Participation	2 hrs	0.25 days	Fri 11/09/15	Fri 11/09/15	281SS	Sw Lead Eng	$350.00
283		Review Participation	2 hrs	0.25 days	Fri 11/09/15	Fri 11/09/15	281SS	Proj Mgmt	$370.00
284		Minute Review Meeting	2 hrs	0.25 days	Fri 11/09/15	Fri 11/09/15	280	Proj Mgmt	$370.00
285		DELIVERABLE - Send Minutes to Cust Rep	0 hrs	0 days	Fri 11/09/15	Fri 11/09/15	284		$0.00
286		Part 4 - Travel From Site	24 hrs	1 day	Fri 11/09/15	Mon 14/09/15	271		$4,440.00
287		Travel From Site (Sys Eng)	8 hrs	1 day	Fri 11/09/15	Mon 14/09/15		Sys Lead Eng	$1,560.00
288		Travel From Site (Sw Eng)	8 hrs	1 day	Fri 11/09/15	Mon 14/09/15	287SS	Sw Lead Eng	$1,400.00
289		Travel From Site (Mgmt)	8 hrs	1 day	Fri 11/09/15	Mon 14/09/15	287SS	Proj Mgmt	$1,480.00
290		Part 4 - Test Report	35 hrs	3.88 days	Mon 14/09/15	Fri 18/09/15	286		$6,415.00
291		Write Test Report / Quality Evidence	28 hrs	3 days	Mon 14/09/15	Thu 17/09/15			$5,380.00
292		Write Report	24 hrs	3 days	Mon 14/09/15	Thu 17/09/15		Sys Lead Eng	$4,680.00
293		Support	4 hrs	0.5 days	Mon 14/09/15	Mon 14/09/15	292SS	Sw Lead Eng	$700.00
294		Internal Review Report	4 hrs	0.5 days	Thu 17/09/15	Thu 17/09/15	291	Test Lead Eng	$500.00
295		Document Presentation Preparation	2 hrs	0.25 days	Fri 18/09/15	Fri 18/09/15	294	Doco Support	$250.00
296		Review - Doco	1 hr	0.13 days	Fri 18/09/15	Fri 18/09/15	295	Proj Mgmt	$185.00
297		DELIVERABLE - Send Site Acceptance Test Report to Cust Rep	0 hrs	0 days	Fri 18/09/15	Fri 18/09/15	296		$0.00
298		MILESTONE - Closeout Report & Invoicing	2 hrs	0.25 days	Fri 18/09/15	Fri 18/09/15	290	Proj Mgmt	$370.00

Figure 107: Scenario's project schedule – Part 4 – Satisfaction Acceptance Test (SAT).

```
For this scenario's project, the Satisfaction Acceptance
Test will be conducted onsite at the customer's location
using the customer's provided equipment and the project's
procured equipment that was also shipped to site; and,
these tests will be conducted in front of the Customer's
Representatives.
```

As stated previously in [Section 3.5.5], **the purpose of the Satisfaction Acceptance Test (SAT) is to retest those features & functionality that were found to be non-conforming during the Functional Acceptance Test (FAT), and to verify that the provided | installed project deliverables are operating "satisfactorily" at the customer's in-situ location**. Thereby, the performing organization's representative can (analogy wise) *"hand over the keys"*, confident in the knowledge that the customer can henceforth successfully *"take it for a drive"* to use the installed deliverables as has been configured.

Thus, it is important to **keep the Customer's Representatives focused on verifying that the project's deliverables are operating "satisfactory" in-situ onsite**; as this, **does NOT mean retesting every feature & functionality as was previously done during the FAT**. However, there may be particular tests from the FAT that need to be repeated so as to

393

verify that the deliverables are operating as expected onsite. Conversely, what the performing organization does not want to happen is for the Customer's Representatives to be reinterpreting the deliverables "fitness for use", and debating *"how it should function"*.

Consequently, the SAT should be performed as prescribed in the approved Acceptance Test Procedures, so that if something does go wrong then it can be reproduced later on for rectification. Therefore, **prior to conducting the SAT**, it is advisable to **hold a 'Commencement Meeting' with the Customer's Representatives**, so that the performing organization can **layout** before them, **both the approved Detailed Specifications and the approved (SAT) Acceptance Test Procedures**, ... and, just in case any concerns arise, have the signed off 'FAT Report' (and OQE) ready to be presented as contrary evidence.

At the finish of the SAT, a **'Test Completion Review' meeting** should be held between the customer & performing organizations' representatives to **discuss & note any concerns with respect to the test outcomes and to "hand over the keys"** (if the SAT was successful). By the end of this meeting, **the Project Manager's objective is to have the Customer's Representatives sign-off on the completion of the SAT.** ... And, once the performing organization's onsite SAT personnel have finished (and returned from site), then the **SAT Report** (including OQE, deviations, waivers, and concessions) would be written up & reviewed prior to being sent to the Customer's Representatives, **then the sign-off on the progress payment for the conclusion of the SAT.** ... As always, try to keep the Progress Payments ahead of the Costs Incurred to date; *e.g. 85% versus 82% respectively*.

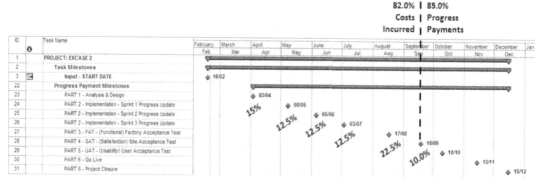

Figure 108: Scenario's project schedule – progress payments to conclusion of SAT.

"Umm, I think there are a few things missing from this schedule; primarily those defect repair activities that would occur at the conclusion of the FAT prior to the occurrence of the SAT. As these will definitely need to be added to the next revision of this project schedule. Oh, and I would also extend the Level Of Effort and duration for the evaluation testing of each development sprint's output. However, for now let's continue on with reviewing this schedule as it is, and then you can update it later, before calculating the budget costing in a spreadsheet."

"Damn, I knew I forgot something; glad it got picked up now, cause we'd definitely would have probably had to do that work."

NOTE TO SELF: need to make the following changes to the current schedule:

- ☐ **Task 128, 135, 142 – Part 2 – Evaluation Testing >> Evaluate Test, Assistance** … need to double the Level Of Effort for these tasks. Noting that it will push back each of the associated Progress Payment invoicing.

- ☐ **Task 167 – Part 2 – Dry Run Test** … need to add new task for preparation of system media prior to the test.

- ☐ **Task 182 – Part 2 – Test Readiness Review** … need to add a parallel group of tasks, for the Project Implementation Team to repair any defects discovered during the 'Dry Run Test'; make this an *"all hands on deck"* Project Implementation Team activity of one (possibly two) week's duration.

- ☐ **Task 197 – Part 3 – FAT – (Functional) Factory Acceptance Test** … need to add new task for preparation of "formal" system media prior to the test.

- ☐ **Task 205 – Part 3 – FAT – (Functional) Factory Acceptance Test >> Conduct FAT** … need to change "Sys Lead Eng" to "Sw Lead Eng".

- ☐ **Task 226 – Part 4 – SAT – (Satisfaction) Site Acceptance Test** … need to add a parallel group of tasks for the *"as needed"* Project Implementation Team to remotely support the onsite Team.

- ☐ **Task 226 – Part 4 – SAT – (Satisfaction) Site Acceptance Test** … need to add group of tasks before here, for the Project Implementation Team to repair any defects discovered during the 'FAT'; so make this an *"all hands on deck"* Project Implementation Team activity of one (possibly two) week's duration.

- ☐ **Task 231, 235, 239 – Part 4 – Media Shipped To Suite** … need to check the Logistics Manager's hourly rates as these appear to be way too high; check all rates costings.

- ☐ **Task 261 – Part 4 – SAT – System Integration** … need to add dependence for personnel having arrived, not just dependent on media & hardware having arrived.

- ☐ **Task 299 – Part 5 – UAT – (Usability) User Acceptance Test** … need to add group of tasks before here, for the Project Implementation Team to repair any defects discovered during the 'SAT'; so make this an *"as needed"* Project Implementation Team activity of a week's duration.

- ☐ **Task 299 – Part 5 – UAT – (Usability) User Acceptance Test** … need to add group of tasks for the preparation of the formal system media that needs to be sent to site; copy Part 4 – SAT tasks 227 through 239.

- ☐ **Task 299 – Part 5 – UAT – (Usability) User Acceptance Test** … need to add a parallel group of tasks for the *"as needed"* Project Implementation Team to remotely support the onsite Team.

- ☐ **Task 347 – Part 6 – Go Live** … need to add group of tasks before here, for the Project Implementation Team to repair any defects discovered during the 'UAT'; so make this an *"as needed"* Project Implementation Team activity. Could be done remotely in parallel with the UAT, if there is insufficient time to return from the UAT.

- ☐ **Task 347 – Part 6 – Go Live** … need to add group of tasks for the preparation of the system media that needs to be sent to site; copy Part 4 – SAT tasks 227 through 239.

- ☐ **Need to ensure that the Go Live mandated milestone date can still be achieved**; may need to add extra [People] & [Resources] during the defect repair tasks after the 'Dry Run Test', 'FAT', 'SAT', and 'UAT' so as to reduce the duration to make the End Date.

Acceptance Testing ... *no such thing as enough testing*

It cannot be over **emphasized, the importance of the 'Evaluation' testing that accompanies the development sprints, and the 'Dry Run' testing**, as these are the **most feasible & affordable opportunities to resolve any issues with non-conforming features & functionality**. Similarly, effective & efficient **"proper" (Factory) Functional Acceptance Testing (FAT) is the last truly affordable opportunity to make corrections of both Defect Repairs & Change Requests to the project's deliverables**; because, by such a late stage as the (Site) Satisfaction Acceptance Testing (SAT), and especially the (User) Usability Acceptance Testing (UAT), making changes & corrections can be a rather costly endeavor.

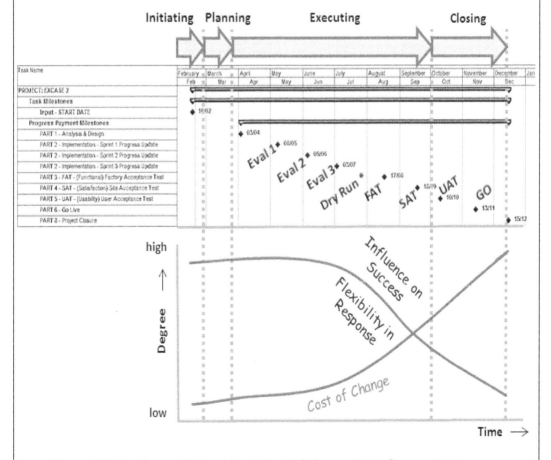

Figure 109: Testings relationship to Cost Of Change Vs. Influence On Success.

Which brings us to the next rule of "Adaptive & Proactive" SDLC Project Management.

RULE 50: There is never enough testing prior to release.

ID	Task Name	Work	Duration	Start	Finish	Predecessors	Resource Names	Cost
299	PART 5 - UAT - (Usability) User Acceptance Test	144.5 hrs	15.94 days	Fri 18/09/15	Sat 10/10/15			$41,452.50
300	Part 5 - Travel To Site	32 hrs	8 days	Fri 18/09/15	Wed 30/09/15			$20,680.00
301	TRAVEL DATE	0 hrs	0 days	Tue 29/09/1	Tue 29/09/1	14		$0.00
302	Travel Coordination	8 hrs	1 day	Fri 18/09/15	Mon 21/09/1	301SS-7 days	Logistics Mgr	$1,240.00
303	Travel To Site (Sys Eng)	8 hrs	1 day	Tue 29/09/1	Wed 30/09/1	301SS	Sw Lead Eng	$1,400.00
304	Travel To Site (Sw Eng)	8 hrs	1 day	Tue 29/09/1	Wed 30/09/1	301SS	Sys Lead Eng	$1,560.00
305	Travel To Site (Mgmt)	8 hrs	1 day	Tue 29/09/1	Wed 30/09/1	301SS	Proj Mgmt	$1,480.00
306	Travel to & from site Cost (Sys Eng)	0 hrs	0 days	Tue 29/09/1	Tue 29/09/1	301SS		$2,000.00
307	Travel to & from site Cost (Sw Eng)	0 hrs	0 days	Tue 29/09/1	Tue 29/09/1	301SS		$2,000.00
308	Travel to & from site Cost (Mgmt)	0 hrs	0 days	Tue 29/09/1	Tue 29/09/1	301SS		$2,000.00
309	Accommodation & Allowance Cost (Sys Eng)	0 hrs	0 days	Tue 29/09/1	Tue 29/09/1	301SS		$3,000.00
310	Accommodation & Allowance Cost (Sw Eng)	0 hrs	0 days	Tue 29/09/1	Tue 29/09/1	301SS		$3,000.00
311	Accommodation & Allowance Cost (Mgmt)	0 hrs	0 days	Tue 29/09/1	Tue 29/09/1	301SS		$3,000.00
312	Part 5 - System Readiness Check	18 hrs	1.25 days	Wed 30/09/15	Thu 01/10/15	300		$3,350.00
313	Application Setup Confirmation	18 hrs	1.25 days	Wed 30/09/15	Thu 01/10/15			$3,350.00
314	Web-Server – Setup	2 hrs	0.25 days	Wed 30/09/1	Wed 30/09/1		Sys Lead Eng	$390.00
315	Database-Server – Setup	2 hrs	0.25 days	Wed 30/09/1	Thu 01/10/1	314	Sys Lead Eng	$390.00
316	Setup Verification & Sanity Testing	2 hrs	0.25 days	Thu 01/10/1	Thu 01/10/1	315	Sys Lead Eng	$390.00
317	BUFFER - Time & Cost	4 hrs	0.5 days	Thu 01/10/1	Thu 01/10/1	316	Sys Lead Eng	$780.00
318	Assist	8 hrs	1 day	Wed 30/09/1	Thu 01/10/1	317FF	Sw Lead Eng	$1,400.00
319	Part 5 - Onsite Test	33.5 hrs	1.56 days	Thu 01/10/15	Mon 05/10/15	312		$6,197.50
320	Commencement Meeting - With Cust Rep Representatives	1.5 hrs	0.06 days	Thu 01/10/15	Thu 01/10/15			$277.50
321	Review Participation	0.5 hrs	0.06 days	Thu 01/10/1	Thu 01/10/1		Sys Lead Eng	$97.50
322	Review Participation	0.5 hrs	0.06 days	Thu 01/10/1	Thu 01/10/1	321SS	Sw Lead Eng	$87.50
323	Review Participation	0.5 hrs	0.06 days	Thu 01/10/1	Thu 01/10/1	321SS	Proj Mgmt	$92.50
324	Conduct UAT	24 hrs	1 day	Thu 01/10/15	Fri 02/10/15	320		$4,440.00
325	Conduct Test	8 hrs	1 day	Thu 01/10/1	Fri 02/10/1		Sys Lead Eng	$1,560.00
326	Support	8 hrs	1 day	Thu 01/10/1	Fri 02/10/1	325SS	Sw Lead Eng	$1,400.00
327	Support & Record Taking	8 hrs	1 day	Thu 01/10/1	Fri 02/10/1	325SS	Proj Mgmt	$1,480.00
328	Review Meeting - Test Completion Review	6 hrs	0.25 days	Fri 02/10/15	Mon 05/10/1	324		$1,110.00
329	Review Participation	2 hrs	0.25 days	Fri 02/10/1	Mon 05/10/1		Sys Lead Eng	$390.00
330	Review Participation	2 hrs	0.25 days	Fri 02/10/1	Mon 05/10/1	329SS	Sw Lead Eng	$350.00
331	Review Participation	2 hrs	0.25 days	Fri 02/10/1	Mon 05/10/1	329SS	Proj Mgmt	$370.00
332	Minute Review Meeting	2 hrs	0.25 days	Mon 05/10/1	Mon 05/10/1	328	Proj Mgmt	$370.00
333	DELIVERABLE - Send Minutes to Cust Rep	0 hrs	0 days	Mon 05/10/1	Mon 05/10/1	332		$0.00
334	Part 5 - Travel From Site	24 hrs	1 day	Mon 05/10/15	Tue 06/10/15	319		$4,440.00
335	Travel From Site (Sys Eng)	8 hrs	1 day	Mon 05/10/1	Tue 06/10/1		Sys Lead Eng	$1,560.00
336	Travel From Site (Sw Eng)	8 hrs	1 day	Mon 05/10/1	Tue 06/10/1	335SS	Sw Lead Eng	$1,400.00
337	Travel From Site (Mgmt)	8 hrs	1 day	Mon 05/10/1	Tue 06/10/1	335SS	Proj Mgmt	$1,480.00
338	Part 5 - Test Report	35 hrs	3.88 days	Tue 06/10/15	Sat 10/10/15	334		$6,415.00
339	Write Test Report / Quality Evidence	28 hrs	3 days	Tue 06/10/1	Fri 09/10/15			$5,380.00
340	Write Report	24 hrs	3 days	Tue 06/10/1	Fri 09/10/1		Sys Lead Eng	$4,680.00
341	Support	4 hrs	0.5 days	Tue 06/10/1	Tue 06/10/1	340SS	Sw Lead Eng	$700.00
342	Internal Review Report	4 hrs	0.5 days	Fri 09/10/1	Fri 09/10/1	339	Test Lead Eng	$600.00
343	Document Presentation Preparation	2 hrs	0.25 days	Fri 09/10/1	Sat 10/10/1	342	Docu Support	$250.00
344	Review - Docu	1 hr	0.13 days	Sat 10/10/1	Sat 10/10/1	343	Proj Mgmt	$185.00
345	DELIVERABLE - Send User Acceptance Test Report to Cust Rep	0 hrs	0 days	Sat 10/10/1	Sat 10/10/1	344		$0.00
346	MILESTONE - Closeout Report & Invoicing	2 hrs	0.25 days	Sat 10/10/1	Sat 10/10/1	338	Proj Mgmt	$370.00

Figure 110: Scenario's project schedule – Part 5 – Usability Acceptance Test (UAT).

For this scenario's project, the Usability Acceptance Test will be conducted sometime after the SAT, and prior to the Go Live. That is, after the defect repairs from the FAT and SAT have been incorporated into the deliverables that will be used for the UAT.

As stated previously in [Section 3.5.5], **the purpose of the Usability Acceptance Test (UAT) is to verify that the deliverables in-situ operation is as was commissioned** (agreed to); **hence, the deliverables are ready for end-user usage.**

While the UAT may occur immediately after the SAT, or some time afterwards; there is a major difference between the UAT and the SAT (and more specifically with the FAT) in that, while the SAT & FAT are well scripted activities with clearly defined Acceptance Test Procedures, the UAT is a more ad-hoc affair that is only bounded by the limitations specified in the Release Notes that accompany the project's deliverables (as provided by the performing organization).

Hence, for the UAT it is highly advisable that the customer organization provide end-users who are knowledgeable as to the purpose of the project's deliverables, and who have some expertise in the field of endeavor where these deliverables are to be used.

DO NOT underestimate the impact of the end-user, because without an appropriate level of work skills, experience, and training, then the end-user can misunderstand, misinterpret, and not know what the deliverables are & are not capable of doing. Consequently, **the end-user could decree the deliverable to be "broken", when in fact they are not using the deliverables as was duly defined (per the approved Detailed Specifications) and as was agreed to be acceptable (per the approved Acceptance Criteria)**.

Conversely, being provided with end-users who have plenty of experience & expertise in the field of endeavor (i.e. persons who carry some level of operational respect within the customer organization) then this can greatly improve the perceived quality & acceptability of the project deliverables; because, these "expert users" can push the application | system beyond what the previous testers have been capable of achieving. ... And potentially, due to the satisfaction of these expert users, they could directly & indirectly influence the Customer's Representatives to sign-off on the deliverables, ASAP.

{déjà vu} ... At the finish of the UAT, a **'Test Completion Review' meeting** should be held between the customer & performing organizations' representatives (including available expert users) to **discuss & note any concerns with respect to the test outcomes and to talk over the potential for the Go Live to commence as was scheduled**. By the end of this meeting, **the Project Manager's objective is to have the Customer's Representatives sign-off on the completion of the UAT.** ... And, once the performing organization's onsite UAT personnel have finished (and returned from site), then the **UAT Report (with a recommendation to proceed or not proceed with the Go Live)** would be written up & reviewed prior to being **sent to the Customer's Representatives, then the sign-off on the associated progress payment for the conclusion of the UAT**. ... As always, try to keep the Progress Payments ahead of the Costs Incurred to date; *e.g. 90% versus 87.5% respectively*.

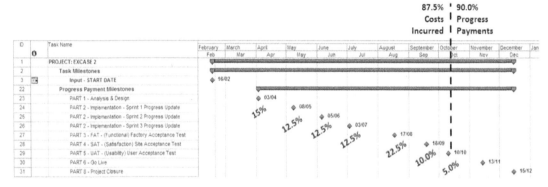

Figure 111: Scenario's project schedule – progress payments to conclusion of UAT.

NOTE: While the Project Manager may have diligently calculated the percentages for **Progress Payments versus Costs Incurred to date**; during the preceding contract's negotiation between the Customer's Representatives and the performing organization's senior management, with the stroke of a pen, these **percentages could be overwritten with greater values being placed on those final parts, and lesser values given to those implementation parts** (where much of the internal costs are incurred). **This is a senior management call**, as they have to **weigh up the cash flow implications versus the need to secure the contracted work**.

ID	Task Name	Work	Duration	Start	Finish	Predecessors	Resource Names	Cost
347	PART 6 - Go Live	174.5 hrs	23 days	Thu 15/10/15	Fri 13/11/15			$51,377.50
348	Part 6 - Travel To Site	32 hrs	8 days	Thu 15/10/15	Mon 26/10/15			$21,880.00
349	TRAVEL DATE	0 hrs	0 days	Mon 26/10/1	Mon 26/10/1	350FS-4 days		$0.00
350	Travel Coordination	8 hrs	1 day	Thu 15/10/1	Thu 15/10/1	349FS-7 days	Logistics Mgr	$1,240.00
351	Travel To Site (Sys Eng)	8 hrs	1 day	Mon 26/10/1	Mon 26/10/1	349SS	Sys Lead Eng	$1,560.00
352	Travel To Site (Sw Eng)	8 hrs	1 day	Mon 26/10/1	Mon 26/10/1	349SS	Sw Lead Eng	$2,600.00
353	Travel To Site (Mgmt)	8 hrs	1 day	Mon 26/10/1	Mon 26/10/1	349SS	Proj Mgmt	$1,480.00
354	Travel to & from site Cost (Sys Eng)	0 hrs	0 days	Mon 26/10/1!	Mon 26/10/1!	349SS		$2,000.00
355	Travel to & from site Cost (Sw Eng)	0 hrs	0 days	Mon 26/10/1!	Mon 26/10/1!	349SS		$2,000.00
356	Travel to & from site Cost (Mgmt)	0 hrs	0 days	Mon 26/10/1!	Mon 26/10/1!	349SS		$2,000.00
357	Accommodation & Allowance Cost (Sys Eng)	0 hrs	0 days	Mon 26/10/1!	Mon 26/10/1!	349SS		$3,000.00
358	Accommodation & Allowance Cost Sw Eng)	0 hrs	0 days	Mon 26/10/1!	Mon 26/10/1!	349SS		$3,000.00
359	Accommodation & Allowance Cost (Mgmt)	0 hrs	0 days	Mon 26/10/1!	Mon 26/10/1!	349SS		$3,000.00
360	Part 6 - System Readiness Check	56 hrs	4 days	Mon 26/10/15	Thu 29/10/15	368FS-8 days		$10,360.00
361	Application Setup Confirmation	56 hrs	4 days	Mon 26/10/15	Thu 29/10/15			$10,360.00
362	Web-Server – Setup	4 hrs	0.5 days	Mon 26/10/1	Mon 26/10/1		Sys Lead Eng	$780.00
363	Database-Server – Setup	4 hrs	0.5 days	Mon 26/10/1	Mon 26/10/1	362	Sys Lead Eng	$780.00
364	Go Live Switchover - Setup	4 hrs	0.5 days	Tue 27/10/1	Tue 27/10/1	363	Sys Lead Eng	$780.00
365	Setup Verification & Sanity Testing	8 hrs	1 day	Tue 27/10/1	Wed 28/10/1	364	Sys Lead Eng	$1,560.00
366	BUFFER - Time & Cost	8 hrs	1 day	Wed 28/10/1!	Thu 29/10/1	365	Sys Lead Eng	$1,560.00
367	Assist	28 hrs	3.5 days	Mon 26/10/1	Thu 29/10/1	363SS	Sw Lead Eng	$4,900.00
368	Part 6 - Go Live	33.5 hrs	3.56 days	Thu 29/10/15	Tue 03/11/15			$7,872.50
369	Commencement Meeting - With Cust Rep Representatives	1.5 hrs	0.06 days	Thu 29/10/1	Fri 30/10/1!			$352.50
370	Review Participation	0.5 hrs	0.06 days	Thu 29/10/1	Thu 29/10/1	373SF	Sys Lead Eng	$97.50
371	Review Participation	0.5 hrs	0.06 days	Thu 29/10/1	Thu 29/10/1	370SS	Sw Lead Eng	$162.50
372	Review Participation	0.5 hrs	0.06 days	Thu 29/10/1	Thu 29/10/1	370SS	Proj Mgmt	$92.50
373	Conduct Go Live activities	24 hrs	1 day	Fri 30/10/1!	Sat 31/10/1!			$6,040.00
374	Conduct	8 hrs	1 day	Fri 30/10/1!	Sat 31/10/1!	377SF	Sys Lead Eng	$1,560.00
375	Support	8 hrs	1 day	Fri 30/10/1!	Sat 31/10/1!	374FF	Sw Lead Eng	$3,000.00
376	Support	8 hrs	1 day	Fri 30/10/1!	Sat 31/10/1!	374FF	Proj Mgmt	$1,480.00
377	**GO LIVE DATE**	0 hrs	0 days	Sat 31/10/1	Sat 31/10/1!	16		$0.00
378	Review Meeting - Completion Review	6 hrs	0.25 days	Tue 03/11/15	Tue 03/11/15	377FS+2 days		$1,110.00
379	Review Participation	2 hrs	0.25 days	Tue 03/11/1	Tue 03/11/1		Sys Lead Eng	$390.00
380	Review Participation	2 hrs	0.25 days	Tue 03/11/1	Tue 03/11/1	379SS	Sw Lead Eng	$350.00
381	Review Participation	2 hrs	0.25 days	Tue 03/11/1	Tue 03/11/1	379SS	Proj Mgmt	$370.00
382	Minute Review Meeting	2 hrs	0.25 days	Tue 03/11/1	Tue 03/11/1	378	Proj Mgmt	$370.00
383	DELIVERABLE - Send Minutes to Cust Rep	0 hrs	0 days	Tue 03/11/1	Tue 03/11/1	382		$0.00
384	Part 6 - Travel From Site	24 hrs	1 day	Tue 03/11/15	Wed 04/11/15	368		$6,040.00
385	Travel From Site (Sys Eng)	8 hrs	1 day	Tue 03/11/1	Wed 04/11/1		Sys Lead Eng	$1,560.00
386	Travel From Site (Sw Eng)	8 hrs	1 day	Tue 03/11/1	Wed 04/11/1	385SS	Sw Lead Eng	$3,000.00
387	Travel From Site (Mgmt)	8 hrs	1 day	Tue 03/11/1	Wed 04/11/1	385SS	Proj Mgmt	$1,480.00
388	Part 6 - Go Live Report	27 hrs	2.88 days	Tue 10/11/15	Fri 13/11/15	384		$4,855.00
389	Write Go Live Report	20 hrs	2 days	Tue 10/11/15	Thu 12/11/15			$3,820.00
390	Write Report	16 hrs	2 days	Tue 10/11/1	Thu 12/11/1		Sys Lead Eng	$3,120.00
391	Support	4 hrs	0.5 days	Tue 10/11/1	Wed 11/11/1	390SS	Sw Lead Eng	$700.00
392	Internal Review Report	4 hrs	0.5 days	Thu 12/11/1	Fri 13/11/1!	389	Test Lead Eng	$600.00
393	Document Presentation Preparation	2 hrs	0.25 days	Fri 13/11/1!	Fri 13/11/1!	392	Doco Support	$250.00
394	Review - Doco	1 hr	0.13 days	Fri 13/11/1!	Fri 13/11/1!	393	Proj Mgmt	$185.00
395	DELIVERABLE - Go Live Report to Cust Rep	0 hrs	0 days	Fri 13/11/1!	Fri 13/11/1!	394		$0.00
396	MILESTONE - Closeout Report & Invoicing	2 hrs	0.25 days	Fri 13/11/1!	Fri 13/11/1!	388	Proj Mgmt	$370.00

Figure 112: Scenario's project schedule – Part 6 – Go Live.

For this scenario's project, the most critical part of the project is the Go Live occurring on a specific date. Where if this doesn't happen then, well, … *"let's just make sure this does happen."*

From the customer's perspective, the Go Live is the most important & critical part of the project, as they could have a lot riding on the timely success of this milestone. Conversely, from the performing organization's perspective, this part of the project marks the **nearing of the end**; to *"screw-up"* now could mean that much needed [People] & [Resources] are *"bogged down"* with completing this project, when it was planned (and is necessary) that they **be available for assigning to other projects and to Business As Usual activities**.

With such a critical milestone as the rollout | switchover to a new or upgraded application | system, then ensure that there is some form of contingency plans in place, just in case the Go Live were to stumble or fail.

Hence, there needs to be a plan & procedures (documented & distributed to those who have to know) for switching back to the previous incumbent application | system, can the installation be uninstalled | rolled back, are there alternate working arrangements while the problems are resolved.

OR, is it "*a wing & a pray*", "*she'll be right mate*" overoptimistic ado.

"Hope for the best, but prepare for the worst."

Which brings us to the next rule of "Adaptive & Proactive" SDLC Project Management.

RULE 51: Failure to plan ... *adequately*, is planning to fail ... *dramatically*.

```
For this scenario's project, there is no incumbent
application | system to switchover from, nor is there a
rollback if it all goes wrong; there is just unplugging it
from the infrastructure, … then the consequential impact
on the customer's other programs of work that depend on
this timely Go Live; e.g. the public launch of the
product, and the associated marketing & advertisement
centred around that particular Go Live date.
```

{*déjà vu*} … As with the FAT, SAT, and UAT, it is ~~advisable~~ **mandatory to hold** a '**Commencement Meeting**' with both the customer & performing organizations' representatives (as well as any directly involved 3rd parties), prior to commencing the Go Live activities. Effectively this Commencement Meeting **is the kick-off meeting to the proceeding Go Live activities, where the persons involved need to be explained the 'Go Live Plan', the timings, and who is responsible for what & when.** This Go Live Plan was possibly prepared by the Customer's Representatives with input from the performing organization's Technical Authority & Management support, as well as any involved 3rd parties, and then distributed prior to this meeting.

And so, the Go Live activities should occur in accordance with the Go Live Plan.

{déjà vu} ... At the finish | halting of the Go Live, a **'Completion Review' meeting** will be held between the customer & performing organizations' representatives (including directly involved 3rd parties) to **discuss & note any concerns with respect to the Go Live outcomes, and if things didn't go as was planned then to initiate alternative arrangements**.

By the end of this meeting, **the Project Manager's objective is to have the Customer's Representatives sign-off on the completion of the Go Live.** ... And, once the performing organization's onsite Go Live personnel have finished (and returned from site), then the **Go Live Report** would be written up & reviewed prior to being **sent to the Customer's Representatives, and then the sign-off on the associated progress payment for the conclusion of the Go Live**.

ID	Task Name	Work	Duration	Start	Finish	Predecessors	Resource Names	Cost
397	PART 7 - Warrantee Support	240 hrs	30 days	Sat 31/10/1!	Thu 10/12/15	377		$38,800.0(
398	Part 7 - Warrantee Support Period	0 hrs	30 days	Sat 31/10/1	Thu 10/12/1			$0.0(
399	Part 7 - Technical Support	240 hrs	7.88 days	Sat 31/10/1!	Tue 10/11/15	398SS		$38,800.0(
400	Sys Lead Eng	40 hrs	5 days	Mon 02/11/1	Tue 10/11/1		Sys Lead Eng	$7,800.0(
401	Sw Eng Lead	40 hrs	5 days	Sat 31/10/1	Mon 09/11/1		Sw Lead Eng	$7,000.0(
402	Sw Eng 1	40 hrs	5 days	Sat 31/10/1	Thu 05/11/1		Sw Eng 1	$6,000.0(
403	Sw Eng 2	40 hrs	5 days	Sat 31/10/1	Thu 05/11/1		Sw Eng 2	$5,600.0(
404	Test Eng	40 hrs	5 days	Sat 31/10/1	Thu 05/11/1		Test Eng	$5,000.0(
405	PM - liaison	40 hrs	5 days	Sat 31/10/1	Mon 09/11/1		Proj Mgmt	$7,400.0(

Figure 113: Scenario's project schedule – Part 7 – Warranty Support.

```
For this scenario's project, once the Go Live occurs then
the Warranty Support commences as per the Terms &
Conditions of the contracted agreement.
```

"Umm, might be a little bit too high with those technical support hours, how about halving those."

While being at the end of the project's life, this **'Warranty Support'** is when the **performing organization can inadvertently get itself into a bit of trouble** (without some pre-thought, or by being "stingy" with spending project money) on these activities. *For example, due to:*

- ☒ **Not retaining / assigning** (either full-time or part-time) **members of the Project Implementation Team to the project's Warranty Support activities**. Hence, when a valid warranty claim is lodged by the customer then the performing organization has to **"scurry about" finding personnel to deal with the claim**; while, **disrupting other projects and Business As Usual activities** in the process of **hurriedly rearranging & reassigning personnel**, or having to default on the timely servicing of the claim.

- ☒ **If there are activities that need to be done to make the Warranty Support viable** (when a warranty claim is made), then proactively **do these activities before** the performing organization **gets caught out by not being prepared**, and has to react hastily due to their **inability to immediately honour such warranty claims.**

- ☒ While the contracted terms stipulate that the Go Live sign-off marks the commencement of the Warranty Support period; however, some of the project deliverables incorporate 3rd party items whose own warranties commenced when these items were received by the performing organization. Thus, what if **there is a gap of time between when the performing organization originally received the item and when the item is handed over to the customer as part of a deliverable**; potentially, a portion of **this time gap may no longer be covered by the manufacturers / vendors / suppliers warrantee**, and subsequently such warrantee has to be **covered out of the performing organization's own pocket.** Therefore, if additional periods of 3rd party warrantee can be purchased then it is **prudent to purchase such an extension.**

- ☒ (This is more of a Senior Management issue to resolve than the Project Manager) ... The customer's imposed contracted terms are such that, once the deliverables are installed into the "pilot" site and the Go Live is completed, the official Warranty Support does not commence then (and nor are the final payments made); rather, the Warranty Support is to commence at the conclusion of the proceeding 'Field Test

Evaluations'. ... Unfortunately, this "*eval-u-a-ting*" seems to drag on almost indefinitely; as though, being treated by the customer as an extended "free trial".

To prevent the project from dragging on, **there needs to be a clear definition of when Warranty Support ends and when (an independent to the project) ongoing maintenance & support contract commences**; as the servicing of Defect Repairs & Change Requests after the scheduled warrantee period has ended, can be a drain on the profitability of the project, and subsequently a drain on resource availability & a financial burden on the performing organization.

ID		Task Name	Work	Duration	Start	Finish	Predecessors	Resource Names	Cost
406		PART 8 - Project Closure	43.5 hrs	2.94 days	Fri 11/12/1!	Tue 15/12/15	397		$7,907.50
407		Project Closure Activities	24 hrs	1 day	Fri 11/12/1!	Fri 11/12/1!			$4,260.00
408		Project Mgmt	8 hrs	1 day	Fri 11/12/1	Fri 11/12/1		Proj Mgmt	$1,480.00
409		Sw Team	8 hrs	1 day	Fri 11/12/1	Fri 11/12/1	408SS	Sw Lead Eng	$1,400.00
410		Test Team	4 hrs	0.5 days	Fri 11/12/1	Fri 11/12/1	408SS	Test Lead Eng	$600.00
411		Sys info	4 hrs	0.5 days	Fri 11/12/1	Fri 11/12/1	408SS	Sys Lead Eng	$780.00
412		Write Report	12 hrs	1 day	Mon 14/12/1!	Mon 14/12/1!	407		$2,260.00
413		Write Report	8 hrs	1 day	Mon 14/12/1	Mon 14/12/1		Proj Mgmt	$1,480.00
414		Support	4 hrs	0.5 days	Mon 14/12/1	Mon 14/12/1	413SS	Sys Lead Eng	$780.00
415		Internal Review Report	4 hrs	0.5 days	Tue 15/12/1	Tue 15/12/1	412	Management Suppt	$800.00
416		Document Presentation Preparation	1 hr	0.13 days	Tue 15/12/1	Tue 15/12/1	415	Doco Support	$125.00
417		DELIVERABLE - Project Close Out Report	0.5 hr	0.06 days	Tue 15/12/1	Tue 15/12/1	416	Proj Mgmt	$92.50
418		MILESTONE - Closeout Report & Invoicing	2 hrs	0.25 days	Tue 15/12/1	Tue 15/12/1	417	Proj Mgmt	$370.00
419		Stage 1 - Management Support (5% Total Hours)	200 hrs	25 days	Mon 16/02/1	Fri 20/03/1	56SS	Management Suppt	$40,000.00
420		Stage 1 - Technical Authority Support (5% Total Hours)	200 hrs	25 days	Mon 16/02/1	Fri 20/03/1	56SS	Technical Authority	$40,000.00

Figure 114: Scenario's project schedule – Part 8 – Project Closure.

```
For this scenario's project, once the Warranty Support
concludes (as per the Terms & Conditions of the contracted
agreement), then the project can finally be closed out.
```

Project Closure, as detailed in [Section 3.7.4], is when it is ensured that "everyone agrees, It is all done".

"Umm, think you might want to double the durations for those closure activities from a day each to a coupla days per person, as there is always some fiddly bits to tie off at the end."

During the project activities, it would be beneficial to hold an internal project closure review to identify (and possible fix) any unresolved issues.

And, once the performing organization's project closure activities are completed, then write up & review the **Project Close Out Report** prior to **sending this off to the Customer's**

Representatives, and then the sign-off on the associated progress payment for the conclusion of the project.

…

Also note, in [Figure 114], external to 'Part 8 – Project Closure', there are a pair of activities that run parallel to the 'Stage 1 – Activity Execution'; these are 'Management Support' and 'Technical Authority Support'. For this scenario's project these are each given a Level Of Effort (slightly less than) 5% of the total work for the 'Stage 1 – Activity Execution'. This allocation of additional hours would be used for unplanned activities that are highly likely to occur, due to:

- Such activities as the Customer's Representatives could require some amount of managerial and/or technical request servicing (aka "handholding").
- The project is "leading edge" where a noticeable amount of management & technical authority support will be consumed dealing with the management & support of the risks associated with such a development; in such a case a 25-33% addition may be necessary to cover the time involved with resolving the known-unknowns of such a project, similarly a "bleeding edge" technology project may need an even higher ratio.

The important thing to remember, is that no matter how detailed the planned schedule (to the point of micro-managing) there will always be managerial and technical issues that will have to be handled in addition to those activities planned for in advance.

…

Update the Reviewed Schedule

Having concluded the internal review of the (proposed) project's schedule with senior management's (i.e. the Program Manager), then update the next revision of this schedule with all of the noted additions & adjustments. Then, follow up with another quick run through to verify that the revised project schedule is viable; before delving into the detailed calculation of the project's [Cost] budget.

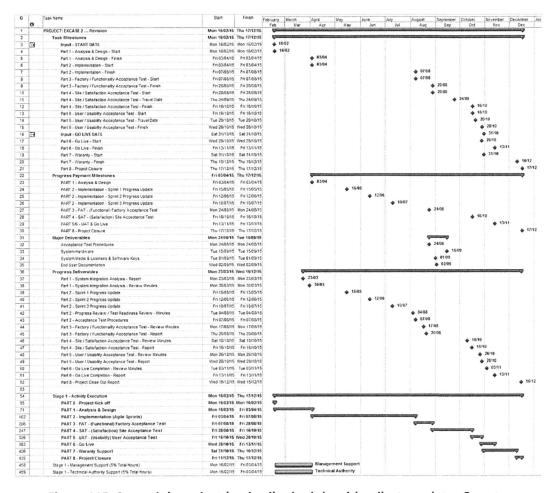

Figure 115: Scenario's project (revised) schedule with milestone dates & costs.

For this scenario's project, having reworked the schedule including all of the review notes, it is found that the schedule has pushed into the Go Live date. … And, if the personnel sent to site were to return between the UAT and the GO Live activities, then it would not be possible for the Go Live to occur on the stipulated date; hence, the proposal is for the UAT to be used as the lead into the Go Live, with the personnel only returning after the Go Live finishes, with remote technical support back at the performing organization to resolve any last minute issues.

ID		Task Name	Work	Duration	Start	Finish	Predecessors	Resource Names	Cost
1		PROJECT: EXCASE 2 ... Revision	5,901 hrs	220.06 days	Mon 16/02/15	Thu 17/12/15			$1,026,262.5
2		Task Milestones	0 hrs	220.06 days	Mon 16/02/15	Thu 17/12/15			$0.00
3	☐	Input - START DATE	0 hrs	0 days	Mon 16/02/1	Mon 16/02/1			$0.00
4		Part 1 - Analysis & Design - Start	0 hrs	0 days	Mon 16/02/1	Mon 16/02/1	71SS		$0.00
5		Part 1 - Analysis & Design - Finish	0 hrs	0 days	Fri 03/04/1	Fri 03/04/1	71FF		$0.00
6		Part 2 - Implementation - Start	0 hrs	0 days	Fri 03/04/1	Fri 03/04/1	102SS		$0.00
7		Part 2 - Implementation - Finish	0 hrs	0 days	Fri 07/08/1	Fri 07/08/1	102FF		$0.00
8		Part 3 - Factory / Functionality Acceptance Test - Start	0 hrs	0 days	Fri 07/08/1	Fri 07/08/1	206SS		$0.00
9		Part 3 - Factory / Functionality Acceptance Test - Finish	0 hrs	0 days	Fri 28/08/1	Fri 28/08/1	206FF		$0.00
10		Part 4 - Site / Satisfaction Acceptance Test - Start	0 hrs	0 days	Fri 28/08/1	Fri 28/08/1	247SS		$0.00
11		Part 4 - Site / Satisfaction Acceptance Test - Travel Date	0 hrs	0 days	Thu 24/09/1	Thu 24/09/1	257FF+7 days		$0.00
12		Part 4 - Site / Satisfaction Acceptance Test - Finish	0 hrs	0 days	Fri 16/10/1	Fri 16/10/1	247FF		$0.00
13		Part 5 - User / Usability Acceptance Test - Start	0 hrs	0 days	Fri 16/10/1	Fri 16/10/1	326SS		$0.00
14		Part 5 - User / Usability Acceptance Test - Travel Date	0 hrs	0 days	Tue 20/10/1	Tue 20/10/1	332SS		$0.00
15		Part 5 - User / Usability Acceptance Test - Finish	0 hrs	0 days	Wed 28/10/1	Wed 28/10/1	326FF		$0.00
16	☐	Input - GO LIVE DATE	0 hrs	0 days	Sat 31/10/1	Sat 31/10/1			$0.00
17		Part 6 - Go Live - Start	0 hrs	0 days	Wed 28/10/1	Wed 28/10/1	383SS		$0.00
18		Part 6 - Go Live - Finish	0 hrs	0 days	Fri 13/11/1	Fri 13/11/1	383FF		$0.00
19		Part 7 - Waranty - Start	0 hrs	0 days	Sat 31/10/1	Sat 31/10/1	426SS		$0.00
20		Part 7 - Waranty - Finish	0 hrs	0 days	Thu 10/12/1	Thu 10/12/1	426FF		$0.00
21		Part 8 - Project Closure	0 hrs	0 days	Thu 17/12/1	Thu 17/12/1	435FF		$0.00
22		Progress Payment Milestones	0 hrs	185.94 days	Fri 03/04/1	Thu 17/12/15			$0.00
23		PART 1 - Analysis & Design	0 hrs	0 days	Fri 03/04/1	Fri 03/04/1	101		$0.00
24		PART 2 - Implementation - Sprint 1 Progress Update	0 hrs	0 days	Fri 15/05/1	Fri 15/05/1	131		$0.00
25		PART 2 - Implementation - Sprint 2 Progress Update	0 hrs	0 days	Fri 12/06/1	Fri 12/06/1	138		$0.00
26		PART 2 - Implementation - Sprint 3 Progress Update	0 hrs	0 days	Fri 10/07/1	Fri 10/07/1	145		$0.00
27		PART 3 - FAT - (Functional) Factory Acceptance Test	0 hrs	0 days	Mon 24/08/1	Mon 24/08/1	238		$0.00
28		PART 4 - SAT - (Satisfaction) Site Acceptance Test	0 hrs	0 days	Fri 16/10/1	Fri 16/10/1	318		$0.00
29		PART 5/6 - UAT & Go Live	0 hrs	0 days	Fri 13/11/1	Fri 13/11/1	425		$0.00
30		PART 8 - Project Closure	0 hrs	0 days	Thu 17/12/1	Thu 17/12/1	457		$0.00
31		Major Deliverables	0 hrs	16 days	Mon 24/08/15	Tue 15/09/15			$0.00
32		Acceptance Test Procedures	0 hrs	0 days	Mon 24/08/1	Mon 24/08/1	237		$0.00
33		System Hardware	0 hrs	0 days	Tue 15/09/1	Tue 15/09/1	265		$0.00
34		System Media & Licenses & Software Keys	0 hrs	0 days	Tue 01/09/1	Tue 01/09/1	252		$0.00
35		End User Documentation	0 hrs	0 days	Wed 02/09/1	Wed 02/09/1	256		$0.00
36		Progress Deliverables	0 hrs	194.56 days	Mon 23/03/15	Wed 16/12/15			$0.00
37		Part 1 - System Integration Analysis - Report	0 hrs	0 days	Mon 23/03/1	Mon 23/03/1	89SS		$0.00
38		Part 1 - System Integration Analysis - Review Minutes	0 hrs	0 days	Mon 30/03/1	Mon 30/03/1	95		$0.00
39		Part 2 - Sprint 1 Progress Update	0 hrs	0 days	Fri 15/05/1	Fri 15/05/1	130		$0.00
40		Part 2 - Sprint 2 Progress Update	0 hrs	0 days	Fri 12/06/1	Fri 12/06/1	137		$0.00
41		Part 2 - Sprint 3 Progress Update	0 hrs	0 days	Fri 10/07/1	Fri 10/07/1	144		$0.00
42		Part 2 - Progress Review / Test Readiness Review - Minutes	0 hrs	0 days	Tue 04/08/1	Tue 04/08/1	199		$0.00
43		Part 2 - Acceptance Test Procedures	0 hrs	0 days	Fri 07/08/1	Fri 07/08/1	204		$0.00
44		Part 3 - Factory / Functionality Acceptance Test - Review Minutes	0 hrs	0 days	Mon 17/08/1	Mon 17/08/1	225		$0.00
45		Part 3 - Factory / Functionality Acceptance Test - Report	0 hrs	0 days	Thu 20/08/1	Thu 20/08/1	232		$0.00
46		Part 4 - Site / Satisfaction Acceptance Test - Review Minutes	0 hrs	0 days	Sat 10/10/1	Sat 10/10/1	305		$0.00
47		Part 4 - Site / Satisfaction Acceptance Test - Report	0 hrs	0 days	Fri 16/10/1	Fri 16/10/1	317		$0.00
48		Part 5 - User / Usability Acceptance Test - Review Minutes	0 hrs	0 days	Mon 26/10/1	Mon 26/10/1	367		$0.00
49		Part 5 - User / Usability Acceptance Test - Report	0 hrs	0 days	Wed 28/10/1	Wed 28/10/1	375		$0.00
50		Part 6 - Go Live Completion - Review Minutes	0 hrs	0 days	Tue 03/11/1	Tue 03/11/1	412		$0.00
51		Part 6 - Go Live Completion - Report	0 hrs	0 days	Fri 13/11/1	Fri 13/11/1	424		$0.00
52		Part 8 - Project Close Out Report	0 hrs	0 days	Wed 16/12/1	Wed 16/12/1	456		$0.00
53									
54		Stage 1 - Activity Execution	5,351 hrs	220.06 days	Mon 16/02/15	Thu 17/12/15			$916,262.50
55		PART 0 - Project Kick off	17 hrs	0.63 days	Mon 16/02/15	Mon 16/02/15	3		$2,700.00
71		PART 1 - Analysis & Design	216 hrs	33.5 days	Mon 16/02/15	Fri 03/04/15	55		$46,460.00
102		PART 2 - Implementation (Agile Sprints)	3,769 hrs	90.38 days	Fri 03/04/15	Fri 07/08/15	71		$612,415.00
206		PART 3 - FAT - (Functional) Factory Acceptance Test	409.5 hrs	15 days	Fri 07/08/15	Fri 28/08/15	102		$62,425.00
247		PART 4 - SAT - (Satisfaction) Site Acceptance Test	405.5 hrs	36.19 days	Fri 28/08/15	Fri 16/10/15	206		$86,207.50
326		PART 5 - UAT - (Usability) User Acceptance Test	231.5 hrs	8.25 days	Fri 16/10/15	Wed 28/10/15	247		$53,707.50
383		PART 6 - Go Live	106 hrs	13.63 days	Wed 28/10/15	Fri 13/11/15			$19,410.00
426		PART 7 - Waranty Support	120 hrs	30 days	Sat 31/10/15	Thu 10/12/15	406		$19,400.00
435		PART 8 - Project Closure	76.5 hrs	4.06 days	Fri 11/12/15	Thu 17/12/15	426		$13,537.50
458		Stage 1 - Management Support (5% Total Hours)	275 hrs	34.38 days	Mon 16/02/1	Fri 03/04/1	54SS	Management Support	$55,000.00
459		Stage 1 - Technical Authority Support (5% Total Hours)	275 hrs	34.38 days	Mon 16/02/1	Fri 03/04/1	54SS	Technical Authority	$55,000.00

Figure 116: Scenario's project (revised) schedule with milestone dates & costs.

For the scenario's project, the Level Of Effort is accumulated to 5901 hours (up from 4940 hours), calendar duration is still 10 months, and the Internal Total Cost of $1,026K (up from $910K) ... which is about a $1.28M to $1.54M Contract Price (up from the $1.14M to $1.37M initially guesstimated) which is a 12% increase.

Which brings us to an addendum to "Adaptive & Proactive" SDLC Project Management.

RULE 51ᴬ: Failing to catch those missing tasks and misplaced dependences can have a dramatic effect on the dates, and the Level Of Effort involved. ... And, on the calculated budget.

NOTE: When viewing the revised schedule on the following pages, there are probably a few changes, additions, subtractions, and rearrangements that you could suggest; hence why, the schedule should be reviewed by various persons in the performing organization prior to being presented in whatever form (summation or detailed) to senior management, ... and, eventually to the Customer's Representatives.

ID	Task Name	Work	Duration	Start	Finish	Predecessors	Resource Names	Cost
55	PART 0 - Project Kick off	17 hrs	0.63 days	Mon 16/02/15	Mon 16/02/15	3		$2,700.00
56	Meeting Preparation / Presentation	4 hrs	0.5 days	Mon 16/02/1	Mon 16/02/1		Proj Mgmt	$740.00
57	Meeting Participation	13 hrs	0.13 days	Mon 16/02/15	Mon 16/02/15	56		$1,960.00
58	Meeting Participation	1 hr	0.13 days	Mon 16/02/1	Mon 16/02/1		Proj Mgmt	$185.00
59	Meeting Participation	1 hr	0.13 days	Mon 16/02/1	Mon 16/02/1	58SS	Sys Lead Eng	$195.00
60	Meeting Participation	1 hr	0.13 days	Mon 16/02/1	Mon 16/02/1	58SS	BA - SA	$200.00
61	Meeting Participation	1 hr	0.13 days	Mon 16/02/1	Mon 16/02/1	58SS	Sw Lead Eng	$175.00
62	Meeting Participation	1 hr	0.13 days	Mon 16/02/1	Mon 16/02/1	58SS	Sw Eng 1	$150.00
63	Meeting Participation	1 hr	0.13 days	Mon 16/02/1	Mon 16/02/1	58SS	Sw Eng 2	$140.00
64	Meeting Participation	1 hr	0.13 days	Mon 16/02/1	Mon 16/02/1	58SS	Sw Eng 3	$130.00
65	Meeting Participation	1 hr	0.13 days	Mon 16/02/1	Mon 16/02/1	58SS	Sw Eng 4	$120.00
66	Meeting Participation	1 hr	0.13 days	Mon 16/02/1	Mon 16/02/1	58SS	Sw Eng 5	$110.00
67	Meeting Participation	1 hr	0.13 days	Mon 16/02/1	Mon 16/02/1	58SS	Test Lead Eng	$150.00
68	Meeting Participation	1 hr	0.13 days	Mon 16/02/1	Mon 16/02/1	58SS	Test Eng	$125.00
69	Meeting Participation	1 hr	0.13 days	Mon 16/02/1	Mon 16/02/1	58SS	Docs Support	$125.00
70	Meeting Participation	1 hr	0.13 days	Mon 16/02/1	Mon 16/02/1	58SS	Logistics Mgr	$155.00
71	PART 1 - Analysis & Design	216 hrs	33.5 days	Mon 16/02/15	Fri 03/04/15	55		$46,460.00
72	Part 1 - System Integration Analysis & Design	214 hrs	33.25 days	Mon 16/02/15	Thu 02/04/15			$46,090.00
73	Travel To Site	16 hrs	2 days	Mon 16/02/15	Wed 18/02/15			$7,800.00
74	Travel Coordination	8 hrs	1 day	Mon 16/02/1	Tue 17/02/1		Logistics Mgr	$1,240.00
75	TRAVEL DATE	0 hrs	0 days	Tue 17/02/1	Tue 17/02/1	74		$0.00
76	Travel To Site	8 hrs	1 day	Tue 17/02/1	Wed 18/02/15	75	Sys Lead Eng	$1,560.00
77	Travel to & from site Cost	0 hrs	0 days	Wed 18/02/1	Wed 18/02/1	76		$2,000.00
78	Accommodation & Allowance Cost	0 hrs	0 days	Tue 17/02/15	Tue 17/02/15	76SS		$3,000.00
79	Analysis of Current Front Office System	24 hrs	3 days	Wed 18/02/1	Mon 23/02/1	73	Sys Lead Eng	$4,680.00
80	Analysis of Current Back Office System	56 hrs	7 days	Mon 23/02/1	Wed 04/03/1	79	Sys Lead Eng	$10,920.00
81	Travel From Site	8 hrs	1 day	Wed 04/03/1	Thu 05/03/1	80	Sys Lead Eng	$1,560.00
82	Write Report	60 hrs	10 days	Thu 05/03/15	Thu 19/03/15	81		$11,800.00
83	Write Report	40 hrs	5 days	Thu 05/03/1	Thu 12/03/1		Sys Lead Eng	$7,800.00
84	Support	20 hrs	2.5 days	Thu 05/03/1	Tue 10/03/1	83SS	BA - SA	$4,000.00
85	BUFFER - Time	0 hrs	5 days	Thu 12/03/1	Thu 19/03/1	83		$0.00
86	Internal Review Report	8 hrs	1 day	Thu 19/03/1	Fri 20/03/1	82	BA - SA	$1,600.00
87	Document Presentation Preparation	4 hrs	0.5 days	Fri 20/03/1	Mon 23/03/1	86	Docs Support	$500.00
88	Review - Docs	1 hr	0.13 days	Mon 23/03/1	Mon 23/03/1	87	Proj Mgmt	$185.00
89	DELIVERABLE - Send Analysis & Design Report to Cust Rep to Ev	0 hrs	5 days	Mon 23/03/1	Mon 30/03/1	88		$0.00
90	Review Meeting - Analysis & Design Report	12 hrs	0.5 days	Mon 30/03/15	Mon 30/03/15	89		$2,320.00
91	Review Participation	4 hrs	0.5 days	Mon 30/03/1	Mon 30/03/1		Sys Lead Eng	$780.00
92	Review Participation	4 hrs	0.5 days	Mon 30/03/1	Mon 30/03/1	91SS	BA - SA	$800.00
93	Review Participation	4 hrs	0.5 days	Mon 30/03/1	Mon 30/03/1	91SS	Proj Mgmt	$740.00
94	Minute Review Meeting	2 hrs	0.25 days	Mon 30/03/1	Mon 30/03/1	90	Proj Mgmt	$370.00
95	DELIVERABLE - Send Minutes to Cust Rep	0 hrs	0 days	Mon 30/03/1	Mon 30/03/1	94		$0.00
96	Update Report	16 hrs	2 days	Tue 31/03/1	Wed 01/04/15	94	Sys Lead Eng	$3,120.00
97	Internal Review Report	4 hrs	0.5 days	Thu 02/04/1	Thu 02/04/1	96	BA - SA	$800.00
98	Document Presentation Preparation	2 hrs	0.25 days	Thu 02/04/1	Thu 02/04/1	97	Docs Support	$250.00
99	Review - Docs	1 hr	0.13 days	Thu 02/04/1	Thu 02/04/1	98	Proj Mgmt	$185.00
100	Send Updated Analysis & Design Report to Cust Rep	0 hrs	0 days	Thu 02/04/1	Thu 02/04/1	99		$0.00
101	MILESTONE - Closeout Report & Invoicing	2 hrs	0.25 days	Thu 02/04/1	Fri 03/04/1	72	Proj Mgmt	$370.00

Figure 117: Revised Schedule ... Part 0 – Project Kick Off and Part 1 – Analysis & Design.

ID	Task Name	Work	Duration	Start	Finish	Predecessors	Resource Names	Cost
102	PART 2 - Implementation (Agile Sprints)	3,769 hrs	90.38 days	Fri 03/04/1!	Fri 07/08/1!	71		$612,415.00
103	Part 2 - Sprint 1	960 hrs	20 days	Fri 03/04/1!	Fri 01/05/1!			$132,000.00
104	Sw Eng Lead	160 hrs	20 days	Fri 03/04/1!	Fri 01/05/1		Sw Lead Eng	$28,000.00
105	Sw Eng 1	160 hrs	20 days	Fri 03/04/1	Fri 01/05/1	104SS	Sw Eng 1	$24,000.00
106	Sw Eng 2	160 hrs	20 days	Fri 03/04/1	Fri 01/05/1	104SS	Sw Eng 2	$22,400.00
107	Sw Eng 3	160 hrs	20 days	Fri 03/04/1	Fri 01/05/1	104SS	Sw Eng 3	$20,800.00
108	Sw Eng 4	160 hrs	20 days	Fri 03/04/1	Fri 01/05/1	104SS	Sw Eng 4	$19,200.00
109	Sw Eng 5	160 hrs	20 days	Fri 03/04/1	Fri 01/05/1	104SS	Sw Eng 5	$17,600.00
110	Part 2 - Sprint 2	960 hrs	20 days	Fri 01/05/1!	Fri 29/05/1!	103		$132,000.00
111	Sw Eng Lead	160 hrs	20 days	Fri 01/05/1	Fri 29/05/1		Sw Lead Eng	$28,000.00
112	Sw Eng 1	160 hrs	20 days	Fri 01/05/1	Fri 29/05/1	111SS	Sw Eng 1	$24,000.00
113	Sw Eng 2	160 hrs	20 days	Fri 01/05/1	Fri 29/05/1	111SS	Sw Eng 2	$22,400.00
114	Sw Eng 3	160 hrs	20 days	Fri 01/05/1	Fri 29/05/1	111SS	Sw Eng 3	$20,800.00
115	Sw Eng 4	160 hrs	20 days	Fri 01/05/1	Fri 29/05/1	111SS	Sw Eng 4	$19,200.00
116	Sw Eng 5	160 hrs	20 days	Fri 01/05/1	Fri 29/05/1	111SS	Sw Eng 5	$17,600.00
117	Part 2 - Sprint 3	960 hrs	20 days	Fri 29/05/1!	Fri 26/06/1!	110		$132,000.00
118	Sw Eng Lead	160 hrs	20 days	Fri 29/05/1	Fri 26/06/1		Sw Lead Eng	$28,000.00
119	Sw Eng 1	160 hrs	20 days	Fri 29/05/1	Fri 26/06/1	118SS	Sw Eng 1	$24,000.00
120	Sw Eng 2	160 hrs	20 days	Fri 29/05/1	Fri 26/06/1	118SS	Sw Eng 2	$22,400.00
121	Sw Eng 3	160 hrs	20 days	Fri 29/05/1	Fri 26/06/1	118SS	Sw Eng 3	$20,800.00
122	Sw Eng 4	160 hrs	20 days	Fri 29/05/1	Fri 26/06/1	118SS	Sw Eng 4	$19,200.00
123	Sw Eng 5	160 hrs	20 days	Fri 29/05/1	Fri 26/06/1	118SS	Sw Eng 5	$17,600.00
124	Part 2 - Evaluation Testing	348 hrs	50.5 days	Fri 01/05/1!	Fri 10/07/1!			$48,780.00
125	Part 2 - Evaluate - Sprint 1	114 hrs	10.25 days	Fri 01/05/1!	Fri 15/05/1!	103		$15,890.00
126	Evaluate Test	80 hrs	10 days	Fri 01/05/1	Fri 15/05/1		Test Eng	$10,000.00
127	Assistance	16 hrs	2 days	Fri 01/05/1	Tue 05/05/1	126SS	Test Lead Eng	$2,400.00
128	Assistance	16 hrs	2 days	Fri 01/05/1	Tue 05/05/1	126SS	Sys Lead Eng	$3,120.00
129	Progress Summary	2 hrs	0.25 days	Fri 15/05/1	Fri 15/05/1	126	Proj Mgmt	$370.00
130	DELIVERABLE - Progress Summary to Cust Rep	0 hrs	0 days	Fri 15/05/1	Fri 15/05/1	125		$0.00
131	MILESTONE - Progress Payment Invoicing	2 hrs	0.25 days	Fri 15/05/1	Fri 15/05/1	130	Proj Mgmt	$370.00
132	Part 2 - Evaluate - Sprint 2	114 hrs	10.25 days	Fri 29/05/1	Fri 12/06/1!	110		$15,890.00
133	Evaluate Test	80 hrs	10 days	Fri 12/06/1	Fri 12/06/1		Test Eng	$10,000.00
134	Assistance	16 hrs	2 days	Fri 29/05/1	Tue 02/06/1	133SS	Test Lead Eng	$2,400.00
135	Assistance	16 hrs	2 days	Fri 29/05/1	Tue 02/06/1	133SS	Sys Lead Eng	$3,120.00
136	Progress Summary	2 hrs	0.25 days	Fri 12/06/1	Fri 12/06/1	133	Proj Mgmt	$370.00
137	DELIVERABLE - Progress Summary to Cust Rep	0 hrs	0 days	Fri 12/06/1	Fri 12/06/1	132		$0.00
138	MILESTONE - Progress Payment Invoicing	2 hrs	0.25 days	Fri 12/06/1	Fri 12/06/1	137	Proj Mgmt	$370.00
139	Part 2 - Evaluate - Sprint 3	114 hrs	10.25 days	Fri 26/06/1!	Fri 10/07/1!	117		$15,890.00
140	Evaluate Test	80 hrs	10 days	Fri 26/06/1	Fri 10/07/1		Test Eng	$10,000.00
141	Assistance	16 hrs	2 days	Fri 26/06/1	Tue 30/06/1	140SS	Test Lead Eng	$2,400.00
142	Assistance	16 hrs	2 days	Fri 26/06/1	Tue 30/06/1	140SS	Sys Lead Eng	$3,120.00
143	Progress Summary	2 hrs	0.25 days	Fri 10/07/1	Fri 10/07/1	140	Proj Mgmt	$370.00
144	DELIVERABLE - Progress Summary to Cust Rep	0 hrs	0 days	Fri 10/07/1	Fri 10/07/1	139		$0.00
145	MILESTONE - Progress Payment Invoicing	2 hrs	0.25 days	Fri 10/07/1	Fri 10/07/1	144	Proj Mgmt	$370.00
146	Part 2 - System Integration	52 hrs	74.5 days	Fri 03/04/1!	Thu 16/07/15			$99,910.00
147	Hardware Acquisition & Delivery	16 hrs	42 days	Fri 03/04/1!	Tue 02/06/15			$12,890.00
148	Request Hardware	8 hrs	1 day	Fri 03/04/1	Mon 06/04/1	86	Sys Lead Eng	$1,560.00
149	Review Hardware Request	2 hrs	0.25 days	Mon 06/04/1	Mon 06/04/1	148	BA - SA	$400.00
150	Hardware Componentry Cost	0 hrs	0 days	Mon 06/04/1!	Mon 06/04/1!	149		$10,000.00
151	Order Hardware Componentry	2 hrs	0.25 days	Mon 06/04/1	Mon 06/04/1	149	Logistics Mgr	$310.00
152	BUFFER - Delivery Time	0 hrs	40 days	Mon 06/04/1	Mon 01/06/1	151		$0.00
153	Handling Hardware Delivery	4 hrs	0.5 days	Mon 01/06/1	Tue 02/06/1	152	Logistics Mgr	$620.00
154	System Integration	36 hrs	32.5 days	Tue 02/06/15	Thu 16/07/15	147		$87,020.00
155	End User Licensing	4 hrs	0.5 days	Tue 02/06/15	Tue 02/06/15			$80,780.00
156	Web-Server – End User License - Cost	0 hrs	0 days	Tue 02/06/1!	Tue 02/06/1!			$15,000.00
157	Database-Server – End User License - Cost	0 hrs	0 days	Tue 02/06/1!	Tue 02/06/1!			$65,000.00
158	Co-ordination	4 hrs	0.5 days	Tue 02/06/1	Tue 02/06/1	156	Sys Lead Eng	$780.00
159	Application Setup	32 hrs	4 days	Fri 10/07/1	Thu 16/07/15	155,124		$6,240.00
160	Web-Server – Setup	8 hrs	1 day	Fri 10/07/1	Mon 13/07/1		Sys Lead Eng	$1,560.00
161	Database-Server – Setup	8 hrs	1 day	Mon 13/07/1	Tue 14/07/1	160	Sys Lead Eng	$1,560.00
162	Setup Verification & Dry Run Testing	8 hrs	1 day	Tue 14/07/1	Wed 15/07/1	161	Sys Lead Eng	$1,560.00
163	BUFFER - Time & Cost	8 hrs	1 day	Wed 15/07/1	Thu 16/07/1	162	Sys Lead Eng	$1,560.00

Figure 118: Revised Schedule ... Part 2 – Implementation (Top Half).

"As an implementer, did you ever stop and think what your day's work cost your employer; and, what a day of your project team's time also costs?"

Well, in the scenario's project case, each development sprint has an internal cost of $132,000 plus $15,890 for each evaluation test, plus system integration $99,910, plus dry run testing $61,330, plus review $6,025, plus management, equals a total of $612,415 in four months. ... *"Or, a rather decent mortgage repayment on a nice house".*

ID	Task Name	Work	Duration	Start	Finish	Predecessors	Resource Names	Cost
164	Part 2 - Dry Run Test	446 hrs	86.5 days	Fri 03/04/15	Mon 03/08/15			$61,330.00
165	Acceptance Test Procedures (1st Pass)	40 hrs	5 days	Fri 03/04/1	Fri 10/04/1	72	Test Eng	$5,000.00
166	Internal Review Acceptance Test Procedures	8 hrs	1 day	Fri 10/04/1	Mon 13/04/1	165	Test Lead Eng	$1,200.00
167	Internal Review Acceptance Test Procedures	8 hrs	1 day	Mon 13/04/1	Tue 14/04/1	166	Sys Lead Eng	$1,560.00
168	Update Acceptance Test Procedures (2nd Pass)	16 hrs	2 days	Tue 14/04/1	Thu 16/04/1	167	Test Eng	$2,000.00
169	System Media	13 hrs	1 day	Thu 16/04/15	Fri 17/04/15	168		$2,245.00
170	Media Preparation	8 hrs	1 day	Thu 16/04/1	Fri 17/04/1		Sys Lead Eng	$1,560.00
171	Presentation Preparation & Packaging	3 hrs	0.38 days	Thu 16/04/1	Thu 16/04/1		Doco Support	$375.00
172	DELIVERABLE - Shipment Coordination / Export Licensing	2 hrs	0.25 days	Thu 16/04/1	Thu 16/04/1		Logistics Mgr	$310.00
173	Dry Run Test	48 hrs	4 days	Thu 16/07/15	Wed 22/07/1	146,169		$6,760.00
174	Conduct Test	24 hrs	3 days	Thu 16/07/1	Tue 21/07/1		Test Eng	$3,000.00
175	Support	8 hrs	1 day	Thu 16/07/1	Fri 17/07/1	174SS	Test Lead Eng	$1,200.00
176	Support	8 hrs	1 day	Thu 16/07/1	Fri 17/07/1	174SS	Sys Lead Eng	$1,560.00
177	BUFFER - Time & Cost	8 hrs	1 day	Tue 21/07/1	Wed 22/07/1	174	Test Eng	$1,000.00
178	Write Test Report / Quality Evidence	33 hrs	3 days	Wed 22/07/1	Mon 27/07/1	173		$4,565.00
179	Write Report	24 hrs	3 days	Wed 22/07/1	Mon 27/07/1		Test Eng	$3,000.00
180	Support	4 hrs	0.5 days	Wed 22/07/1	Thu 23/07/1	179SS	Test Lead Eng	$600.00
181	Support	4 hrs	0.5 days	Wed 22/07/1	Thu 23/07/1	179SS	Sys Lead Eng	$780.00
182	Review Doco	1 hr	0.13 days	Thu 23/07/1	Thu 23/07/1	181	Proj Mgmt	$185.00
183	Send Invitation to Cust Rep for Test Readiness Review	0 hrs	5 days	Mon 27/07/1	Mon 03/08/1	178		$0.00
184	Defect Repair	280 hrs	6.25 days	Wed 22/07/1	Thu 30/07/15	173		$38,000.00
185	Sw Eng Lead	40 hrs	5 days	Wed 22/07/1	Thu 30/07/1		Sw Lead Eng	$7,000.00
186	Sw Eng 1	40 hrs	5 days	Wed 22/07/1	Wed 29/07/1	185SS	Sw Eng 1	$6,000.00
187	Sw Eng 2	40 hrs	5 days	Wed 22/07/1	Wed 29/07/1	185SS	Sw Eng 2	$5,600.00
188	Sw Eng 3	40 hrs	5 days	Wed 22/07/1	Wed 29/07/1	185SS	Sw Eng 3	$5,200.00
189	Sw Eng 4	40 hrs	5 days	Wed 22/07/1	Wed 29/07/1	185SS	Sw Eng 4	$4,800.00
190	Sw Eng 5	40 hrs	5 days	Wed 22/07/1	Wed 29/07/1	185SS	Sw Eng 5	$4,400.00
191	Test Eng	40 hrs	5 days	Wed 22/07/1	Wed 29/07/1	185SS	Test Eng	$5,000.00
192	Part 2 - Test Readiness Review	41 hrs	3.63 days	Mon 03/08/1	Fri 07/08/1	164		$6,025.00
193	Review Meeting - Test Readiness Review	16 hrs	0.5 days	Mon 03/08/15	Tue 04/08/15			$2,620.00
194	Review Participation	4 hrs	0.5 days	Mon 03/08/1	Tue 04/08/1		Test Eng	$500.00
195	Review Participation	4 hrs	0.5 days	Mon 03/08/1	Tue 04/08/1	194SS	Test Lead Eng	$600.00
196	Review Participation	4 hrs	0.5 days	Mon 03/08/1	Tue 04/08/1	194SS	Sys Lead Eng	$780.00
197	Review Participation	4 hrs	0.5 days	Mon 03/08/1	Tue 04/08/1	194SS	Proj Mgmt	$740.00
198	Minute Review Meeting	2 hrs	0.25 days	Tue 04/08/1	Tue 04/08/1	193	Proj Mgmt	$370.00
199	DELIVERABLE - Send Minutes to Cust Rep	0 hrs	0 days	Tue 04/08/1	Tue 04/08/1	198		$0.00
200	Update Acceptance Test Procedures	16 hrs	2 days	Tue 04/08/1	Thu 06/08/1	199	Test Eng	$2,000.00
201	Internal Review Report	4 hrs	0.5 days	Thu 06/08/1	Thu 06/08/1	200	Test Lead Eng	$600.00
202	Document Presentation Preparation	2 hrs	0.25 days	Thu 06/08/1	Fri 07/08/1	201	Doco Support	$250.00
203	Review Doco	1 hr	0.13 days	Fri 07/08/1	Fri 07/08/1	202	Proj Mgmt	$185.00
204	DELIVERABLE - Send Acceptance Test Procedures to Cust Rep	0 hrs	0 days	Fri 07/08/1	Fri 07/08/1	203		$0.00
205	MILESTONE - Closeout Report & Invoicing	2 hrs	0.25 days	Fri 07/08/1	Fri 07/08/1	192	Proj Mgmt	$370.00

Figure 119: Revised Schedule ... Part 2 – Implementation (Bottom Half).

ID	Task Name	Work	Duration	Start	Finish	Predecessors	Resource Names	Cost
206	PART 3 - FAT - (Functional) Factory Acceptance Test	409.5 hrs	15 days	Fri 07/08/1	Fri 28/08/1	102		$62,425.00
207	Part 3 - System Media	9.5 hrs	1.19 day	Fri 07/08/15	Mon 10/08/15			$1,747.50
208	Media Preparation	8 hrs	1 day	Fri 07/08/1	Mon 10/08/1		Sys Lead Eng	$1,560.00
209	Presentation Preparation	1.5 hrs	0.19 day	Mon 10/08/1	Mon 10/08/1	208	Doco Support	$187.50
210	Part 3 - Commencement Meeting - With Cust Rep Representatives	2 hrs	0.06 days	Mon 10/08/15	Mon 10/08/1	207		$327.50
211	Review Participation	0.5 hr	0.06 days	Mon 10/08/1	Mon 10/08/1		Test Eng	$62.50
212	Review Participation	0.5 hr	0.06 day	Mon 10/08/1	Mon 10/08/1	211SS	Test Lead Eng	$75.00
213	Review Participation	0.5 hr	0.06 days	Mon 10/08/1	Mon 10/08/1	211SS	Sys Lead Eng	$97.50
214	Review Participation	0.5 hr	0.06 days	Mon 10/08/1	Mon 10/08/1	211SS	Proj Mgmt	$92.50
215	Part 3 - Conduct FAT	56 hrs	4 days	Mon 10/08/15	Fri 14/08/1	210		$8,760.00
216	Conduct Test	24 hrs	3 days	Mon 10/08/1	Thu 13/08/1		Test Eng	$3,000.00
217	Support	8 hrs	1 day	Thu 13/08/1	Thu 13/08/1	216SS	Sys Lead Eng	$1,560.00
218	Support	24 hrs	3 days	Mon 10/08/1	Thu 13/08/1	216SS	Sw Lead Eng	$4,200.00
219	Part 3 - Review Meeting - Test Completion Review	10 hrs	0.5 days	Fri 14/08/15	Mon 17/08/15	215		$1,680.00
220	Review Participation	2 hrs	0.25 days	Fri 14/08/1	Fri 14/08/1		Test Eng	$250.00
221	Review Participation	2 hrs	0.25 days	Fri 14/08/1	Fri 14/08/1	220SS	Test Lead Eng	$300.00
222	Review Participation	2 hrs	0.25 days	Fri 14/08/1	Fri 14/08/1	220SS	Sys Lead Eng	$390.00
223	Review Participation	2 hrs	0.25 days	Fri 14/08/1	Fri 14/08/1	220SS	Proj Mgmt	$370.00
224	Minute Review Meeting	2 hrs	0.25 days	Mon 17/08/1	Mon 17/08/1	220	Proj Mgmt	$370.00
225	DELIVERABLE - Send Minutes to Cust Rep	0 hrs	0 days	Mon 17/08/1	Mon 17/08/1	224		$0.00
226	Part 3 - Write Test Report / Quality Evidence	50 hrs	5.25 days	Mon 17/08/1	Mon 24/08/15	224		$6,940.00
227	Write Report	24 hrs	3 days	Mon 17/08/1	Thu 20/08/1		Test Eng	$3,000.00
228	Support	4 hrs	0.5 days	Mon 17/08/1	Mon 17/08/1	227SS	Sw Lead Eng	$700.00
229	Support	4 hrs	0.5 days	Mon 17/08/1	Mon 17/08/1	227SS	Sys Lead Eng	$780.00
230	Internal Review Report	4 hrs	0.5 days	Thu 20/08/1	Thu 20/08/1	227	Test Lead Eng	$600.00
231	Document Presentation Preparation	2 hrs	0.25 days	Thu 20/08/1	Thu 20/08/1	230	Doco Support	$250.00
232	DELIVERABLE - Send Factory Acceptance Test Report to Cust Rep	0 hrs	0 days	Thu 20/08/1	Thu 20/08/1	231		$0.00
233	Update Acceptance Test Procedures	8 hrs	1 day	Fri 21/08/1	Fri 21/08/1	224,232	Test Eng	$1,000.00
234	Internal Review Report	2 hrs	0.25 days	Mon 24/08/1	Mon 24/08/1	233	Test Lead Eng	$300.00
235	Document Presentation Preparation	1 hr	0.13 days	Mon 24/08/1	Mon 24/08/1	234	Doco Support	$125.00
236	Review - Doco	1 hr	0.13 days	Mon 24/08/1	Mon 24/08/1	235	Proj Mgmt	$185.00
237	DELIVERABLE - Send Updated Acceptance Test Procedures to C	0 hrs	0 days	Mon 24/08/1	Mon 24/08/1	236		$0.00
238	MILESTONE - Closeout Report & Invoicing	2 hrs	0.25 days	Mon 24/08/1	Mon 24/08/1	226	Proj Mgmt	$370.00
239	Part 3 - Defect Repair	280 hrs	6.25 days	Thu 20/08/15	Fri 28/08/1	226SS+3 days		$42,600.00
240	Sw Eng Lead	40 hrs	5 days	Thu 20/08/1	Thu 28/08/1		Sw Lead Eng	$7,000.00
241	Sw Eng 1	40 hrs	5 days	Thu 20/08/1	Thu 27/08/1	240SS	Sw Eng 1	$6,000.00
242	Sw Eng 2	40 hrs	5 days	Thu 20/08/1	Thu 27/08/1	240SS	Sw Eng 2	$5,600.00
243	Sw Eng 3	40 hrs	5 days	Thu 20/08/1	Thu 27/08/1	240SS	Sw Eng 3	$5,200.00
244	Sys Lead Eng	40 hrs	5 days	Thu 20/08/1	Thu 27/08/1	240SS	Sys Lead Eng	$7,800.00
245	Test Lead Eng	40 hrs	5 days	Thu 20/08/1	Thu 27/08/1	240SS	Test Lead Eng	$6,000.00
246	Test Eng	40 hrs	5 days	Thu 20/08/1	Thu 27/08/1	240SS	Test Eng	$5,000.00

Figure 120: Revised Schedule ... Part 3 – Functional Acceptance Test (FAT).

ID	Task Name	Work	Duration	Start	Finish	Predecessors	Resource Names	Cost
247	PART 4 - SAT - (Satisfaction) Site Acceptance Test	405.5 hrs	36.19 days	Fri 28/08/15	Fri 16/10/15	206		$86,207.50
248	Part 4 - Media Shipped To Site	25 hrs	3.13 days	Fri 28/08/15	Wed 02/09/15			$3,775.00
249	System Media	13 hrs	1.63 days	Fri 28/08/15	Tue 01/09/15			$2,245.00
250	Media Preparation	8 hrs	1 day	Fri 28/08/1	Mon 31/08/1		Sys Lead Eng	$1,560.00
251	Presentation Preparation & Packaging	3 hrs	0.38 days	Mon 31/08/1	Mon 31/08/1	250	Doco Support	$375.00
252	DELIVERABLE - Shipment Coordination / Export Licensing	2 hrs	0.25 days	Mon 31/08/1	Tue 01/09/1	251	Logistics Mgr	$310.00
253	End User Documentation	12 hrs	1.5 days	Tue 01/09/15	Wed 02/09/15	249		$1,530.00
254	Doco Preparation	8 hrs	1 day	Tue 01/09/1	Wed 02/09/1		Doco Support	$1,000.00
255	Presentation Preparation & Packaging	3 hrs	0.38 days	Wed 02/09/1	Wed 02/09/1	254	Doco Support	$375.00
256	DELIVERABLE - Shipment Coordination	1 hr	0.13 days	Wed 02/09/1	Wed 02/09/1	255	Logistics Mgr	$155.00
257	Part 4 - Hardware Shipped To Site	34 hrs	12 days	Fri 28/08/15	Tue 15/09/15			$7,850.00
258	Packaging Materials	0 hrs	0 days	Fri 28/08/1	Fri 28/08/1			$250.00
259	Package for Shipping	18 hrs	1 day	Fri 28/08/15	Mon 31/08/15	258		$2,870.00
260	Package	8 hrs	1 day	Fri 28/08/1	Mon 31/08/1		Sys Lead Eng	$1,560.00
261	Assist	8 hrs	1 day	Fri 28/08/1	Mon 31/08/1	260SS	Test Eng	$1,000.00
262	Assist	2 hrs	2 hrs	Mon 31/08/1	Mon 31/08/1	260SS	Logistics Mgr	$310.00
263	Shipment Coordination / Export Licensing	8 hrs	1 day	Fri 28/08/1	Mon 31/08/1	259SS	Logistics Mgr	$1,240.00
264	Hardware Dispatch	8 hrs	1 day	Mon 31/08/1	Tue 01/09/1	263,259	Logistics Mgr	$1,240.00
265	DELIVERABLE - Shipment Of Hardware	0 hrs	10 days	Tue 01/09/1	Tue 15/09/1	264		$2,250.00
266	Part 4 - Travel To Site	32 hrs	12 days	Tue 15/09/15	Thu 01/10/15			$20,680.00
267	TRAVEL DATE	0 hrs	0 days	Thu 24/09/1	Thu 24/09/1	11		$0.00
268	Travel Coordination	8 hrs	1 day	Tue 15/09/1	Wed 16/09/1	267SS-7 days	Logistics Mgr	$1,240.00
269	Travel To Site (Sys Eng)	8 hrs	1 day	Thu 24/09/1	Thu 01/10/1	267SS	Sys Lead Eng	$1,560.00
270	Travel To Site (Sw Eng)	8 hrs	1 day	Thu 24/09/1	Thu 01/10/1	267SS	Sw Lead Eng	$1,400.00
271	Travel To Site (Mgmt)	8 hrs	1 day	Mon 28/09/1	Mon 28/09/1	267SS	Proj Mgmt	$1,480.00
272	Travel to & from site Cost (Sys Eng)	0 hrs	0 days	Thu 24/09/1	Thu 24/09/1	267SS		$2,000.00
273	Travel to & from site Cost (Sw Eng)	0 hrs	0 days	Thu 24/09/1	Thu 24/09/1	267SS		$2,000.00
274	Travel to & from site Cost (Mgmt)	0 hrs	0 days	Thu 24/09/1	Thu 24/09/1	267SS		$2,000.00
275	Accommodation & Allowance Cost (Sys Eng)	0 hrs	0 days	Thu 24/09/1	Thu 24/09/1	267SS		$3,000.00
276	Accommodation & Allowance Cost (Sw Eng)	0 hrs	0 days	Thu 24/09/1	Thu 24/09/1	267SS		$3,000.00
277	Accommodation & Allowance Cost (Mgmt)	0 hrs	0 days	Thu 24/09/1	Thu 24/09/1	267SS		$3,000.00
278	Part 4 - System Integration	52 hrs	3.5 days	Thu 01/10/15	Tue 06/10/15	257,248,266		$9,660.00
279	Hardware Setup	16 hrs	1 day	Thu 01/10/15	Fri 02/10/15	257		$2,960.00
280	Handling Hardware Delivery	8 hrs	1 day	Thu 01/10/1	Fri 02/10/1		Sys Lead Eng	$1,560.00
281	Assist	8 hrs	1 day	Thu 01/10/1	Fri 02/10/1	280FF	Sw Lead Eng	$1,400.00
282	Application Setup	36 hrs	2.5 days	Fri 02/10/15	Tue 06/10/15	279		$6,700.00
283	Web-Server -- Setup	4 hrs	0.5 days	Fri 02/10/1	Fri 02/10/1		Sys Lead Eng	$780.00
284	Database-Server -- Setup	4 hrs	0.5 days	Mon 05/10/1	Mon 05/10/1	283	Sys Lead Eng	$780.00
285	Setup Verification & Sanity Testing	4 hrs	0.5 days	Mon 05/10/1	Mon 05/10/1	284	Sys Lead Eng	$780.00
286	BUFFER - Time & Cost	8 hrs	1 day	Tue 06/10/1	Tue 06/10/1	285	Sys Lead Eng	$1,560.00
287	Assist	16 hrs	2 days	Mon 05/10/1	Tue 06/10/1	286FF	Sw Lead Eng	$2,800.00
288	Part 4 - Onsite Test	81.5 hrs	3.56 days	Wed 07/10/15	Sat 10/10/15	278		$14,317.50
289	Commencement Meeting - With Cust Rep Representatives	1.5 hrs	0.06 days	Wed 07/10/1	Wed 07/10/15			$277.50
290	Review Participation	0.5 hrs	0.06 days	Wed 07/10/1	Wed 07/10/1		Sys Lead Eng	$97.50
291	Review Participation	0.5 hrs	0.06 days	Wed 07/10/1	Wed 07/10/1	290SS	Sw Lead Eng	$87.50
292	Review Participation	0.5 hrs	0.06 days	Wed 07/10/1	Wed 07/10/1	290SS	Proj Mgmt	$92.50
293	Conduct SAT	56 hrs	3 days	Wed 07/10/15	Sat 10/10/15	289		$10,360.00
294	Conduct Test	24 hrs	3 days	Wed 07/10/1	Sat 10/10/1		Sys Lead Eng	$4,680.00
295	Assist	24 hrs	3 days	Wed 07/10/1	Sat 10/10/1	294SS	Sw Lead Eng	$4,200.00
296	Support & Record Taking	8 hrs	1 day	Wed 07/10/1	Thu 08/10/1	294SS	Proj Mgmt	$1,480.00
297	Remote Support	16 hrs	1 day	Wed 07/10/15	Thu 08/10/15	293SS		$2,200.00
298	Eng Support	8 hrs	1 day	Wed 07/10/1	Thu 08/10/1		Sw Eng 1	$1,200.00
299	Test Support	8 hrs	1 day	Wed 07/10/1	Thu 08/10/1	298SS	Test Eng	$1,000.00
300	Review Meeting - Test Completion Review	6 hrs	0.25 days	Sat 10/10/1	Sat 10/10/1	293		$1,110.00
301	Review Participation	2 hrs	0.25 days	Sat 10/10/1	Sat 10/10/1		Sys Lead Eng	$390.00
302	Review Participation	2 hrs	0.25 days	Sat 10/10/1	Sat 10/10/1	301SS	Sw Lead Eng	$350.00
303	Review Participation	2 hrs	0.25 days	Sat 10/10/1	Sat 10/10/1	301SS	Proj Mgmt	$370.00
304	Minute Review Meeting	2 hrs	0.25 days	Sat 10/10/1	Sat 10/10/1	300	Proj Mgmt	$370.00
305	DELIVERABLE - Send Minutes to Cust Rep	0 hrs	0 days	Sat 10/10/1	Sat 10/10/1	304		$0.00
306	Part 4 - Travel From Site	24 hrs	1 day	Sat 10/10/15	Mon 12/10/15	288		$4,440.00
307	Travel From Site (Sys Eng)	8 hrs	1 day	Sat 10/10/1	Mon 12/10/1		Sys Lead Eng	$1,560.00
308	Travel From Site (Sw Eng)	8 hrs	1 day	Sat 10/10/1	Mon 12/10/1	307SS	Sw Lead Eng	$1,400.00
309	Travel From Site (Mgmt)	8 hrs	1 day	Sat 10/10/1	Mon 12/10/1	307SS	Proj Mgmt	$1,480.00
310	Part 4 - Test Report	35 hrs	3.88 days	Fri 02/10/1	Fri 16/10/1	306		$6,415.00
311	Write Test Report / Quality Evidence	28 hrs	3 days	Mon 12/10/15	Thu 15/10/15			$5,380.00
312	Write Report	24 hrs	3 days	Mon 12/10/1	Thu 15/10/1		Sys Lead Eng	$4,680.00
313	Support	4 hrs	0.5 days	Mon 12/10/1	Tue 13/10/1	312SS	Sw Lead Eng	$700.00
314	Internal Review Report	4 hrs	0.5 days	Thu 15/10/1	Fri 16/10/1	311	Test Lead Eng	$600.00
315	Document Presentation Preparation	2 hrs	0.25 days	Fri 16/10/1	Fri 16/10/1	314	Doco Support	$250.00
316	Review - Doco	1 hr	0.13 days	Fri 16/10/1	Fri 16/10/1	315	Proj Mgmt	$185.00
317	DELIVERABLE - Send Site Acceptance Test Report to Cust Rep	0 hrs	0 days	Fri 16/10/1	Fri 16/10/1	316		$0.00
318	MILESTONE - Closeout Report & Invoicing	2 hrs	0.25 days	Fri 16/10/1	Fri 16/10/1	310	Proj Mgmt	$370.00
319	Part 4 - Defect Repair	120 hrs	3.75 days	Mon 12/10/15	Fri 16/10/1	310SS		$18,700.00
320	Sw Eng Lead	20 hrs	2.5 days	Mon 12/10/1	Fri 16/10/1		Sw Lead Eng	$3,500.00
321	Sw Eng 1	20 hrs	2.5 days	Mon 12/10/1	Thu 15/10/1	320SS	Sw Eng 1	$3,000.00
322	Sw Eng 2	20 hrs	2.5 days	Mon 12/10/1	Thu 15/10/1	320SS	Sw Eng 2	$2,800.00
323	Sys Lead Eng	20 hrs	2.5 days	Mon 12/10/1	Thu 15/10/1	320SS	Sys Lead Eng	$3,900.00
324	Test Lead Eng	20 hrs	2.5 days	Mon 12/10/1	Thu 15/10/1	320SS	Test Lead Eng	$3,000.00
325	Test Eng	20 hrs	2.5 days	Mon 12/10/1	Thu 15/10/1	320SS	Test Eng	$2,500.00

Figure 121: Revised Schedule ... Part 4 – Satisfaction Acceptance Test (SAT).

ID	Task Name	Work	Duration	Start	Finish	Predecessors	Resource Names	Cost
326	PART 5 - UAT - (Usability) User Acceptance Test	231.5 hrs	8.25 days	Fri 16/10/1!	Wed 28/10/1!	247		$53,707.5(
327	Part 5 - Media Shipped To Site	9 hrs	1.13 days	Fri 16/10/1!	Mon 19/10/1!			$1,465.0(
328	Media Preparation	4 hrs	0.5 days	Fri 16/10/1!	Mon 19/10/1		Sys Lead Eng	$780.0(
329	Presentation Preparation & Packaging	3 hrs	0.38 days	Mon 19/10/1	Mon 19/10/1	328	Doco Support	$375.0(
330	DELIVERABLE - Shipment Coordination / Export Licensing	2 hrs	0.25 days	Mon 19/10/1	Mon 19/10/1	329	Logistics Mgr	$310.0(
331	Part 5 - Travel To Site	32 hrs	3.13 days	Fri 16/10/1!	Wed 21/10/1!			$20,680.0(
332	TRAVEL DATE	0 hrs	0 days	Tue 20/10/1	Tue 20/10/1	327FF+1 day		$0.0(
333	Travel Coordination	8 hrs	1 day	Fri 16/10/1!	Mon 19/10/1	332SS-7 days	Logistics Mgr	$1,240.0(
334	Travel To Site (Sys Eng)	8 hrs	1 day	Tue 20/10/1	Wed 21/10/1	332SS	Sw Lead Eng	$1,400.0(
335	Travel To Site (Sw Eng)	8 hrs	1 day	Tue 20/10/1	Wed 21/10/1	332SS	Sys Lead Eng	$1,560.0(
336	Travel To Site (Mgmt)	8 hrs	1 day	Tue 20/10/1	Wed 21/10/1	332SS	Proj Mgmt	$1,480.0(
337	Travel to & from site Cost (Sys Eng)	0 hrs	0 days	Tue 20/10/1!	Tue 20/10/1!	332SS		$2,000.0(
338	Travel to & from site Cost (Sw Eng)	0 hrs	0 days	Tue 20/10/1!	Tue 20/10/1!	332SS		$2,000.0(
339	Travel to & from site Cost (Mgmt)	0 hrs	0 days	Tue 20/10/1!	Tue 20/10/1!	332SS		$2,000.0(
340	Accommodation & Allowance Cost (Sys Eng)	0 hrs	0 days	Tue 20/10/1!	Tue 20/10/1!	332SS		$3,000.0(
341	Accommodation & Allowance Cost (Sw Eng)	0 hrs	0 days	Tue 20/10/1!	Tue 20/10/1!	332SS		$3,000.0(
342	Accommodation & Allowance Cost (Mgmt)	0 hrs	0 days	Tue 20/10/1!	Tue 20/10/1!	332SS		$3,000.0(
343	Part 5 - System Readiness Check	18 hrs	1.25 days	Wed 21/10/1!	Fri 23/10/1!	331		$3,350.0(
344	Application Setup Confirmation	18 hrs	1.25 days	Wed 21/10/1!	Fri 23/10/1!			$3,350.0(
345	Web-Server – Setup	2 hrs	0.25 days	Wed 21/10/1!	Thu 22/10/1		Sys Lead Eng	$390.0(
346	Database-Server – Setup	2 hrs	0.25 days	Thu 22/10/1	Thu 22/10/1	345	Sys Lead Eng	$390.0(
347	Setup Verification & Sanity Testing	2 hrs	0.25 days	Thu 22/10/1	Thu 22/10/1	346	Sys Lead Eng	$390.0(
348	BUFFER - Time & Cost	4 hrs	0.5 days	Thu 22/10/1	Fri 23/10/1	347	Sys Lead Eng	$780.0(
349	Assist	8 hrs	1 day	Thu 22/10/1	Fri 23/10/1	348FF	Sw Lead Eng	$1,400.0(
350	Part 5 - Onsite Test	49.5 hrs	1.56 days	Fri 23/10/1!	Mon 26/10/1!	343		$8,397.5(
351	Commencement Meeting - With Cust Rep Representatives	1.5 hrs	0.06 days	Fri 23/10/1!	Fri 23/10/1!			$277.5(
352	Review Participation	0.5 hrs	0.06 days	Fri 23/10/1	Fri 23/10/1		Sys Lead Eng	$97.5(
353	Review Participation	0.5 hrs	0.06 days	Fri 23/10/1	Fri 23/10/1	352SS	Sw Lead Eng	$87.5(
354	Review Participation	0.5 hrs	0.06 days	Fri 23/10/1	Fri 23/10/1	352SS	Proj Mgmt	$92.5(
355	Conduct UAT	24 hrs	1 day	Fri 23/10/1!	Mon 26/10/1!	351		$4,440.0(
356	Conduct Test	8 hrs	1 day	Fri 23/10/1	Mon 26/10/1		Sys Lead Eng	$1,560.0(
357	Support	8 hrs	1 day	Fri 23/10/1	Mon 26/10/1	356SS	Sw Lead Eng	$1,400.0(
358	Support & Record Taking	8 hrs	1 day	Fri 23/10/1	Mon 26/10/1	356SS	Proj Mgmt	$1,480.0(
359	Remote Support	16 hrs	1 day	Fri 23/10/1!	Mon 26/10/1!	355SS		$2,200.0(
360	Eng Support	8 hrs	1 day	Fri 23/10/1	Mon 26/10/1		Sw Eng 1	$1,200.0(
361	Test Support	8 hrs	1 day	Fri 23/10/1	Mon 26/10/1	360SS	Test Eng	$1,000.0(
362	Review Meeting - Test Completion Review	6 hrs	0.25 days	Mon 26/10/15	Mon 26/10/1!	355		$1,110.0(
363	Review Participation	2 hrs	0.25 days	Mon 26/10/1	Mon 26/10/1		Sys Lead Eng	$390.0(
364	Review Participation	2 hrs	0.25 days	Mon 26/10/1	Mon 26/10/1	363SS	Sw Lead Eng	$350.0(
365	Review Participation	2 hrs	0.25 days	Mon 26/10/1	Mon 26/10/1	363SS	Proj Mgmt	$370.0(
366	Minute Review Meeting	2 hrs	0.25 days	Mon 26/10/1	Mon 26/10/1	362	Proj Mgmt	$370.0(
367	DELIVERABLE - Send Minutes to Cust Rep	0 hrs	0 days	Mon 26/10/1	Mon 26/10/1	366		$0.0(
368	Part 5 - Test Report	27 hrs	2.88 days	Mon 26/10/15	Wed 28/10/1!	350SS+1 day		$4,855.0(
369	Write Test Report / Quality Evidence	20 hrs	2 days	Mon 26/10/15	Wed 28/10/1			$3,820.0(
370	Write Report	16 hrs	2 days	Mon 26/10/1	Wed 28/10/1		Sys Lead Eng	$3,120.0(
371	Support	4 hrs	0.5 days	Mon 26/10/1	Mon 26/10/1	370SS	Sw Lead Eng	$700.0(
372	Internal Review Report	4 hrs	0.5 days	Wed 28/10/1	Wed 28/10/1	369	Test Lead Eng	$600.0(
373	Document Presentation Preparation	2 hrs	0.25 days	Wed 28/10/1	Wed 28/10/1	372	Doco Support	$250.0(
374	Review - Doco	1 hr	0.13 days	Wed 28/10/1	Wed 28/10/1	373	Proj Mgmt	$185.0(
375	DELIVERABLE - Send User Acceptance Test Report to Cust Rep	0 hrs	0 days	Wed 28/10/1	Wed 28/10/1	374		$0.0(
376	Part 5 - Defect Repair	96 hrs	2 days	Mon 26/10/15	Wed 28/10/1!	355SS+1 day		$14,960.0(
377	Sw Eng Lead	16 hrs	2 days	Mon 26/10/1	Wed 28/10/1		Sw Lead Eng	$2,800.0(
378	Sw Eng 1	16 hrs	2 days	Mon 26/10/1	Wed 28/10/1	377SS	Sw Eng 1	$2,400.0(
379	Sw Eng 2	16 hrs	2 days	Mon 26/10/1	Wed 28/10/1	377SS	Sw Eng 2	$2,240.0(
380	Sys Lead Eng	16 hrs	2 days	Mon 26/10/1	Wed 28/10/1	377SS	Sys Lead Eng	$3,120.0(
381	Test Lead Eng	16 hrs	2 days	Mon 26/10/1	Wed 28/10/1	377SS	Test Lead Eng	$2,400.0(
382	Test Eng	16 hrs	2 days	Mon 26/10/1	Wed 28/10/1	377SS	Test Eng	$2,000.0(

Figure 122: Revised Schedule ... Part 5 – Usability Acceptance Test (UAT).

"Gee, how the hours and dollars keep adding up; ... that block of tasks there is the cost of a brand new family car, ... that chunk could pay for the renovation of the bathrooms & kitchen, ... there's a holiday trip."

Yes, that is **a fair amount of money (capital)** that both the customer & performing organizations **will be investing in the conduct of this proposed project**. ... So will it really provide a **worthwhile Return On Investment** for both of these organizations?

ID	Task Name	Work	Duration	Start	Finish	Predecessors	Resource Names	Cost
383	PART 6 - Go Live	106 hrs	13.63 days	Wed 28/10/15	Fri 13/11/15			$19,410.00
384	Part 6 - Media Shipped To Site	9 hrs	1.13 days	Wed 28/10/15	Thu 29/10/15	376		$1,565.00
385	Media Preparation	4 hr	0.5 days	Wed 28/10/1	Wed 28/10/1		Sys Lead Eng	$780.00
386	Presentation Preparation & Packaging	3 hrs	0.38 days	Wed 28/10/1	Wed 28/10/1	385	Doco Support	$375.00
387	DELIVERABLE - Shipment Coordination / Export Licensing	2 hrs	0.25 days	Thu 29/10/1	Thu 29/10/1	386	Logistics Mgr	$310.00
388	Courier Media to Site	0 hrs	0 days	Thu 29/10/1	Thu 29/10/1	387		$100.00
389	Part 6 - System Readiness Check	12 hrs	1 day	Thu 29/10/15	Fri 30/10/15	384		$2,260.00
390	Application Setup Confirmation	12 hrs	1 day	Thu 29/10/15	Fri 30/10/15			$2,260.00
391	Web-Server – Setup	1 hr	0.13 days	Thu 29/10/1	Thu 28/10/1		Sys Lead Eng	$195.00
392	Database-Server – Setup	1 hr	0.13 days	Thu 29/10/1	Thu 29/10/1	391	Sys Lead Eng	$195.00
393	Go Live Switchover - Setup	1 hr	0.13 days	Thu 29/10/1	Thu 29/10/1	392	Sys Lead Eng	$195.00
394	Mini SAT - Setup Verification & Sanity Testing	2 hrs	0.25 days	Thu 29/10/1	Thu 29/10/1	393	Sys Lead Eng	$390.00
395	BUFFER - Time & Cost	3 hrs	0.38 days	Thu 29/10/1	Fri 30/10/1	394	Sys Lead Eng	$585.00
396	Assist	4 hrs	0.5 days	Fri 30/10/1	Fri 30/10/1	395FF	Sw Lead Eng	$700.00
397	Part 6 - Go Live	32 hrs	3.5 days	Thu 29/10/15	Tue 03/11/15			$5,920.00
398	Commencement Meeting - With Cust Rep Representatives	3 hrs	0.13 days	Thu 29/10/15	Fri 30/10/15			$555.00
399	Review Participation	1 hr	0.13 days	Thu 29/10/1	Fri 30/10/1	402SF	Sys Lead Eng	$195.00
400	Review Participation	1 hr	0.13 days	Thu 29/10/1	Thu 29/10/1	399SS	Sw Lead Eng	$175.00
401	Review Participation	1 hr	0.13 days	Thu 29/10/1	Thu 29/10/1	399SS	Proj Mgmt	$185.00
402	Conduct Go Live activities	24 hrs	1 day	Fri 30/10/1	Sat 31/10/15			$4,440.00
403	Conduct	8 hrs	1 day	Fri 30/10/1	Sat 31/10/1	406SF	Sys Lead Eng	$1,560.00
404	Support	8 hrs	1 day	Fri 30/10/1	Sat 31/10/1	403FF	Sw Lead Eng	$1,400.00
405	Support	8 hrs	1 day	Fri 30/10/1	Sat 31/10/1	403FF	Proj Mgmt	$1,480.00
406	GO LIVE DATE	0 hrs	0 days	Sat 31/10/1	16			$0.00
407	Review Meeting - Completion Review	3 hrs	0.13 days	Tue 03/11/15	Tue 03/11/15	406FS+2 days		$555.00
408	Review Participation	1 hr	0.13 days	Tue 03/11/1	Tue 03/11/1		Sys Lead Eng	$195.00
409	Review Participation	1 hr	0.13 days	Tue 03/11/1	Tue 03/11/1	408SS	Sw Lead Eng	$175.00
410	Review Participation	1 hr	0.13 days	Tue 03/11/1	Tue 03/11/1	408SS	Proj Mgmt	$185.00
411	Minute Review Meeting	2 hrs	0.25 days	Tue 03/11/1	Tue 03/11/1	407	Proj Mgmt	$370.00
412	DELIVERABLE - Send Minutes to Cust Rep	0 hrs	0 days	Tue 03/11/1	Tue 03/11/1	411		$0.00
413	Part 6 - Travel From Site	24 hrs	1 day	Tue 03/11/15	Wed 04/11/15	397		$4,440.00
414	Travel From Site (Sys Eng)	8 hrs	1 day	Tue 03/11/1	Wed 04/11/1		Sys Lead Eng	$1,560.00
415	Travel From Site (Sw Eng)	8 hrs	1 day	Tue 03/11/1	Wed 04/11/1	414SS	Sw Lead Eng	$1,400.00
416	Travel From Site (Mgmt)	8 hrs	1 day	Tue 03/11/1	Wed 04/11/1	414SS	Proj Mgmt	$1,480.00
417	Part 6 - Go Live Report	27 hrs	2.88 days	Tue 10/11/15	Fri 13/11/15	413		$4,855.00
418	Write Go Live Report	20 hrs	2 days	Tue 10/11/15	Thu 12/11/15			$3,820.00
419	Write Report	16 hrs	2 days	Tue 10/11/1	Thu 12/11/1		Sys Lead Eng	$3,120.00
420	Support	4 hrs	0.5 days	Tue 10/11/1	Wed 11/11/1	419SS	Sw Lead Eng	$700.00
421	Internal Review Report	4 hrs	0.5 days	Thu 12/11/1	Fri 13/11/1	418	Test Lead Eng	$600.00
422	Document Presentation Preparation	2 hrs	0.25 days	Fri 13/11/1	Fri 13/11/1	421	Doco Support	$250.00
423	Review - Doco	1 hr	0.13 days	Fri 13/11/1	Fri 13/11/1	422	Proj Mgmt	$185.00
424	DELIVERABLE - Go Live Report to Cust Rep	0 hrs	0 days	Fri 13/11/1	Fri 13/11/1	423		$0.00
425	MILESTONE - Closeout Report & Invoicing	2 hrs	0.25 days	Fri 13/11/1	Fri 13/11/1	417	Proj Mgmt	$370.00

Figure 123: Revised Schedule ... Part 6 – Go Live.

ID	Task Name	Work	Duration	Start	Finish	Predecessors	Resource Names	Cost
426	PART 7 - Warranty Support	120 hrs	30 days	Sat 31/10/1	Thu 10/12/15	406		$19,400.00
427	Part 7 - Warrantee Support Period	0 hrs	30 days	Sat 31/10/1	Thu 10/12/1			$0.00
428	Part 7 - Technical Support	120 hrs	5.38 days	Sat 31/10/1	Fri 06/11/15	427SS		$19,400.00
429	Sys Lead Eng	20 hrs	2.5 days	Mon 02/11/1	Fri 06/11/1		Sys Lead Eng	$3,900.00
430	Sw Eng Lead	20 hrs	2.5 days	Sat 31/10/1	Wed 04/11/1		Sw Lead Eng	$3,500.00
431	Sw Eng 1	20 hrs	2.5 days	Sat 31/10/1	Tue 03/11/1		Sw Eng 1	$3,000.00
432	Sw Eng 2	20 hrs	2.5 days	Sat 31/10/1	Tue 03/11/1		Sw Eng 2	$2,800.00
433	Test Eng	20 hrs	2.5 days	Sat 31/10/1	Tue 03/11/1		Test Eng	$2,500.00
434	PM - liaison	20 hrs	2.5 days	Sat 31/10/1	Wed 04/11/1		Proj Mgmt	$3,700.00
435	PART 8 - Project Closure	76.5 hrs	4.06 days	Fri 11/12/1	Thu 17/12/15	426		$13,537.50
436	Project Closure Activities	48 hrs	2 days	Fri 11/12/1	Mon 14/12/15			$8,520.00
437	Project Mgmt	16 hrs	2 days	Fri 11/12/1	Mon 14/12/1		Proj Mgmt	$2,960.00
438	Sw Team	16 hrs	2 days	Fri 11/12/1	Mon 14/12/1	437SS	Sw Lead Eng	$2,800.00
439	Test Team	8 hrs	1 day	Fri 11/12/1	Fri 11/12/1	437SS	Test Lead Eng	$1,200.00
440	Sys info	8 hrs	1 day	Fri 11/12/1	Fri 11/12/1	437SS	Sys Lead Eng	$1,560.00
441	Write Report	21 hrs	1.13 days	Tue 15/12/15	Wed 16/12/15	436		$3,630.00
442	Internal Review	9 hrs	0.13 days	Tue 15/12/15	Tue 15/12/15			$1,370.00
443	Meeting Participation	1 hr	0.13 days	Tue 15/12/1	Tue 15/12/1		Proj Mgmt	$185.00
444	Meeting Participation	1 hr	0.13 days	Tue 15/12/1	Tue 15/12/1	443SS	Sys Lead Eng	$195.00
445	Meeting Participation	1 hr	0.13 days	Tue 15/12/1	Tue 15/12/1	443SS	Sw Lead Eng	$175.00
446	Meeting Participation	1 hr	0.13 days	Tue 15/12/1	Tue 15/12/1	443SS	Sw Eng 1	$150.00
447	Meeting Participation	1 hr	0.13 days	Tue 15/12/1	Tue 15/12/1	443SS	Sw Eng 2	$140.00
448	Meeting Participation	1 hr	0.13 days	Tue 15/12/1	Tue 15/12/1	443SS	Sw Eng 3	$130.00
449	Meeting Participation	1 hr	0.13 days	Tue 15/12/1	Tue 15/12/1	443SS	Sw Eng 4	$120.00
450	Meeting Participation	1 hr	0.13 days	Tue 15/12/1	Tue 15/12/1	443SS	Test Lead Eng	$150.00
451	Meeting Participation	1 hr	0.13 days	Tue 15/12/1	Tue 15/12/1	443SS	Test Eng	$125.00
452	Write Report	8 hrs	1 day	Tue 15/12/1	Wed 16/12/1	442	Proj Mgmt	$1,480.00
453	Support	4 hrs	0.5 days	Tue 15/12/1	Tue 15/12/1	452SS	Sys Lead Eng	$780.00
454	Internal Review Report	4 hrs	0.5 days	Wed 16/12/1	Wed 16/12/1	441	Management Supp	$800.00
455	Document Presentation Preparation	1 hr	0.13 days	Wed 16/12/1	Wed 16/12/1	454	Doco Support	$125.00
456	DELIVERABLE - Project Close Out Report	0.5 hr	0.06 days	Wed 16/12/1	Wed 16/12/1	455	Proj Mgmt	$92.50
457	MILESTONE - Closeout Report & Invoicing	2 hrs	0.25 days	Wed 16/12/1	Thu 17/12/1	456	Proj Mgmt	$370.00
458	Stage 1 - Management Support (5% Total Hours)	275 hrs	34.38 days	Mon 16/02/1	Fri 03/04/1	54SS	Management Supp	$55,000.00
459	Stage 1 - Technical Authority Support (5% Total Hours)	275 hrs	34.38 days	Mon 16/02/1	Fri 03/04/1	54SS	Technical Authority	$55,000.00

Figure 124: Revised Schedule ... Part 7 – Warranty Support and Part 8 – Project Closure.

Integrating the Project Schedule Into The Program Roadmap

With the revised schedule having been re-reviewed, then the Project Manager will proceed with producing the [Cost] calculation spreadsheet; while the Program Manager takes a copy of that schedule and works this as a timeline representation of a summation schedule into the **Integrated Master Schedule (IMS) | Master Program Schedule (MPS) | Program Roadmap**, … or, various other names used depending on the organization and industry.

The Program Manager will use information extracted from this (proposed) project's schedule to determine:

- How does the project's flow of overarching work activities (with major milestones marked out) compare to the performing organization's other (proposed & actual) projects that are also included in the "roadmap"? **Are there multiple projects vying** for (almost simultaneous) **access to limited resources** such as; scarce equipment (*e.g. the thingamajig analyser*), restrained work-stream facilities (*e.g. testing labs*), specialist work groups (*e.g. circuit layout*), and individuals (*e.g. chief mathematician*)?

- Are there multiple projects whose combined (estimated) **cash outflows for a given period** (monthly | quarterly) **would exceed the performing organization's potential to cover such a financial burden**? When are the **expected cash inflows** to occur?

- What is the ~~resource loading~~ **people loading** across the various projects, are there **shortfalls in availability** (where additional personnel would need to be engaged), are there times when **better arranging could result in optimal utilization**?

> Which brings us to the next rule of "Adaptive & Proactive" SDLC Project Management.
>
> **RULE 52:** No project is an island, entirely isolated to itself. … A project is one of many in a stream of past, present and future; where each affects each other's chances of success.

The Project Schedule's Accumulated Level Of Effort as a S-Curve

Having revised the project's schedule, if the Level of Effort for the "Management Support" and the "Technical Authority Support" activities were spread evenly across the entire duration of the project, and then **the weekly totals for the Level Of Effort for all** of the project's **activities were accumulated then the S-Curve**, in [Figure 125], **would result**.

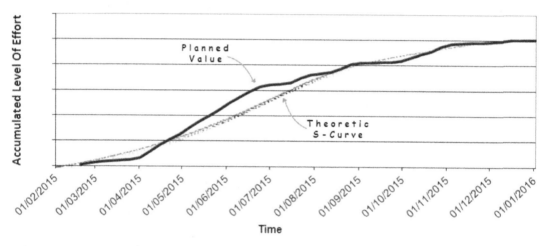

Figure 125: Scenario's project – accumulated Level Of Effort over the project's duration.

Now, **this "internal costs incurred" S-Curve [Cost Baseline] is yet to include the Safety Margin, Management Reserve, and Profit Margin** as these also **need to be included** so as to **determine a "viable" Contracted Price**. But these additions, in difference to the side diagram, probably won't be included as clearly defined drop-ins on top; rather, these will be **incorporated into the "Sell Price" of the Labour Rates and Material Prices charged**. Hence, there is a **need for a dedicated calculation spreadsheet** (aka "Cost Account Authorization") to **produce an overall budget Contract Price**.

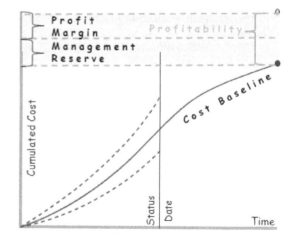

6.3.4. What Will It Cost?

Simplified Cost Formula

There is an old saying that, *"time is money"*, and, *"you break it, you buy it"*, which from a project [Resources] perspective is *"you used it, you pay for what you used"*; essentially, a **"Pay As You Go" philosophy**, where:

$$Costs = (People\ ^{Labour\ Rates} * Time\ ^{Used}) +$$
$$(Resource\ ^{Unit\ Prices} * Amount\ ^{Used}) +$$
$$(Fixed\ Operating\ Costs\ ^{To\ Date})$$

But, what to use for the **marked-up "Sell Price"** for the [People] Labour Rates and the [Resource] Unit Prices, and those Fixed Operating Costs, when there **needs to be included Escalated Costs** (*exchange rates & inflation*), **Cost Contingencies** (*warranty repairs & latent defects*), **Risk Contingencies** (*type of work being conducted*), **Overhead Burden**, **Management Reserve**, and **Profit Margin** … which all have to be incorporated into the [Cost] calculation.

$$People\ ^{Labour\ Rates} = Base\ Labour\ Rates * 1.??$$
$$1.?? = Percentage\ for\ Escalated\ Costs +$$
$$Percentage\ for\ Cost\ Contingencies +$$
$$Percentage\ for\ Risk\ Contingencies +$$
$$Percentage\ for\ Overhead\ Burden +$$
$$Percentage\ for\ Management\ Reserve +$$
$$Percentage\ for\ Profit\ Margin$$

The [People] **Labour Rates will be different depending on the type of work conducted**, where some relatively simple activities have a small Risk Contingency, whereas other activities (such as R&D) have a much greater Risk Contingency. Hence, in the cost calculation spreadsheet, **different [People] will be assigned different categories of work**.

DO NOT allow these Sell Price percentage combinations for Escalated Costs, Cost Contingencies, Risk Contingencies, Overhead Burden, Management Reserve, and Profit Margin **to be known by the Customer's Representative** (or competitors); because, to have this information revealed will greatly **damage** the performing organization's **capability to "profitably" negotiate** with the customer. … And, to enable the undercutting by competitors.

"So, don't accidently go emailing the costing spreadsheet to the customer instead of a separately prepared quotation. … DOPE!!"

Similarly for [Resources], there would be **different Sale Price markups based on the type of materials & services that will be provided**. For example; travel, accommodation, server hardware, software licenses, packaging & freight.

Contract Price – Overall Budget Calculation

Hence, **using the performing organization's "standardized" Cost Account Authorization spreadsheet**, start constructing the budget; as was done previously in [Section 5.4.2].

Labour Categories	Labour Rates	
Description	Resource Code	Base Value
BA - SA	RES1	$ 200.00
Sys Lead Eng	RES2	$ 195.00
Doco Support	RES3	$ 125.00
Proj Mgmt	RES4	$ 185.00
Sw Lead Eng	RES5	$ 175.00
Sw Eng 1	RES6	$ 150.00
Sw Eng 2	RES7	$ 140.00
Sw Eng 3	RES8	$ 130.00
Sw Eng 4	RES9	$ 120.00
Sw Eng 5	RES10	$ 110.00
Logistics Mgr	RES11	$ 155.00
Test Eng	RES12	$ 125.00
Test Lead Eng	RES13	$ 150.00
Technical Authority	RES14	$ 200.00
Management Support	RES15	$ 200.00

Labour SOC		"Bums On" Seat Operational Cost	
Description	Category	Base Value	
Software Engineer	SOC1	$	50.00
IT Engineer	SOC2	$	50.00
Graphics / Web Designer	SOC3	$	50.00
System Engineer	SOC4	$	50.00
Exec Mgmt/Senior Management	SOC5	$	50.00
EA / DOC	SOC6	$	50.00
Researcher	SOC7	$	50.00

Burdens & Margins							
Description	Category	Mgmt	Wnty	Latent	Contgy	OHead	Profit
Fixed Price Contract - Development	L0	10.0%	1.5%	1.5%	5.0%	20.0%	12.5%
Internal Contract – Development	L1	5.0%			1.5%	20.0%	
T&M Contract – Development	L2					20.0%	12.5%
R&D Contract - Development	L3	10.0%			10.0%	20.0%	25.0%
Commercialization	L4	10.0%	1.5%	1.5%	5.0%	20.0%	
Subcontractors (Fixed)	L5	10.0%	1.5%	1.5%	5.0%	5.0%	12.5%
Subcontractors (T&M)	L6	10.0%			10.0%	10.0%	12.5%
Material, Other	L7					5.0%	
Packaging and Freight	L8					5.0%	
Leasing	L9	5.0%			1.5%	5.0%	
S/W Licences	L10					5.0%	

What is 'Ohead' Overhead in the Burdens & Margins? This is to cover the overhead cost of printers & servers, I.T. infrastructure, I.T. support staff, office administration, executive management, staff training, recruitment, operating permits, ... etc.

With the revised schedule, for each of the major parts of the project (*e.g. Part 0 – Project Kick Off, Part 1 – Analysis & Design, Part 2 – Implementation, ... Part 8 – Project Closure*) accumulate the hours | Level Of Effort per [People] working on that particular part.

Then, in the **Cost Account Authorization spreadsheet, in a worksheet for the Work Breakdown Structure enter the work to be done by** each of the involved performing organization's **[People], and the [Resources] to be utilized**. ... As, the **spreadsheet automatically calculates the Sell Price percentage combination markups per entered item**; thereby, **determining the project's accumulated Contract Price** (to be offered).

Work Breakdown Structure						Cost Estimate				Sell
Activity	Description	Cat.	Hours	Rate	SOC	Materials	Other	Travel	Qty	Price
PART 0 - Project Kick off	Tasks Conducted By	L0								$ -
	BA - SA	L0	1.0	RES1	SOC7					$ 398
	Sys Lead Eng	L0	1.0	RES2	SOC4					$ 390
	Doco Support	L0	1.0	RES3	SOC6					$ 279
	Proj Mgmt	L0	5.0	RES4	SOC5					$ 1,872
	Sw Lead Eng	L0	1.0	RES5	SOC1					$ 358
	Sw Eng 1	L0	1.0	RES6	SOC1					$ 319
	Sw Eng 2	L0	1.0	RES7	SOC1					$ 303
	Sw Eng 3	L0	1.0	RES8	SOC1					$ 287
	Sw Eng 4	L0	1.0	RES9	SOC1					$ 271
	Sw Eng 5	L0	1.0	RES10	SOC1					$ 255
	Logistics Mgr	L0	1.0	RES11	SOC7					$ 327
	Test Eng	L0	1.0	RES12	SOC7					$ 279
	Test Lead Eng	L0	1.0	RES13	SOC7					$ 319
	Technical Authority	L0	0.0	RES14	SOC7					$ -
	Management Support	L0	0.0	RES15	SOC5					$ -
		L0								$ -
										$ 5,655

Figure 126: Cost Calculation for ... Part 0 – Project Kick Off.

Work Breakdown Structure						Cost Estimate				Sell
Activity	Description	Cat.	Hours	Rate	SOC	Materials	Other	Travel	Qty	Price
PART 1 - Analysis & Design	Tasks Conducted By	L0								$ -
	BA - SA	L0	36.0	RES1	SOC7					$ 14,337
	Sys Lead Eng	L0	156.0	RES2	SOC4		3,000	2,000	1	$ 68,849
	Doco Support	L0	6.0	RES3	SOC6					$ 1,673
	Proj Mgmt	L0	10.0	RES4	SOC5					$ 3,744
	Sw Lead Eng	L0	0.0	RES5	SOC1					$ -
	Sw Eng 1	L0	0.0	RES6	SOC1					$ -
	Sw Eng 2	L0	0.0	RES7	SOC1					$ -
	Sw Eng 3	L0	0.0	RES8	SOC1					$ -
	Sw Eng 4	L0	0.0	RES9	SOC1					$ -
	Sw Eng 5	L0	0.0	RES10	SOC1					$ -
	Logistics Mgr	L0	8.0	RES11	SOC7					$ 2,613
	Test Eng	L0	0.0	RES12	SOC7					$ -
	Test Lead Eng	L0	0.0	RES13	SOC7					$ -
	Technical Authority	L0	0.0	RES14	SOC7					$ -
	Management Support	L0	0.0	RES15	SOC5					$ -
		L0								$ -
										$ 91,215

Figure 127: Cost Calculation for ... Part 1 – Analysis & Design.

Work Breakdown Structure						Cost Estimate				Sell
Activity	Description	Cat.	Hours	Rate	SOC	Materials	Other	Travel	Qty	Price
PART 2 - Implementation (Agile Sprints)	Tasks Conducted By	L0								$ -
	BA - SA	L0	2.0	RES1	SOC7					$ 797
	Sys Lead Eng	L0	124.0	RES2	SOC4					$ 48,395
	Doco Support	L0	5.0	RES3	SOC6					$ 1,394
	Proj Mgmt	L0	22.0	RES4	SOC5					$ 8,236
	Sw Lead Eng	L0	520.0	RES5	SOC1					$ 186,281
	Sw Eng 1	L0	520.0	RES6	SOC1					$ 165,672
	Sw Eng 2	L0	520.0	RES7	SOC1					$ 157,368
	Sw Eng 3	L0	520.0	RES8	SOC1					$ 149,105
	Sw Eng 4	L0	520.0	RES9	SOC1					$ 140,821
	Sw Eng 5	L0	520.0	RES10	SOC1					$ 132,538
	Logistics Mgr	L0	8.0	RES11	SOC7					$ 2,613
	Test Eng	L0	412.0	RES12	SOC7					$ 114,855
	Test Lead Eng	L0	76.0	RES13	SOC7					$ 24,214
	Technical Authority	L0	0.0	RES14	SOC7					$ -
	Management Support	L0	0.0	RES15	SOC5					$ -
	Hardware Componentry Cost	L7				10,000			1	$ 18,500
	Web-Server – End User License - Cost	L10				15,000			1	$ 15,750
	Database-Server – End User License - Cost	L10				65,000			1	$ 68,250
		L0								$ -
										$ 1,226,908

Figure 128: Cost Calculation for ... Part 2 – Implementation.

Work Breakdown Structure			Cost Estimate							Sell
Activity	Description	Cat.	Hours	Rate	SOC	Materials	Other	Travel	Qty	Price
PART 3 - FAT - (Functional) Factory Acceptance Test	Tasks Conducted By	L0								$ -
	BA - SA	L0	0.0	RES1	SOC7					$ -
	Sys Lead Eng	L0	62.5	RES2	SOC4					$ 24,393
	Doco Support	L0	4.5	RES3	SOC6					$ 1,254
	Proj Mgmt	L0	7.5	RES4	SOC5					$ 2,808
	Sw Lead Eng	L0	68.0	RES5	SOC1					$ 24,373
	Sw Eng 1	L0	40.0	RES6	SOC1					$ 12,744
	Sw Eng 2	L0	40.0	RES7	SOC1					$ 12,107
	Sw Eng 3	L0	40.0	RES8	SOC1					$ 11,470
	Sw Eng 4	L0	0.0	RES9	SOC1					$ -
	Sw Eng 5	L0	0.0	RES10	SOC1					$ -
	Logistics Mgr	L0	0.0	RES11	SOC7					$ -
	Test Eng	L0	98.5	RES12	SOC7					$ 27,459
	Test Lead Eng	L0	48.5	RES13	SOC7					$ 15,452
	Technical Authority	L0	0.0	RES14	SOC7					$ -
	Management Support	L0	0.0	RES15	SOC5					$ -
		L0								$ -
										$ 132,060

Figure 129: Cost Calculation for ... Part 3 – Functional Acceptance Test (FAT).

Work Breakdown Structure			Cost Estimate							Sell
Activity	Description	Cat.	Hours	Rate	SOC	Materials	Other	Travel	Qty	Price
PART 4 - SAT - (Satisfaction) Site Acceptance Test	Tasks Conducted By	L0								$ -
	BA - SA	L0	0.0	RES1	SOC7					$ -
	Sys Lead Eng	L0	130.5	RES2	SOC4		3,000	2,000	1	$ 58,897
	Doco Support	L0	16.0	RES3	SOC6					$ 4,460
	Proj Mgmt	L0	31.5	RES4	SOC5		3,000	2,000	1	$ 19,757
	Sw Lead Eng	L0	90.5	RES5	SOC1		3,000	2,000	1	$ 40,402
	Sw Eng 1	L0	28.0	RES6	SOC1					$ 8,921
	Sw Eng 2	L0	20.0	RES7	SOC1					$ 6,053
	Sw Eng 3	L0	0.0	RES8	SOC1					$ -
	Sw Eng 4	L0	0.0	RES9	SOC1					$ -
	Sw Eng 5	L0	0.0	RES10	SOC1					$ -
	Logistics Mgr	L0	29.0	RES11	SOC7					$ 9,470
	Test Eng	L0	36.0	RES12	SOC7					$ 10,036
	Test Lead Eng	L0	24.0	RES13	SOC7					$ 7,646
	Technical Authority	L0	0.0	RES14	SOC7					$ -
	Management Support	L0	0.0	RES15	SOC5					$ -
	Packaging Materials	L8				250			1	$ 263
	Shipment Of Hardware	L8				2,250			1	$ 2,363
		L0								$ -
										$ 168,269

Figure 130: Cost Calculation for ... Part 4 – Satisfaction Acceptance Test (SAT).

Work Breakdown Structure			Cost Estimate							Sell
Activity	Description	Cat.	Hours	Rate	SOC	Materials	Other	Travel	Qty	Price
PART 5 - UAT - (Usability) User Acceptance Test	Tasks Conducted By	L0								$ -
	BA - SA	L0	0.0	RES1	SOC7					$ -
	Sys Lead Eng	L0	64.5	RES2	SOC4		3,000	2,000	1	$ 33,138
	Doco Support	L0	5.0	RES3	SOC6					$ 1,394
	Proj Mgmt	L0	21.5	RES4	SOC5		3,000	2,000	1	$ 16,014
	Sw Lead Eng	L0	46.5	RES5	SOC1		3,000	2,000	1	$ 24,632
	Sw Eng 1	L0	24.0	RES6	SOC1					$ 7,648
	Sw Eng 2	L0	16.0	RES7	SOC1					$ 4,843
	Sw Eng 3	L0	0.0	RES8	SOC1					$ -
	Sw Eng 4	L0	0.0	RES9	SOC1					$ -
	Sw Eng 5	L0	0.0	RES10	SOC1					$ -
	Logistics Mgr	L0	10.0	RES11	SOC7					$ 3,265
	Test Eng	L0	24.0	RES12	SOC7					$ 6,691
	Test Lead Eng	L0	20.0	RES13	SOC7					$ 6,372
	Technical Authority	L0	0.0	RES14	SOC7					$ -
	Management Support	L0	0.0	RES15	SOC5					$ -
		L0								$ -
										$ 103,995

Figure 131: Cost Calculation for ... Part 5 – Usability Acceptance Test (UAT).

Work Breakdown Structure						Cost Estimate				Sell
Activity	Description	Cat.	Hours	Rate	SOC	Materials	Other	Travel	Qty	Price
PART 6 - Go Live	Tasks Conducted By	L0								$ -
	BA - SA	L0	0.0	RES1	SOC7					$ -
	Sys Lead Eng	L0	46.0	RES2	SOC4					$ 17,953
	Doco Support	L0	5.0	RES3	SOC6					$ 1,394
	Proj Mgmt	L0	23.0	RES4	SOC5					$ 8,610
	Sw Lead Eng	L0	26.0	RES5	SOC1					$ 9,319
	Sw Eng 1	L0	0.0	RES6	SOC1					$ -
	Sw Eng 2	L0	0.0	RES7	SOC1					$ -
	Sw Eng 3	L0	0.0	RES8	SOC1					$ -
	Sw Eng 4	L0	0.0	RES9	SOC1					$ -
	Sw Eng 5	L0	0.0	RES10	SOC1					$ -
	Logistics Mgr	L0	2.0	RES11	SOC7					$ 853
	Test Eng	L0	0.0	RES12	SOC7					$ -
	Test Lead Eng	L0	4.0	RES13	SOC7					$ 1,274
	Technical Authority	L0	0.0	RES14	SOC7					$ -
	Management Support	L0	0.0	RES15	SOC5					$ -
	Courier Media To Site	L8				100			1	$ 105
		L0								$ -
	Interim Total - Tasks Conducted By									$ 39,309

Figure 132: Cost Calculation for ... Part 6 – Go Live.

Work Breakdown Structure						Cost Estimate				Sell
Activity	Description	Cat.	Hours	Rate	SOC	Materials	Other	Travel	Qty	Price
PART 7 - Warranty Support	Tasks Conducted By	L0								$ -
	BA - SA	L0	0.0	RES1	SOC7					$ -
	Sys Lead Eng	L0	20.0	RES2	SOC4					$ 7,806
	Doco Support	L0	0.0	RES3	SOC6					$ -
	Proj Mgmt	L0	20.0	RES4	SOC5					$ 7,487
	Sw Lead Eng	L0	20.0	RES5	SOC1					$ 7,169
	Sw Eng 1	L0	20.0	RES6	SOC1					$ 6,372
	Sw Eng 2	L0	20.0	RES7	SOC1					$ 6,053
	Sw Eng 3	L0	0.0	RES8	SOC1					$ -
	Sw Eng 4	L0	0.0	RES9	SOC1					$ -
	Sw Eng 5	L0	0.0	RES10	SOC1					$ -
	Logistics Mgr	L0	0.0	RES11	SOC7					$ -
	Test Eng	L0	20.0	RES12	SOC7					$ 5,576
	Test Lead Eng	L0	0.0	RES13	SOC7					$ -
	Technical Authority	L0	0.0	RES14	SOC7					$ -
	Management Support	L0	0.0	RES15	SOC5					$ -
		L0								$ -
	Interim Total - Tasks Conducted By									$ 40,482

Figure 133: Cost Calculation for ... Part 7 – Warranty Support.

Work Breakdown Structure						Cost Estimate				Sell
Activity	Description	Cat.	Hours	Rate	SOC	Materials	Other	Travel	Qty	Price
PART 8 - Project Closure	Tasks Conducted By	L0								$ -
	BA - SA	L0	0.0	RES1	SOC7					$ -
	Sys Lead Eng	L0	13.0	RES2	SOC4					$ 5,074
	Doco Support	L0	1.0	RES3	SOC6					$ 279
	Proj Mgmt	L0	27.5	RES4	SOC5					$ 10,295
	Sw Lead Eng	L0	17.0	RES5	SOC1					$ 6,093
	Sw Eng 1	L0	1.0	RES6	SOC1					$ 319
	Sw Eng 2	L0	1.0	RES7	SOC1					$ 303
	Sw Eng 3	L0	1.0	RES8	SOC1					$ 287
	Sw Eng 4	L0	1.0	RES9	SOC1					$ 271
	Sw Eng 5	L0	0.0	RES10	SOC1					$ -
	Logistics Mgr	L0	0.0	RES11	SOC7					$ -
	Test Eng	L0	1.0	RES12	SOC7					$ 279
	Test Lead Eng	L0	9.0	RES13	SOC7					$ 2,867
	Technical Authority	L0	0.0	RES14	SOC7					$ -
	Management Support	L0	4.0	RES15	SOC5					$ 1,593
		L0								$ -
	Interim Total - Tasks Conducted By									$ 27,658

Figure 134: Cost Calculation for ... Part 8 – Project Closure.

Work Breakdown Structure			Cost Estimate							Sell
Activity	Description	Cat.	Hours	Rate	SOC	Materials	Other	Travel	Qty	Price
Stage 1 - Management & Technical Authority Support	Tasks Conducted By	L0								$ -
	Technical Authority	L0	275.0	RES14	SOC7					$ 109,519
	Management Support	L0	275.0	RES15	SOC5					$ 109,519
		L0								$ -
	Interim Total - Tasks Conducted By									$ 219,038

Figure 135: Cost Calculation for … Management & Technical Authority Support.

Thus, for the scenario's project the Internal Costs versus the Sell Price are as follows:

	LOE Hours	Internal Cost	Sell Price
PART 0 - Project Kick off	17.0	$ 3,550	$ 5,655
PART 1 - Analysis & Design	216.0	$ 57,260	$ 91,215
PART 2 - Implementation (Agile Sprints)	3,769.0	$ 800,865	$ 1,226,908
PART 3 - FAT - (Functional) Factory Acceptance Test	409.5	$ 82,900	$ 132,060
PART 4 - SAT - (Satisfaction) Site Acceptance Test	405.5	$ 106,483	$ 168,269
PART 5 - UAT - (Usability) User Acceptance Test	231.5	$ 65,283	$ 103,995
PART 6 - Go Live	106.0	$ 24,710	$ 39,309
PART 7 - Warranty Support	120.0	$ 25,400	$ 40,462
PART 8 - Project Closure	76.5	$ 17,363	$ 27,658
Stage 1 - Management & Technical Authority Support	550.0	$ 137,500	$ 219,038
TOTALS	5,901.0	$ 1,321,313	$ 2,054,569

Hence, for the scenario's project the performing organization's senior management would provide the customer's representatives with a $2.055M Offered Price.

Contract Price – Negotiations

During the contract negotiations between the performing & customer organizations, there may be some concerns raised that the **offered Contract Price is "*way too high*"**.

With no reduction in the agreed [Scope] of the project, then … **instead of slashing the Level Of Effort hours [Time]**, … the performing organization could **consider** the following options:

- **Reducing the Base Labour Rates** per [People] involved with the project; noting that, this base rate would include some "fat" to cover the cost of the project's Initiating Phase work which occurred prior to the contract's sign off, and to cover the costs involved with those proposed projects that were not won.   … "*Also, $200 per hour for a permanent salaried worker at 40 hours/week x 52 weeks / year = $416,000 per annum; but rest assured that, most probably, not one of the project team members are on a salary of half that rate. Similarly, a junior software engineer at $110 per hour, definitely ain't getting a yearly salary of $229K; though, they would probably be satisfied with a third of that*".

    For example; a 25% "discount" on Labour Rates would result in an Internal Cost of $1,173,788 (vs. $1,321,313) and a Contract Sell Price of $1,819,562 (vs. $2,054,569). … Still not good enough, then how about a 50% "discount" for an Internal Cost of $1,026,263 (vs. $1,321,313) and a Contract Sell Price of $1,584,554 (vs. $2,054,569).

- **Reducing the Burdens & Margins** (i.e. the "magic number" percentages) for Cost Contingencies (*warranty repairs & latent defects*), Risk Contingencies (*type of work conducted*), Overhead Burden, Management Reserve, and Profit Margin.

    For example; if the Profit Margin was reduced from 12.5% to 5% (without any other changes) then this would result in the Internal Cost remaining the same ($1,321,313), but a Contract Sell Price of $1,924,080 (vs. $2,054,569). Thus, what downwards adjustments could be achieved by tweaking others of those combination percentages?

ONLY A FOOL reduces the Contract Price by reducing the Level Of Effort without a corresponding reduction in the Scope Of Deliverables.

Please refer to [Section 5.4.3] for "pricing the project to a certain death"; because, one of the battles that the Project Manager will face, is when Sales & Business Development (and even senior management) reduces the project's Contracted Price by cutting back on the Level Of Effort without reducing the Scope Of Deliverables, **as this puts the "actual burden" for the project's profitability onto the members of the Project Implementation Team, … as unrecorded overtime hours that these [People] will have to work**, if this project is to have any chance of being completed on [Time] schedule within the stipulated [Cost] budget, … which can result in these [People] being unable to focus on producing "satisfactory" [Quality] [Scope] of deliverables.

Which brings us to the next rule of "Adaptive & Proactive" SDLC Project Management.

RULE 53: Level Of Effort is to [Scope], … as [Time] is to [Cost].

COFFEE BREAK DISCUSSION … *Like to take a razor to that goat.*

It really gets on my goat, when you spend a lot of time & effort preparing a realistic project schedule and diligently working out a viable project budget; then some fool in Sales & Business Development (or even senior management), who has zero (or no recent) experience with actual hands-on system development, they come through and willy-nilly slice chunks of hours off the durations for tasks without having the faintest idea as to the complexity of what is actually involved with those implementation tasks.

"Do you realize that we've bitched about this problem twice before in coffee break discussions. … Maybe we need some counselling."

"Yah, lay back on the couch, and tell me, all about it".

7. PM you ... Field Test 3

7.1. Overview

This field test examines the following topics with respect to real-world project management:

- Monitoring the project's performance using the project schedule.
    - The relationship of the Project Schedule to the Level Of Effort (S-Curve), and the [People] Head Count over the duration of the project.
- Monitoring the project's performance using agile's Task Board and Burn Down Chart.
- Monitoring the project's progress using Earned Value Performance Measures (EVPM).
    - Interpretation of the project's S-Curve, the Schedule Performance Index (SPI), the Cost Performance Index (CPI), the Estimate At Completion (EAC), Head Count utilization, and the Percentage of Schedule Progress Variance.
- Using these above listed performance measures as Key Performance Indicators (KPIs) for the monitoring of a program of projects.
- The relationship between project implementation, management, and governance.

7.2. Introduction

That last field test concentrated on the intricacies of the Initiating & Planning Phases for a "Fixed Price" project; whereas, this current field test will focus on the particulars involved with the Monitoring & Control Phase of a project.

7.2.1. The Scenario

```
This project is an in-house development, while not as
complicated as the previous project, it does have some
agile sprints embedded within a waterfall project layout.
```

7.2.2. Start At The Beginning

As with the previous projects, start from a clean slate and answer these opening questions:

Initiating Phase [Section 3.3]

- ☐ What do **THEY** (the customer) **WANT**?
- ✓ Produced ... **Customer Requirements**.
- ☐ What will **WE** (the implementer) **GIVE** (to the customer)?
- ☐ What will **WE** (the implementer) **NOT** be **GIVING** (to the customer)?
- ☐ Approximately **HOW LONG** do **WE** (the implementer) **EXPECT** it to **TAKE**?
- ☐ What will it **COST US** (the implementer) **TO GIVE** it (to the customer)?
- ☐ What will **WE** (the implementer) **GET IN RETURN** (from the customer)?
- ☐ What is **AT STAKE FOR US** (the implementer)?
- ✓ Signed off ... **Project Charter**.

Planning Phase   [Section 3.4]

- ☐ **WE** (the implementer) **NEED WHAT**, when, and how much of it?

- ✓ Signed off ... [Scope Baseline]   **Detailed Specifications**.
- ✓ Produced ...   **Work Breakdown Structure**.
- ✓ Produced ...   **Requirements Traceability Matrix**.
- ✓ Produced ...   **Acceptance Criteria**.

- ☐ **WE** (the implementer) **NEED** what, **WHEN**, and how much of it?

- ✓ Signed off ... [Time Baseline]   **Project Schedule**.
- ✓ Produced ...   **Milestone List**.

- ☐ **WE** (the implementer) **NEED** what, when, and **HOW MUCH** of it?

- ✓ Signed off ... [Cost Baseline]   **Project Budget**.

- ☐ How will **WE** (the implementer) **KNOW WE GOT IT RIGHT**?

- ✓ Signed off ... [Quality Baseline]   **Acceptance Criteria**.

- ☐ What do **WE** (the implementer) **THINK CAN GO WRONG**?

- ✓ Produced ...   **Risks & Issues Register**.
- ✓ Produced ...   **Baseline Change Request process**.
- ✓ Produced ...   **Defect Reporting process**.

- ☐ Planning Phase Completion
- ✓ Signed off   **Gate Review**.

Executing Phase   [Section 3.5]

- ✓ Conduct ...   **Kick-Off meeting**.

- ☐ **How** exactly **WILL** the Project Implementation **TEAM DO IT**?

- ☐ Is the Project Implementation **TEAM DOING IT**?

- ☐ Is the Project Implementation **TEAM CHECKING IT**?

- ☐ Is the Project Implementation **TEAM DONE YET**?

429

7.2.3. Monitoring & Control, Today's Concern

Given that the scenario's project has passed through the Gates (i.e. "sanity checkpoints") for the Initiating Phase and the Planning Phase, and given that the Executing Phase is now underway, then the Project Manager's concern turns primarily towards the Monitoring & Control of the project, as was previously detailed in [Section 3.6]. That is, keeping one eye on the road, keeping the other eye on the gauges, and both hands on the steering wheel.

Clock	RPM	Speedo	FUEL
TIME DURATION	**PEOPLE & RESOURCES**	**SCOPE COVERAGE**	**COST BUDGET**

[Scope] Monitoring & Control [Section 3.6.3]

[Scope] Monitoring & Control **needs to continually provide answers to** the following questions, … in a form that can be easily understood & readily utilized by all of the involved & interested project stakeholders.

- ☐ **What's in & what's out**?

- ☐ **What's done, what's doing, what's yet to do**?

- ☐ **What's being added & what's being subtracted**?

- ☐ **What's in doubt & what's next to discuss**?

By using such devices as background colour coding (i.e. BLACK – out of scope, WHITE – yet to do, GREY – completed previously, GREEN – just finished, AMBER | YELLOW – work in progress, RED – blocking issue) the entries in the **Requirements Traceability Matrix (RTM) depicts the coverage of the project's approved Detailed Specifications**, and the **Work Breakdown Structure (WBS) depicts the status of each package of work**.

[Time] Monitoring & Control [Section 3.6.4]

[Time] Monitoring & Control **needs to continually provide answers to** the following questions:

- ☐ **What** tasks **[Scope] should have been completed** by now?
- ☐ **How far should** these tasks **[Scope] have progressed** by now?
- ☐ **What** tasks **[Scope], [People], and [Resources] should** currently be **engaged**?
- ☐ **What** tasks **[Scope], [People], and [Resources] are** currently **engaged**?
- ☐ **What** tasks **[Scope], [People], and [Resources] should** be **engaged next**?
- ☐ **What progress has been made through** the **assigned** tasks **[Scope]** that these **[People] & [Resources]** are **currently engaged** with?
- ☐ **Is the reality** of what & when tasks [Scope], [People], and [Resources] are being engaged **deviating away from those approved [Baselines]**?
- ☐ **How to realign** tasks [Scope], [People], and [Resources] **with the [Baselines]**?
- ☐ **Given the realities** of the current situation & prevailing circumstances, **does "The Plan" need to be re-baselined** to deal with the risks & issues at hand?

[Cost] Monitoring & Control [Section 3.6.5]

[Cost] Monitoring & Control **needs to continually provide answers to** the following questions:

- ☐ **Planned Value (PV)** – what was the **planned spend to date**, to get the project to **where it was planned to be by now** (instead of where it currently is)?
- ☐ **Actual Cost (AC)** – what was **actually spent to date** (when the measurements were taken), to get the project to **where it really is now**?
- ☐ **Earned Value (EV)** – what was **planned to have been spent** to date (when the measurements were taken), to get the project to **where it really is now**?

- ☐ **Cost Variance (CV)** – **how far above or below** the [**Cost Baseline**] budget to get the project to the point **where it really is now**?

- ☐ **Schedule Variance (SV)** – **how far ahead or behind** the [**Time Baseline**] schedule of **where** it **really is now**, **against where** it was **planned to be by now**?

- ☐ **Cost Performance Index (CPI)** – **how good or bad** is the project's **performance** when **compared** to the [**Cost Baseline**] budget (in terms of labour hours | monetary units), to get the project to **where it really is now**?

- ☐ **Schedule Performance Index (SPI)** – **how good or bad** is the project's **performance** when **compared** to the [**Time Baseline**] schedule (in terms of labour hours | monetary units), where the project **really is now**, **against where** it was **planned to be by now**?

- ☐ **Budget At Completion (BAC)** – what was the **total Planned Value when** the project is **finished**?

- ☐ **Estimate To Complete (ETC)** – what **has to be spend from here on** to get this project **across the finish line**?

- ☐ **Estimate At Completion (EAC)** – What do we now calculate as the **total cost when** this project **finishes**?

Consequential Decisions From The Triple Constraints

Given the realities of the current situation & prevailing circumstances:

- ☐ **Does the scheduled tasks [Scope] need to be refactored** … expanded or reduced?

- ☐ **Does the schedule [Time] duration need to be compressed or extended**?

- ☐ **Does the [Planned Costs] need to be re-baselined**?

And, …

- ☐ **Do additional [People] & [Resources] need to be allocated**?

- ☐ **Does the utilization percentages for the existing ones need to be changed**?

[Quality] Monitoring & Control [Section 3.6.6]

[Quality] Monitoring & Control **needs to continually provide answers to** the following questions:

- ☐ Are the project's **tangible & measurable deliverables appropriate** when **compared to** the **agreed [Scope]** and the **agreed Acceptance Criteria**?

- ☐ Is there an **alignment between** the customers and the other stakeholders' **perceptions & expectations** when **compared** to what is in the **approved Detailed Specifications**?

- ☐ ~~Does the project's deliverables satisfy the needs for which these were intended~~?

- ☐ Does the project's **deliverables satisfy the purpose for which these were specified**?

- ☐ Are the **appropriate Quality Assurance and Quality Control** processes & procedures **being followed**, … consistently & sensibly?

- ☐ Is **[Quality]** being considered as **today's issue**, or is "*qual-it-tee*" being put off as a tack-on that will need to be resolved someday after tomorrow?

"Satisfaction" Versus "Performance"

As stated previously in [Section 2.8.2]:

> **RULE 11:** Project success is opinionated, based on the "satisfaction"
> of   [Scope] & [Quality]
> versus
> the relative "performance"
> of   [Time] & [Cost]

Where the [People] & [Resources] utilized will be significant factors influencing whether the project will be able to satisfy the customer, while striving to achieve the project's performance measures. See [Figure 21], reproduced on the following page.

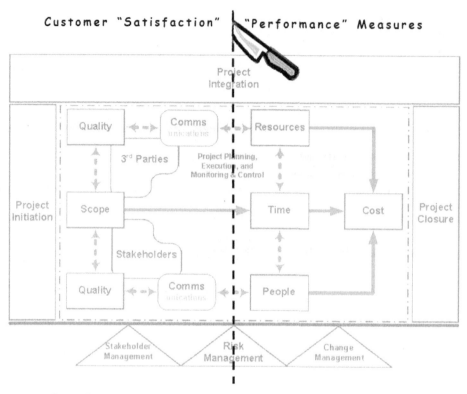

Figure 21: "Satisfaction" versus "Performance" division of the 'Project Management Process'.

Which brings us to an addendum to "Adaptive & Proactive" SDLC Project Management.

RULE 11[B]: Project success is a multi-faceted balancing act of monitoring & controlling the project variables & project constraints, the risks, and changes; while, simultaneously satisfying the needs & wants, concerns, expectations, perceptions, and opinions of the project's stakeholders.

7.3. Monitoring & Control

7.3.1. The Project Schedule

This (in-house) project has been broken up into the following stages of; initiation, requirements gathering, planning, design, development, evaluation testing, defect repair & retest, customer acceptance testing, release, warranty support, and closure.

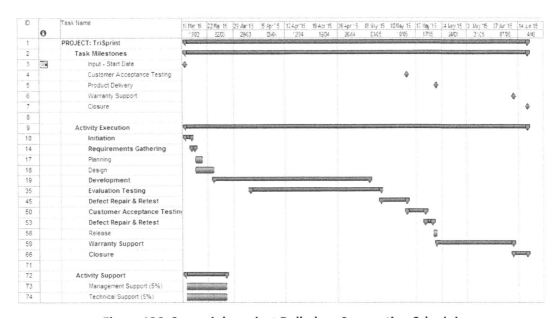

Figure 136: Scenario's project Rolled-up Summation Schedule.

While this project does have monetary value per hour worked, where each hour worked is worth N dollars; what is of importance for now is the Level Of Effort (i.e. work hours of 1590), and the time span of 4 months duration.

"So, if we spitballed the bums-on-seats labour rates as $100 per hour, gives $159K, so 20-25% ballparking for a Rough Order of Magnitude quotation, is about a $200K investment for this project".

But given, this is an in-house project (such as delivery of an application for in-house use) then the "dollars" (with Profit Margin) ain't as interesting as the Level Of Effort involved.

7.3.2. Level Of Effort (S-Curve) & Head Count

Having reviewed the schedule; if the weekly totals for the Level Of Effort for all of the project's activities were accumulated then the S-Curve, in [Figure 137], would result in a **Total Level Of Effort**, also referred to as **Budget At Completion (BAC)**, of 1590 hours.

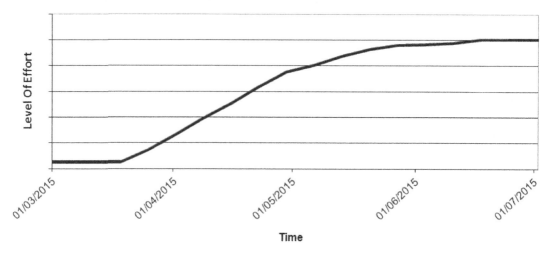

Figure 137: Scenario's project – accumulated Level Of Effort over the project's duration.

And, the expected **Weekly Head Count = (PV) / (40hr x 83% Utilization)**.

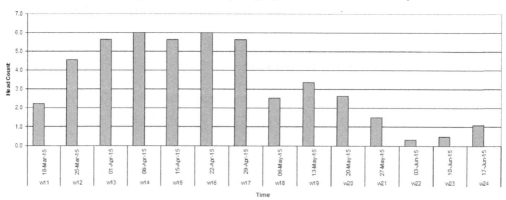

Figure 138: Scenario's project – planned weekly Head Count over the project's duration.

7.3.3. Schedule Performance Measures

Make a "work-in-progress" copy of the approved [Time Baselined] Project Schedule; with this Work-In-Progress Schedule (version controlled for the day or week of concern), track & record the project's real-world progress, so as to pass judgement on the comparison of "actual progress" against the approved "Plan".

1.. **Routinely** at least once a week, preferably on a consistent day of the week (*e.g. every Monday*), and if possible at approximately the same time of day, go **visit strategic members of the Project Implementation Team and/or the Project Working Group** and individually **discuss with them the progress that their area has made** with respect to their **assigned tasks**, as displayed on a visible (screen or physical paper copy) of the Work-In-Progress Schedule.

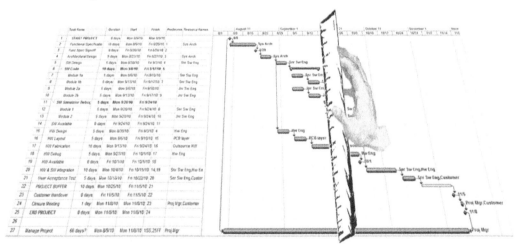

Figure 139: Marking up the Work-In-Progress Schedule.

2.. As per [Figure 139], ruler align on today's date and boldly mark the date line, then ask:

- ☐ **What scheduled tasks have they or their team members been working on**?

    It is advantageous to know in advance what tasks they should be answering for, i.e. to have noted the task numbers that you would expect them to talk about.

If they are found to be working on tasks that they were not scheduled to be working on (i.e. tasks a lot later in the schedule or tasks that do not appear on the schedule) then enquiries should be made as to, "why?"

Decide whether they have to be **steered back on course** to do only those tasks that they have been assigned, or **does the schedule need to be changed to reflect the realities of the current situation & prevailing circumstances**?

☐ **How many days do they still require to complete the work-in-progress tasks?** Not how far have they progressed with the completion of their assigned tasks?

Any of their assigned tasks (or an agile sprint) that appears in whole on the left side of the ruler bounded schedule, [Figure 139], should have been completed by now, any assigned tasks that cross under the ruler should currently be worked on, and assigned tasks that are right of the ruler probably shouldn't be worked on yet.

If each task (sprint) was limited to a maximum of 4 weeks duration, then each week would correspond to a 25% progress marker, and each day would correspond to 5% progress. So, if they respond that they will need a few extra days then start reducing the recorded percentage complete for the corresponding assigned tasks; because, as the Project Manager, **what you don't want is tasks stuck at 80-90% complete, where more effort is being put in but no progress (or rather no slippage) is being represented**.

If the task's percentage complete is decreasing, then this raises questions of:

- o **What is preventing this assigned task from being completed in the allocated time**; i.e. are there **blocking issues** that need to be resolved?

- o **Is scope creep occurring?**

- o **Should this assigned task be moved to a more appropriate / opportune moment in the schedule**, or transferred to another milestone / release?

- o **Does this assigned task have to be dissected into multiple tasks**?
- o **Are there extra tasks needing to be added to the schedule**?

Note down this information for recording into the Work-In-Progress Schedule.

- ☐ Are they **experiencing any blocking issues** hindering their progress?
- ☐ Are there **<Risks> to the project** that could prevent tasks (and the project) from being completed successfully?
- ☐ Are there **tasks** that **could be better arranged** to occur at more opportune timings?
- ☐ Are there **tasks [Scope], [People], and/or [Resources] availability** that are evidently **missing** from the schedule?
- ☐ Are there **tasks [Scope], [People], and/or [Resources]** that are evidently **no longer necessary** for the project to be completed successfully?
- ☐ Are there **baseline tasks [Scope], [Time], [Cost], [Quality], [People], and/or [Resources]** that evidently **will need to be adjusted** for the project to be completed successfully? Will **a Baseline Change Request be required**?

Where & whenever possible, **involve the Project Team members active participation** in the project's scheduling, and **semi-directly with the decision making process** for the project. Especially, make inquiries as to *"what can I do to help you succeed"* to bring those late tasks back on schedule?

"Project accountant" type managers will expect (even demand) that the project's schedule be followed as though gospel; this is critically flawed, because **in reality things hardly ever turn out exactly as was planned.**

It is best to view the schedule as a guide to the advised path to implement the project; knowing that sometimes parts of "The Plan" will have to be adapted, revised, and reapplied to better suit the current situation & prevailing circumstances.

"Yep, it is a foolish manager who has the misguided belief that, if the schedule is zealously abided by, then 'all will be fine'; and consequently, they act as though any deviation (whatsoever) away from 'Their Plan' is a cardinal sin for which the wrongdoer must be punished or reprimanded. Yet, they will proclaim that the project is being done as 'aj-jile'; having no real idea that agile means adapting to changing understandings."

Which brings us to the next rule of "Adaptive & Proactive" SDLC Project Management.

RULE 54:   Don't foolishly expect everything to work out exactly as was planned; cause, it never happens that way.

7.3.4.   Agile Performance Measures

For the scenario's project, there is a portion of the project that will be conducted using agile techniques; i.e. the 'Development' and 'Evaluation Testing' activities.

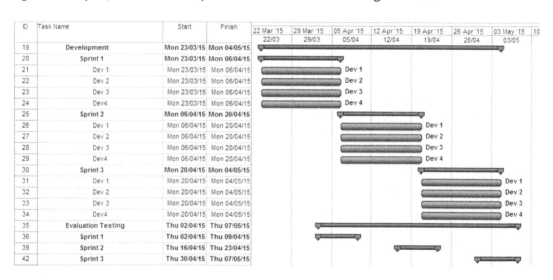

Figure 140: 'Development' activities of the project planned as an agile implementation.

As detailed in [Section 3.5.3] through to [Section 3.5.6], and in [Section 3.6.10], the Project Implementation Team members will have selected those 'To Do Items' to be put into the

'Sprint Backlog' for the current sprint; and then, over the proceeding weeks & days of that sprint, The Team will develop & self-test each 'To Do Item' as they move its corresponding 'card' across the Task Board, as per [Figure 141], ... and, thereby reducing the daily count of the work left to be done. So that by the expiry of the current sprint's allocated time duration, then ideally all of those selected 'To Do Items' would be done with none left over.

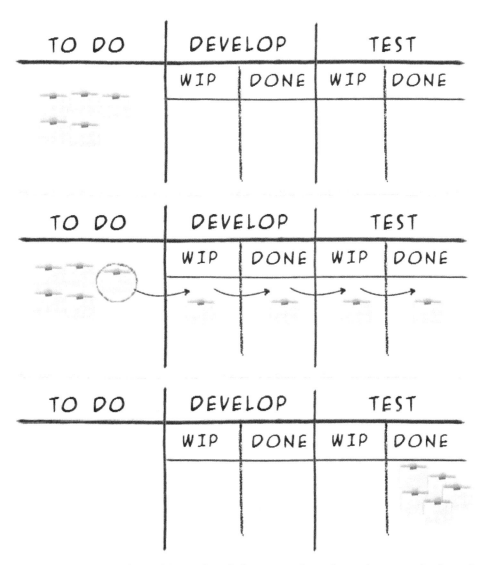

Figure 141: An agile Task Board and the traversing of a task across the board.

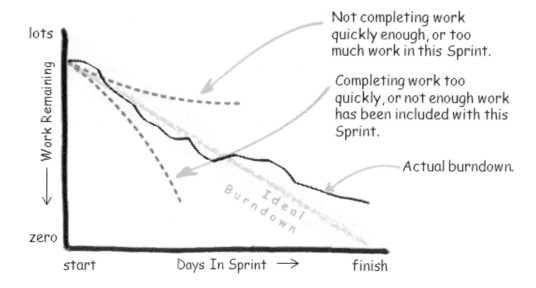

Figure 79: Agile Burn Down Chart with up-to-date data.

While the Project Implementation Team would (ideally) use an agile tracking application to plot the Burn Down Chart on a daily basis, see [Figure 79]; the Project Manager would use the following equation to calculate the '% Work Complete' for the current sprint's summation activity that is to be entered into the Work-In-Progress Schedule.

$$Sprint\ \%\ Work\ Complete = \left(1 - \frac{(STDi - HVADi)}{STDi}\right) * 100$$

Where STDi is the 'Starting To Do' item total,
HVADi is the 'Have Verified As Done' item total.
And, where the item total could be based on a weighting of the technical difficulty – complexity and/or the priority of each 'To Do' item's card.

As an aside, in agile there is a term "velocity" which denotes how many tasks | use case cards are being completed per sprint; i.e. how much work is being done per sprint.

$$Time = \frac{Scope}{Velocity} \approx \frac{Distance}{Speed}$$

> If the velocity is known for how much work is being done per sprint, and given that these sprints are of a constant duration, then one should be able to extrapolate exactly how long (and how many sprints) it is going to take to get through the entire amount of the [Scope] of work to be done; and subsequently, how long it will take to finish off the agile development portion of the project.

Similarly to that for the 'Development' activities, the scenario project's 'Evaluation Testing' activities will also be conducted using agile techniques, and thus use the same methods for determining the scheduled progress being made. Though, for Quality Control best practice reasons, these 'Evaluation Testing' activities will be conducted independently of the 'Development' activities by an impartial (as possible) test team.

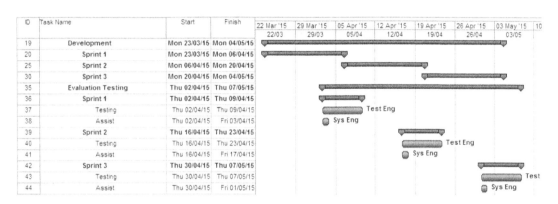

Figure 142: 'Evaluation testing' activities planned as an agile implementation.

Based, on the "velocity" of [Scope] covered for both 'Development' and 'Evaluation Testing', it may become apparent that insufficient [Scope] will be delivered at the completion of these sprints; hence, it may be necessary to reassess:

☐ Given the current loading of [People] & [Resources], **how many sprints would be required to produce all of the agreed [Scope]. ... Are additional sprints necessary**?

   If **more sprints need to be added** to the project, then this would **result in changes to** the approved **[Time Baseline]** Project Schedule; hence, a **Baseline Change Request** assessment would **need to be conducted**.

443

- [ ] Alternatively, **should additional [People] & [Resources] be added to the existing sprints** so as to cover more of the agreed [Scope] in the available [Time]?

    However, the addition of extra [People] & [Resources] will most probably affect the approved [Cost Baseline] budget; hence, a **Baseline Change Request** assessment would need to be conducted.

Though, if additional sprints are required, and/or additional [People] & [Resources] are needed then this could be an indication that there are underlying problems with the project; *e.g. under estimations of the amount of work involved, the occurrence of scope creep, technically difficult blocking issues, inefficient processes & procedures limiting how much work can get done in the given amount of time, and imbalances of ??.*

Suggest that **the Project Implementation Team concentrate firstly on doing those high priority (i.e. high Return On Investment) features & functionality**, followed by the lower priority ones in later sprints. In this way, those unforeseen problems & issues encountered later in the project's life would be associated with those lower priority features & functionality.

Subsequently, it becomes a negotiated discussion as to whether these remaining 'To Do Items' really have to be done, given that these are of relatively lower priority; and, does **the Return On Investment for those remaining 'To Do Items' really justify the [Time] & [Costs] involved**?

Stakeholders who are used to the traditional waterfall methodology often have a real problem with the concept of dropping those low priority features & functionality; as though on pain of death, everything must be there, irrespective of what it is practically worth to them (and to the end-users).

7.3.5. Earned Value Performance Measures

Previously in [Section 7.3.2], the "Planned Hours of Effort" per week (i.e. the **Planned Value – PV**, also known as the **Budget Cost of Work Scheduled – BCWS**) were accumulated over the duration of the project, then this would equal the "planned total hours" to conduct the project (i.e. the **Budget At Completion – BAC**), see the S-Curve in [Figure 137].

			Week														Total
			1	2	3	4	5	6	7	8	9	10	11	12	13	14	
Budget Cost of Work Scheduled																	
	(weekly) Planned Value		74	152	188	200	188	200	188	84	112	88	50	10	16	36	1590
BCWS	Planned Value	PV	74	227	415	615	803	1004	1192	1276	1389	1477	1527	1537	1554	1590	BAC
	Planned % Complete	PV / BAC	5%	14%	26%	39%	51%	63%	75%	80%	87%	93%	96%	97%	98%	100%	

Based on the "Accumulated Planned Hours" it can be determined via the Budget At Completion (*e.g. 1590 hours*) what is the **"Planned Percentage Complete"** per week.

$$\text{Planned \% Complete} = PV / BAC$$

Now that the project is underway and work is being done, then each week the Project Team members can diligently submit their timesheets (i.e. the "Actual Hours Worked") against the project's assigned job code. The Project Manager would then accumulate the timesheet recorded actual hours worked on the project. If these "Actual Hours Worked" per week (i.e. the **Actual Cost – AC**) were accumulated over the entire duration of the project then this would equal the "actual total hours" to conduct the project (i.e. the **Estimate At Completion – EAC**).

			Week														Total
			1	2	3	4	5	6	7	8	9	10	11	12	13	14	
Actual Cost of Work Performed																	
	(weekly) Actual Cost		68	178	198	238	134	282	175	85	99	87	45	10	11	49	1658
ACWP	Actual Cost	AC	68	246	443	682	815	1098	1273	1358	1457	1544	1589	1599	1610	1658	EAC

On a weekly basis, the Project Manager can obtain the **"Actual % Complete"** from the Work-In-Progress Schedule and **multiply** this value **by** the Budget At Completion to **derive** the **Earned Value – EV** (also known as the **Budget Cost of Work Performed – BCWP**).

$$EV = BAC \times \text{Actual \% Complete}$$

			Week															Total
			1	2	3	4	5	6	7	8	9	10	11	12	13	14	
Budget Cost of Work Scheduled																	
	(weekly) Planned Value		74	152	188	200	188	200	188	84	112	88	50	10	16	36	1590
BCWS	Planned Value	PV	74	227	415	615	803	1004	1192	1276	1389	1477	1527	1537	1554	1590	BAC
	Planned % Complete	PV / BAC	5%	14%	26%	39%	51%	63%	75%	80%	87%	93%	96%	97%	98%	100%	
Actual Cost of Work Performed																	
	(weekly) Actual Cost		68	178	198	238	134	282	175	85	99	87	45	10	11	49	1658
ACWP	Actual Cost	AC	68	246	443	682	815	1098	1273	1358	1457	1544	1589	1599	1610	1658	EAC
Budget Cost of Work Performed																	
	Actual % Complete (Schedule)		7%	15%	22%	31%	37%	53%	71%	77%	83%	90%	95%	95%	97%	100%	
BCWP	Earned Value	EV	104.0	238.0	356.6	490.8	594.4	848.7	1127.4	1216.7	1322.3	1436.2	1507.1	1518.0	1539.8	1590.0	

And, then on a regular basis (*e.g. weekly*), this correlated data can be plotted as the Level Of Effort versus Time (i.e. S-Curve).

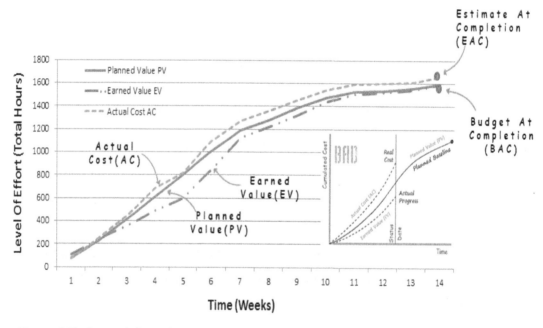

Figure 143: Scenario's project – S-Curve actual performance over the project's duration.

See [Section 3.6.5] … As evident from the scenario project's S-Curve actual performance, in [Figure 143], **for most of the project's duration**, the accumulated **Actual Cost (AC)** of the work undertaken was **greater than the Planned Value (PV)** of the work that was planned to be done; i.e. **the project has been "Over Budget"** for most of the time.

And, the **Earned Value (EV) has been less than the Planned Value (PV)** which means that **the project has been "behind schedule"** for most of the time.

NOTICE how the Earned Value at the end of the project is equal to the BAC (*eg. 1590 hrs*) and not the EAC (*eg. 1658 hrs*). This is because, the Earned Value maps where the project currently is (i.e. the "real" Actual % Complete) against the equivalent point in the planned baseline; hence, **when 100% complete then EV = PV**.

On a regular basis (*e.g. weekly*), the **Schedule Performance Index – SPI** for [Time] and the **Cost Performance Index – CPI** for [Costs] would be calculated.

$$SPI = (EV_{accumulative}) / (PV_{accumulative})$$

$$CPI = (EV_{accumulative}) / (AC_{accumulative})$$

			Week													Total
			1	2	3	4	5	6	7	8	9	10	11	12	13	14	
BCWS	Planned Value	PV	74	227	415	615	803	1004	1192	1276	1389	1477	1527	1537	1554	1590	BAC
ACWP	Actual Cost	AC	68	246	443	682	815	1098	1273	1358	1457	1544	1589	1599	1610	1658	EAC
BCWP	Earned Value	EV	104.0	238.0	356.6	490.8	594.4	848.7	1127.4	1216.7	1322.3	1436.2	1507.1	1518.0	1539.8	1590.0	
Performance Indexes																	
	Schedule Performance Index	SPI	1.40	1.05	0.86	0.80	0.74	0.85	0.95	0.95	0.95	0.97	0.99	0.99	0.99	1.00	SPI = EV / PV
	Cost Performance Index	CPI	1.54	0.97	0.80	0.72	0.73	0.77	0.89	0.90	0.91	0.93	0.95	0.95	0.96	0.96	CPI = EV / AC

And, then on a regular basis (*e.g. weekly*), this correlated data can be plotted as the Performance Index versus Time.

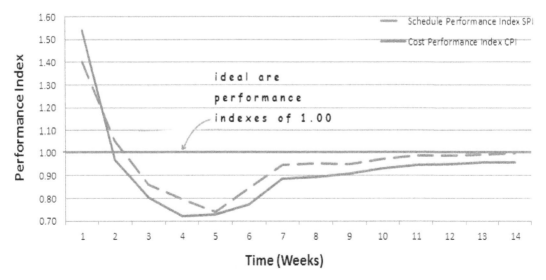

Figure 144: Scenario's project – actual performance indexes over the project's duration.

NOTE: The **SPI** may be **GREATER than OR LESS than ONE** depending on how early or late the project is delivering; whereas, the **CPI** may be **GREATER than OR LESS than ONE** depending on how "profitable" an endeavour is the project.

NOTE: For the first couple of weeks, don't be overly concerned if either the SPI or the CPI values are outside their respective "Performance Index Windows" of being near ONE, see [Figure 150] and [Figure 151], because not enough time has passed and not enough work effort has been accumulated to provide a really meaningful indication about the long term progress being made on the project. However, what is **of interest is the general direction of the "trend" as each performance index moves towards or away from the ideal of ONE**.

For a Fixed Price contract DO NOT present the CPI to the customer's representative (and DO NOT provide the AC data), as this would give an insight into what are the performing organization's margins for this project.

Because, if the CPI is much greater than ONE (a "very profitable endeavor"), then this could be interpreted by the Customer's Representative as, *"sucker, we are charging you a lot more (in Level Of Effort hours) for this project than what is really necessary"*. This realization would probably result in an acrimonious relationship between the Customer's Representatives and the performing organization.

NOTICE how in [Figure 144] the SPI and the CPI both trend downwards very steeply for the first few weeks, then about Week 5, the project makes a noticeable turn around "recovery" as both of these indexes trend upwards to ONE. ... Such, a recovery could be due to; a technical break through having been made, some major blocking issue has been resolved, more effective & efficient work practices are in effect, or the composition of the Project Team has been changed with more appropriately skilled | experienced [People] and/or [Resources] replacing not so effective ones.

Remembering that, **[Cost] at this point is being expressed as Level of Effort labour hours expended and not the actual monetary amounts paid for those [People] & [Resources] utilized.** Hence, **if the improvement in CPI and/or SPI was due to replacing less expensive [People] & [Resources] with higher priced [People] & [Resources] then this will not be reflected in these performance measures.** As any such increases in financial expenditure would be determined in a month or so when the contractors & service providers' invoices are being processed, and when fortnightly & monthly salaries are paid.

For this scenario project there appears to have been a concerted effort to recover the project from an escalating estimated final delivery Level Of Effort (and increasing delivery date), as reflected in the weekly plot of the **Estimate At Completion**, see [Figure 145].

$$EAC = BAC - EV + AC$$

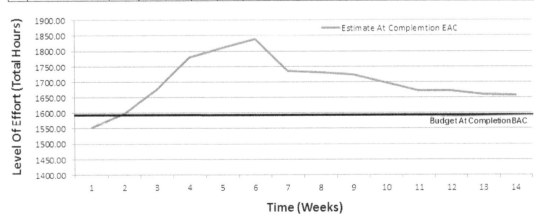

			\multicolumn{14}{c}{Week}	
			1	2	3	4	5	6	7	8	9	10	11	12	13	14	Total
BCWS	Planned Value	PV	74	227	415	615	803	1004	1192	1276	1389	1477	1527	1537	1554	1590	BAC
ACWP	Actual Cost	AC	68	246	443	682	815	1098	1273	1358	1457	1544	1589	1599	1610	1658	EAC
BCWP	Earned Value	EV	104.0	238.0	356.6	490.8	594.4	848.7	1127.4	1216.7	1322.3	1436.2	1507.1	1518.0	1539.8	1590.0	
Estimates																	
	Estimate At Complemtion	EAC	1553.60	1597.76	1676.84	1780.99	1811.12	1839.16	1735.60	1731.39	1724.65	1698.16	1672.01	1671.19	1659.95	1658.39	

Figure 145: Scenario's project – Estimate At Completion over the project's duration.

Based on the plots for SPI, CPI, and EAC, in [Figure 144] and [Figure 145], it could be assumed that the close timing of the positive turnaround of these three performance measures during the 5th and 6th weeks was possibly due to some form of breakthrough or process improvement rather than just putting in extra overtime hours worked (given that the costs CPI also improved). ... However ??

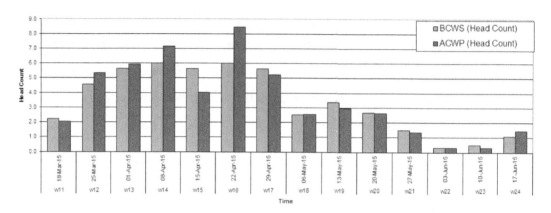

Figure 146: Scenario's project – actual weekly Head Count over the project's duration.

From the actual weekly [People] head count bar chart, [Figure 146], there has been a few weeks (i.e. Week 4 and Week 6) where the project has required noticeably more [People] than was planned. If only those assigned [People] were used on this project and no one else recorded [Time] against the project's timesheet job codes, then this chart indicates the amount of "overtime" that these assigned [People] had to work extra on the project.

"Though, discrepancies with timesheet reporting can transform into inflation or deflation of the project's Actual Cost Values, and this can result in misaligned Earned Value Performance Measures which dilutes the tracking capability. So, be aware if project team members are white-washing the project's timesheet job code with a constant daily number of hours per work day; and, especially watch out for people using the project's job code as a catch all for whatever is their daily activities."

Which brings us to the next rule of "Adaptive & Proactive" SDLC Project Management.

RULE 55: People should only book for the time that they spend on the project, and not the time that they think is expected to be spent.

Performance Measuring When The Time Worked Is Not Recorded

At some performing organizations, timesheets are not utilized and the **hours worked are not recorded against the project**; hence, in such cases **there would be no data available to determine the Actual Costs (AC), and thus the Earned Value Performance Measurement (EVPM) equations for CPI and EAC would be incomplete**. ... And, some organizations don't use EVPM calculations at all; therefore, **some other way is required to judge the project's progress when compared to the approved [Baselines]**.

Even though, EVPM is not being calculated, the baselined schedule can still be used to determine the **Planned Value (PV)**, and this PV **can be accumulated to obtained the Level Of Effort – Budget At Completion (BAC)**. Subsequently, the **Planned % Complete** for each measurement time period (*e.g. weekly*) **can be calculated**. Then on a routine basis, the **Project Implementation Team** can be **interviewed to determine** what **progress** has been made on **those scheduled tasks**; and thus, **determine the Actual % Complete** per time period. If on a routine basis, the **Planned % Complete** was **subtracted from the Actual % Complete** then this would produce a **variance in the scheduled progress being made**.

			\multicolumn{14}{c}{Week}	
			1	2	3	4	5	6	7	8	9	10	11	12	13	14	Total
BCWS	(weekly) Planned Value		74	152	188	200	188	200	188	84	112	88	50	10	16	36	1590
	Planned Value	PV	74	227	415	615	803	1004	1192	1276	1389	1477	1527	1537	1554	1590	BAC
	Planned % Complete	PV / BAC	5%	14%	26%	39%	51%	63%	75%	80%	87%	93%	96%	97%	98%	100%	
	Actual % Complete (Schedule)		7%	15%	22%	31%	37%	53%	71%	77%	83%	90%	95%	95%	97%	100%	
	Actual - Planned % Complete		2%	1%	-4%	-8%	-13%	-10%	-4%	-4%	-4%	-3%	-1%	-1%	-1%	0%	

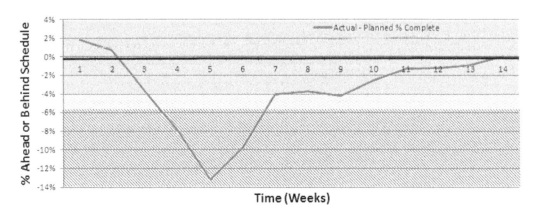

Figure 147: Scenario's project – % schedule variance over the project's duration.

As per [Figure 147], the performing organization could have **predefined numerical descriptors for acceptable & unacceptable variance** between the Planned % Complete and the Actual % Complete, where the **chosen numbers reflect the organization's risk tolerance**. Thus, providing "judgemental reference markers" to be presented in some form of "Project Status Report".

> Which brings us to the next rule of "Adaptive & Proactive" SDLC Project Management.

> **RULE 56:   Project management is blind without repeatable consistent ways of judging progress.**

Integrating Performance Measures Into The Program's KPIs

Having **routinely produced these performance measures** for the project, on a weekly basis, the summation performance data would be **integrated into the Key Performance Indicator (KPI) reporting for the Program Of Work** that contains this project as a constituent part (aka "work stream"), see [Section 3.6.5].

Of interest with the plots of the Performance Indexes versus Time (weekly) graph in [Figure 144] is the weekly trends for both the CPI [Figure 148] and the SPI [Figure 149].

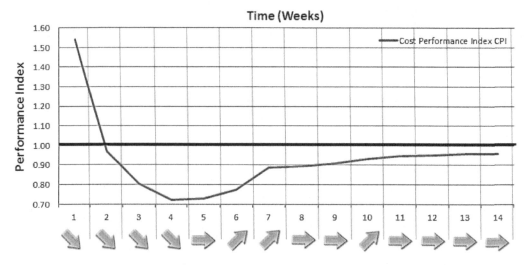

Figure 148: Scenario's project – CPI trend over the project's duration.

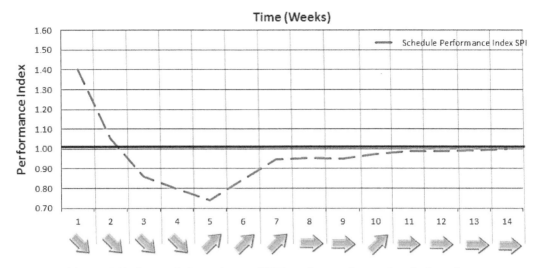

Figure 149: Scenario's project – SPI Trend over the project's duration.

For this project, is the **direction of** the respective performance index **trends on the rise or on the fall or are these stabilising** (hopefully to an index equal to ONE)?

Also of interest is whether the **performance indexes** are **staying within** their respective "Performance Index Windows" around the ideal value of ONE.

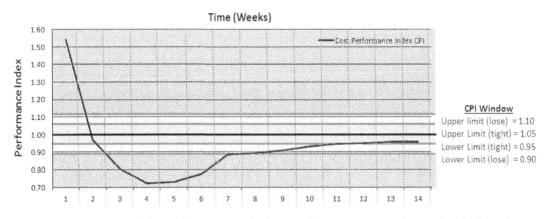

Figure 150: Scenario's project – CPI relative performance over the project's duration.

See [Figure 150], ... as long as the current index value is within the prescribed Performance Index Window ("green zone" per se) then all is well, but outside of this window in the "amber zone" then all is not well, and when in the "red zone" then ... *"please explain; what's your plan, to turn this situation around, as this is not good."*

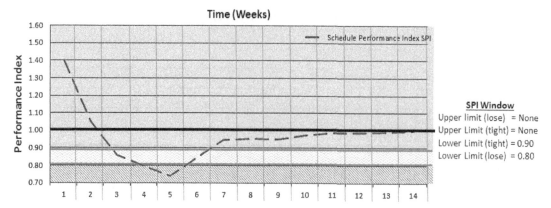

Figure 151: Scenario's project – SPI relative performance over the project's duration.

NOTE: The Performance Index Windows used for SPI and CPI don't necessarily have to have exactly the same limits. Depending on what is more important, [Time] progress or [Cost] spend then either the SPI or the CPI window could be tighter or looser than the other's performance window.

NOTE: As per [Figure 151], **the Performance Index Window doesn't necessarily have to have an upper limit**, for it could be decreed that as long as the value is near ONE or greater then all is good with that performance index's aspect of the project.

However, **there must always be a lower limit for both of these Performance Index Windows**, and what **value is set for this lower limit will represent the** "**risk appetite**" of the primary stakeholders; where the **closer to ONE** then the more "**risk adverse**", whereas the **further from ONE** then the more "**risk tolerant**".

Which brings us to an addendum to "Adaptive & Proactive" SDLC Project Management.

RULE 56^A: **Project management is illuminated by utilizing visual ways of summarizing the progress made, what's going on, and what's wrong.**

7.3.6. Key Performance Indicators and Program Reporting

Each reporting cycle (*e.g. weekly*) the Project Manager would take the most recent slice of performance measuring information for the project and enter this into the performing organization's "standardized" **Project Status Report – PSR | Weekly Progress Report – WPR**. ... *"Ideally, a project summed up in detail as a Project On A Page."*

This Project Status Report template would include fields for such information as:

- ☑ **Identification of which project** is being reported, the **time period** being reported, who is the **Project Manager – Technical Lead – Program Manager – Project Sponsor**, a "reminder" **short description of the project**, and the **type of project** (i.e. Fixed Price, Time & Material, Internal, Research & Development R&D, or Commercialization).

- ☑ **What SDLC phase is the project in**; i.e. Initiating, Planning, Executing, Closing.

- ☑ **Cost Performance Index** (CPI), **Schedule Performance Index** (SPI), **Trend arrows**, and, depending on what else has also been stipulated for inclusion in the report, the baselined **Budget At Completion** (BAC), the current **Estimate At Completion** (EAC), and possibly the EVPM plots for the **S-Curve, SPI & CPI, EAC**, and **% Progress Variance**.

> *"It is very important that consistent reporting periods and data cut off days are used for the reports related to a particular project & program of projects; as to use varying reporting durations would be the analogous equivalent of comparing apples to oranges, while both are fruits, these ain't the same thing to use to determine if you're having a good apple growing season.*
>
> *Also, there needs to be a 'single source of truth' for information as differing sources could produce contradictory information, with the result being that different parts of the organization could end up making conflicting decisions. ... All because, we used this Monday's data, but they used Wednesday's last week."*

- ☑ **List of the Key** (delivery) **Milestones**, including the **percentage** progress towards **completion**, the **baseline delivery dates**, the currently **forecasted delivery dates**, a "RAG Status" **Red-Amber-Green indicator** per milestone as to whether the achievement of the baseline target is at risk, an indicator as to whether a decision is required for **Baseline Change Requests** or a **re-baselining has recently occurred**, and **any comments** about each milestone's change of status (or why the current Amber or Red irrespective of its status in the previous report).

    "If the project is re-baselined and subsequently the RAG Status changes from last week's RED / AMBER to this week's GREEN then definitely include a comment as to the occurrence of the re-baselining; because, the last thing that the Project Manager needs is interrogation by the Steering Committee as to the reasons why the sudden change in status, ... especially if there are a large number of projects & programs being reported on. Remembering that, the Steering Committee probably only has last times Project Status Report or Program Update Report as a frame of reference to go by."

- ☑ If **key milestones** are **not** going to be **achieved then; what is the impact of** such a **delay, what is the proposed Recovery Plan**, and **what are the recommendations** to resolve any problems with this project.

    "Remembering that, the Project Manager's function is to manage the project; whereas, the Project Steering Committee provides strategic governance decisions as to how the project & program can best benefit the performing organization and the customer organization."

- ☑ Summarize the **key achievements during the reporting period**, and what are the **key planned activities for the next reporting period**.

- ☑ **List of Risks**, a short description of each risk, the **date when the risk was identified**, the **date when the risk was last updated**, each risk's **estimated impact on** the project's **schedule & budget**, a **Red-Amber-Green indicator** per risk as to the **likelihood** of its occurrence, and a reference to the associated **Recovery Plan**.

- ☑ **List of Issues**, a **short description of each issue**, the **date when the issue occurred**, the **date when the issue was last updated**, each issue's **estimated impact on** the project's **schedule & budget**, a **Red-Amber-Green indicator** per issue as to the **severity** of its happening, and a reference to the associated **Recovery Plan**.

    "As the Project Manager (with the assistance of the Program Manager and utilizing input from the Project Implementation Team), you should be the one who is putting forwards the proposed Recovery Plans; NOT, leaving it to the Steering Committee to initiate proceedings to recover an ailing project. ... As this is the Project Manager's (and the Program Manager's) responsibility to do that; else, what have they been employed to do? ... Produce fancy graphs, bar charts, and coloured boxes."

- ☑ Are there any [People] & [Resources] **utilization & availability concerns & conflicts**, including a **Red-Amber-Green indicator** per **the urgency** and **the impact** on the project, ... possibly including the weekly head count bar chart to depict the under staffing of the project or the overtime work of those persons assigned to the project.

- ☑ **List of Decision Points** (including a short description of each decision required), by **whom** the particular **decision needs to be made**, and **by** when a particular **date**.

This **Project Status Report** would then be **submitted** (by a particular day of the week) to the **Program Manager for review** (and further information consultation). The Program Manager would in turn use this latest slice of project information, **combined with** similar information from the other **constituent projects in the Program Of Work** (that they are responsible for) to **produce** a "standardized" (weekly) **Program Update Report | Project Dashboard Report**, see [Figure 152]. ... And, based on this accumulated information, the Program Manager (and the Project Steering Committee) can visually determine which projects in the program need further investigation into the causes of the problems that are being encountered for; [Cost] overruns, being behind [Time] schedule, and at <Risk> of not being delivered to the agreed milestones.

The project & program **Progress Reports need to include some way of comparing the current report's progress against the previous period's report** (without the need to have any previous reports at hand); hence, the inclusion of such things as the CPI & SPI & EAC graphs and the trend arrows.

Also, **each Progress Report needs to be for a consistent time duration and be synchronized from one reporting period to the next**. ... *"Not ad-hoc."*

Indicator	Project 1	Project 2	Project 3	Project 4	Project 5	Project 6	Project 7
Project Manager	<PM Name>	<PM Name>	<PM Name>	<PM Name>	<PM Name>	<PM Name>	<PM Name>
Technical Lead	<TL Name>	<TL Name>	<TL Name>	<TL Name>	<TL Name>	<TL Name>	<TL Name>
Contract Type	R&D	Fixed Price	Commercialisation	Internal	Fixed Price	Fixed Price	Internal
State	Kick-Off	Execution	Execution	Execution	Execution	Execution	Closure
CPI (Hours)	0.96	1.01	0.90	0.98	1.00	0.82	0.89
SPI (Hours)	1.11	0.97	0.96	1.07	1.00	0.91	1.45
Milestones	GREEN	YELLOW	RED	GREEN	GREEN	GREEN	GREEN
CPI Trend	→	↗	↗	↗	→	↗	→
SPI Trend	→	→	↘	↘	↘	↗	→

Figure 152: Program Key Performance Indicator Dashboard.

As depicted in [Figure 152], Project 3 is in some amount of trouble with respect to [Cost] overruns and is at risk of not achieving its next key milestone; similarly, Project 6 and Project 7 are also incurring [Cost] overruns. If the Program Manager (prior to the Steering Committee meeting) was to observe that these troubled projects all happen to have the same Project Manager and/or Technical Team Lead then questions definitely **need to** be asked **as to the reasons for the difficulties** that are (personally) being **encountered**.

Indicator	Project 1	Project 2	Project 3	Project 4	Project 5	Project 6	Project 7
COST	0.96	1.01	0.90	0.98	1.00	0.82	0.89
SCHEDULE	1.11	0.97	0.96	1.07	1.00	0.91	1.45
ON TARGET	GREEN	YELLOW	RED	GREEN	GREEN	GREEN	GREEN
Cost TREND	→	↗	↗	↗	→	↗	→
Time TREND	→	→	↘	↘	↘	↗	→

Figure 153: More focus view of the KPI Dashboard.

Don't Tell Me Everything, Only Tell Me What I Really Need To Know

For a performing organization with multiple projects and programs (all occurring at the same time), then **those stakeholders higher up the management structure** usually **don't have the time available to take in the copious amounts of information about** the inner workings of **each project | program**. Instead, they **need** such information in a **standardized regimented concise summation of the key facts, the percentage progress made towards delivery milestones** (not activities done), **the noteworthy risks & issues, and the recommendations**; so that, they can **quickly understand the current situation for each of the projects & programs within their domain of responsibility**. Thereby, **enabling** them to **make timely rational decisions** with respect to each project or for the program, and to be able to ask for more in-depth information on the specific areas of concern to them. Hence, **the less quantity of information that they require per project | program, but the better the filtered quality of the information that they need**.

Thus, for Key Performance Indicators(KPIs) the important things to remember are:

- ☑ **Base KPIs on tangible & measurable objectives that relate to what is to be achieved**, such as; the progress of [Scope] coverage, the advancement towards specific milestone dates [Time], the current budgetary spend [Cost], ... and, the defect rate & clearance rates [Quality], plus the levels of [People] & [Resources] being utilized.

- ☑ **KPIs that can be repeatably quantifiably measured, over consistent time durations, from consistent points of reference.**

"The trick with selecting KPIs are choosing those which are the most effective at providing an insight into the project's current situation. That is, KPIs that are S.M.A.R.T.

S ... Specific to the understanding of a particular aspect of the project,
M ... Measurable based quantifiable data which relates to that aspect,
A ... Attainable & agreed goals which relate to that aspect,
R ... Relevant to what is trying to be achieved for that aspect, and
T ... Timeboxed period over which that aspect is of importance."

Alternate Universe View Of The Same Project ... *as demonstrated*

			Week																			
			1	2	3	4	5	6	7	8	9	10	11	12	13	14	15	16	17	18	19	20
Budget Cost of Work Scheduled																						
	(weekly) Planned Value		74	152	188	200	188	200	188	84	112	88	50	10	16	36	0	0	0	0	0	0
BCWS	Planned Value	PV	74	227	415	615	803	1004	1192	1276	1389	1477	1527	1537	1554	1590	1590	1590	1590	1590	1590	1590
	Planned % Complete	PV/BAC	5%	14%	26%	39%	51%	63%	75%	80%	87%	93%	96%	97%	98%	100%	100%	100%	100%	100%	100%	100%
Actual Cost of Work Performed																						
	(weekly) Actual Cost		68	178	198	238	134	282	175	85	99	87	45	10	11	49	0	0	0	0	0	0
ACWP	Actual Cost	AC	68	246	443	682	815	1098	1273	1358	1457	1544	1589	1599	1610	1658	1658	1658	1658	1658	1658	1658
Budget Cost of Work Performed																						
	Actual % Complete (Schedule)		7%	15%	22%	31%	37%	53%	71%	77%	83%	90%	95%	95%	97%	100%	100%	100%	100%	100%	100%	100%
BCWP	Earned Value	EV	104.0	238.0	356.6	490.8	594.4	848.7	1127.4	1216.7	1322.3	1436.2	1507.1	1518.0	1539.8	1590.0	1590.0	1590.0	1590.0	1590.0	1590.0	1590.0
Performance Indexes																						
	Schedule Performance Index	SPI	1.40	1.05	0.86	0.80	0.74	0.85	0.95	0.95	0.95	0.97	0.99	0.99	0.99	1.00	1.00	1.00	1.00	1.00	1.00	1.00
	Cost Performance Index	CPI	1.54	0.97	0.80	0.72	0.73	0.77	0.89	0.90	0.91	0.93	0.95	0.95	0.96	0.96	0.96	0.96	0.96	0.96	0.96	0.96
Estimates																						
	Estimate At Complemtion	EAC	1553.60	1597.76	1676.84	1780.99	1811.12	1839.16	1735.60	1731.39	1724.65	1698.16	1672.01	1671.19	1659.95	1658.39	1658.39	1658.39	1658.39	1658.39	1658.39	1658.39
	To Complete Performance Index	TCPI	0.98	1.01	1.08	1.21	1.29	1.51	1.46	1.61	2.01	3.37	92.42	-7.84	-2.54	0.00	0.00	0.00	0.00	0.00	0.00	0.00
Progress Measurs																						
	Actual - Planned % Complete		2%	1%	-4%	-8%	-13%	-10%	-4%	-4%	-4%	-3%	-1%	-1%	-1%	0%	0%	0%	0%	0%	0%	0%
HEAD COUNT																						
	Planned Head Count		2.24	4.59	5.67	6.03	5.67	6.03	5.67	2.54	3.38	2.66	1.51	0.31	0.49	1.09	-	-	-	-	-	-
	Actual Head Hount		2.04	5.37	5.95	7.18	4.03	8.51	5.27	2.56	2.98	2.63	1.35	0.30	0.32	1.46	-	-	-	-	-	-

Figure 154: Earned Value Performance Measures – for the Demonstrated Project.

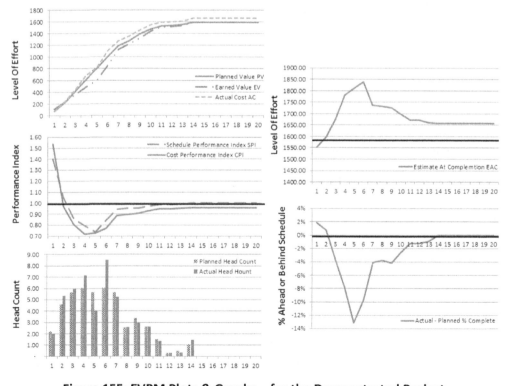

Figure 155: EVPM Plots & Graphs – for the Demonstrated Project.

Alternate Universe View Of The Same Project ... "slipping away"

			Week																			
			1	2	3	4	5	6	7	8	9	10	11	12	13	14	15	16	17	18	19	20
Budget Cost of Work Scheduled																						
	(weekly) Planned Value		74	152	188	200	188	200	188	84	112	88	50	10	16	36	0	0	0	0	0	0
BCWS	Planned Value	PV	74	227	415	615	803	1004	1192	1276	1389	1477	1527	1537	1554	1590	1590	1590	1590	1590	1590	1590
	Planned % Complete	PV/BAC	5%	14%	26%	39%	51%	63%	75%	80%	87%	93%	96%	97%	98%	100%	100%	100%	100%	100%	100%	100%
Actual Cost of Work Performed																						
	(weekly) Actual Cost		91	212	173	144	273	156	134	100	89	100	60	13	19	28	44	43	38	0	0	0
ACWP	Actual Cost	AC	91	303	476	620	893	1049	1183	1283	1372	1472	1532	1545	1564	1592	1636	1679	1717	1717	1717	1717
Budget Cost of Work Performed																						
	Actual % Complete (Schedule)		3%	13%	26%	33%	48%	63%	73%	77%	83%	87%	89%	90%	90%	93%	95%	96%	99%	100%	100%	100%
BCWP	Earned Value	EV	47.7	206.7	413.4	524.7	763.2	1001.7	1160.7	1224.3	1319.7	1383.3	1415.1	1431.0	1431.0	1478.7	1510.5	1526.4	1574.1	1590.0	1590.0	1590.0
Performance Indexes																						
	Schedule Performance Index	SPI	0.64	0.91	1.00	0.85	0.95	1.00	0.97	0.96	0.95	0.94	0.93	0.93	0.92	0.93	0.95	0.96	0.99	1.00	1.00	1.00
	Cost Performance Index	CPI	0.52	0.68	0.87	0.85	0.85	0.95	0.98	0.95	0.96	0.94	0.92	0.93	0.91	0.93	0.92	0.91	0.92	0.93	0.93	0.93
Estimates																						
	Estimate At Complemtion	EAC	1633.30	1686.30	1652.60	1685.30	1719.80	1637.30	1612.30	1648.70	1642.30	1678.70	1706.90	1704.00	1723.00	1703.30	1715.50	1742.60	1732.90	1717.00	1717.00	1717.00
	To Complete Performance Index	TCPI	1.03	1.07	1.06	1.10	1.19	1.09	1.05	1.19	1.24	1.75	3.02	3.53	6.12	-53.63	-1.73	-0.71	-0.13	0.00	0.00	0.00
Progress Measurs																						
	Actual - Planned % Complete		-2%	-1%	0%	-6%	-3%	0%	-2%	-3%	-4%	-6%	-7%	-7%	-8%	-7%	-5%	-4%	-1%	0%	0%	0%
HEAD COUNT																						
	Planned Head Count		2.24	4.59	5.67	6.03	5.67	6.03	5.67	2.54	3.38	2.66	1.51	0.31	0.49	1.09						
	Actual Head Hount		2.74	6.39	5.21	4.34	8.22	4.70	4.04	3.01	2.68	3.01	1.81	0.39	0.57	0.84	1.33	1.30	1.14			

Figure 156: Earned Value Performance Measures – "slipping away".

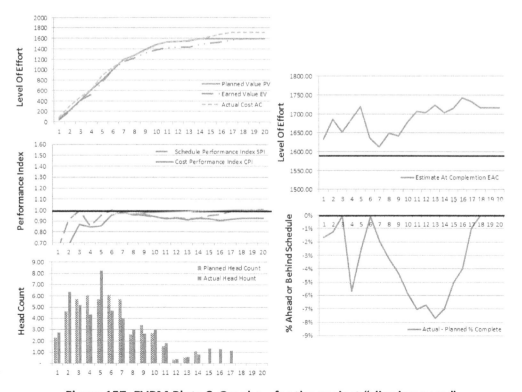

Figure 157: EVPM Plots & Graphs – for the project "slipping away".

Alternate Universe View Of The Same Project ... *with an early finish*

			Week																		
			1	2	3	4	5	6	7	8	9	10	11	12	13	14	15	16	17	18	19	20
Budget Cost of Work Scheduled																						
	(weekly) Planned Value		74	152	188	200	188	200	188	84	112	88	50	10	16	36	0	0	0	0	0	0
BCWS	Planned Value	PV	74	227	415	615	803	1004	1192	1276	1389	1477	1527	1537	1554	1590	1590	1590	1590	1590	1590	1590
	Planned % Complete	PV/BAC	5%	14%	26%	39%	51%	63%	75%	80%	87%	93%	96%	97%	98%	100%	100%	100%	100%	100%	100%	100%
Actual Cost of Work Performed																						
	(weekly) Actual Cost		102	181	196	140	158	268	269	83	162	0	0	0	0	0	0	0	0	0	0	0
ACWP	Actual Cost	AC	102	283	479	619	777	1045	1314	1397	1559	1559	1559	1559	1559	1559	1559	1559	1559	1559	1559	1559
Budget Cost of Work Performed																						
	Actual % Complete (Schedule)		3%	11%	23%	40%	57%	66%	81%	88%	98%	100%	100%	100%	100%	100%	100%	100%	100%	100%	100%	100%
BCWP	Earned Value	EV	47.7	174.9	365.7	636.0	906.3	1049.4	1287.9	1399.2	1558.2	1590.0	1590.0	1590.0	1590.0	1590.0	1590.0	1590.0	1590.0	1590.0	1590.0	1590.0
Performance Indexes																						
	Schedule Performance Index	SPI	0.64	0.77	0.88	1.03	1.13	1.05	1.08	1.10	1.12	1.08	1.04	1.03	1.02	1.00	1.00	1.00	1.00	1.00	1.00	1.00
	Cost Performance Index	CPI	0.47	0.62	0.76	1.03	1.17	1.00	0.98	1.00	1.00	1.02	1.02	1.02	1.02	1.02	1.02	1.02	1.02	1.02	1.02	1.02
Estimates																						
	Estimate At Completion	EAC	1644.30	1698.10	1703.30	1573.00	1460.70	1585.60	1616.10	1587.80	1590.80	1559.00	1559.00	1559.00	1559.00	1559.00	1559.00	1559.00	1559.00	1559.00	1559.00	1559.00
	To Complete Performance Index	TCPI	1.04	1.08	1.10	0.98	0.84	0.99	1.09	0.99	1.03	0.00	0.00	0.00	0.00	0.00	0.00	0.00	0.00	0.00	0.00	0.00
Progress Measurs																						
	Actual - Planned % Complete		-2%	-3%	-3%	1%	6%	3%	6%	8%	11%	7%	4%	3%	2%	0%	0%	0%	0%	0%	0%	0%
HEAD COUNT																						
	Planned Head Count		2.24	4.59	5.67	6.03	5.67	6.03	5.67	2.54	3.38	2.66	1.51	0.31	0.49	1.09	.	.	.	.	.	.
	Actual Head Hount		3.07	5.45	5.90	4.22	4.76	8.07	8.10	2.50	4.88	.	.	.	.	.	.	.	.	.	.	.

Figure 158: Earned Value Performance Measures – with an early finish.

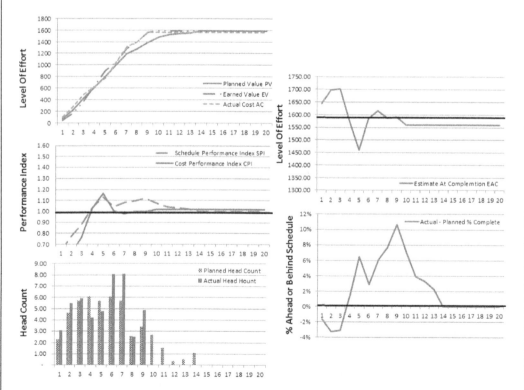

Figure 159: EVPM Plots & Graphs – for the project with an early finish.

Alternate Universe View Of The Same Project ... *with a late finish*

			Week																		
			1	2	3	4	5	6	7	8	9	10	11	12	13	14	15	16	17	18	19	20
Budget Cost of Work Scheduled																						
	(weekly) Planned Value		74	152	188	200	186	200	188	84	112	88	50	10	16	36	0	0	0	0	0	0
BCWS	Planned Value	PV	74	227	415	615	803	1004	1192	1276	1389	1477	1527	1537	1554	1590	1590	1590	1590	1590	1590	1590
	Planned % Complete	PV / BAC	5%	14%	26%	39%	51%	63%	75%	80%	87%	93%	96%	97%	98%	100%	100%	100%	100%	100%	100%	100%
Actual Cost of Work Performed																						
	(weekly) Actual Cost		51	140	250	142	209	148	158	94	130	89	68	9	20	38	22	30	24	40	44	0
ACWP	Actual Cost	AC	51	191	441	583	792	940	1098	1192	1322	1411	1479	1488	1508	1546	1568	1598	1622	1662	1706	1706
Budget Cost of Work Performed																						
	Actual % Complete (Schedule)		2%	11%	20%	31%	40%	51%	61%	68%	73%	81%	84%	84%	85%	87%	90%	93%	94%	95%	96%	100%
BCWP	Earned Value	EV	31.8	174.9	318.0	492.9	636.0	810.9	969.9	1081.2	1160.7	1287.9	1335.6	1335.6	1351.5	1383.3	1431.0	1478.7	1494.6	1510.5	1526.4	1590.0
Performance Indexes																						
	Schedule Performance Index	SPI	0.43	0.77	0.77	0.80	0.79	0.81	0.81	0.85	0.84	0.87	0.87	0.87	0.87	0.87	0.90	0.93	0.94	0.95	0.96	1.00
	Cost Performance Index	CPI	0.62	0.92	0.72	0.85	0.80	0.86	0.88	0.91	0.88	0.91	0.90	0.90	0.90	0.89	0.91	0.93	0.92	0.91	0.89	0.93
Estimates																						
	Estimate At Completion	EAC	1609.20	1606.10	1713.00	1680.10	1746.00	1719.10	1718.10	1700.80	1751.30	1713.10	1733.40	1742.40	1746.50	1752.70	1727.00	1709.30	1717.40	1741.50	1769.60	1706.00
	To Complete Performance Index	TCPI	1.01	1.01	1.11	1.09	1.20	1.20	1.26	1.28	1.60	1.69	2.29	2.49	2.91	4.70	7.23	-13.91	-2.98	-1.10	-0.55	0.00
Progress Measurs																						
	Actual - Planned % Complete		-3%	-3%	-6%	-8%	-11%	-12%	-14%	-12%	-14%	-12%	-12%	-13%	-13%	-13%	-10%	-7%	-6%	-5%	-4%	0%
HEAD COUNT																						
	Planned Head Count		2.24	4.59	5.67	6.03	5.67	6.03	5.67	2.54	3.38	2.66	1.51	0.31	0.49	1.09	.	.	.	.	.	.
	Actual Head Hount		1.54	4.22	7.53	4.28	6.30	4.46	4.76	2.83	3.92	2.68	2.05	0.27	0.60	1.14	0.66	0.90	0.72	1.20	1.33	

Figure 160: Earned Value Performance Measures – with a late finish.

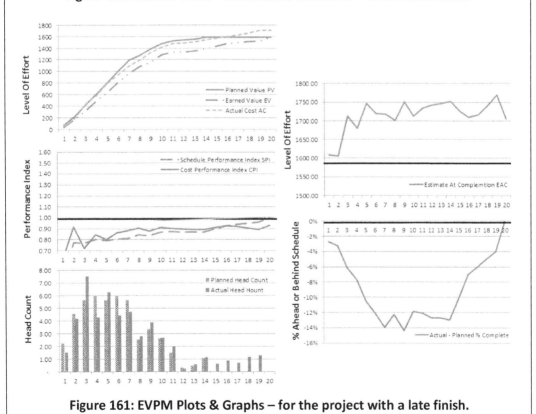

Figure 161: EVPM Plots & Graphs – for the project with a late finish.

463

Don't Tell Me How Hard You're Working, If There Is Nothing To Show

Heard a good one-liner the other day;

"DON'T CONFUSE MOTION WITH PROGRESS."

Which brings us to an addendum to "Adaptive & Proactive" SDLC Project Management.

RULE 56[B]: Don't equate the Level Of Effort exerted as a truthful representation of the progress that has been made towards the next milestone.

Analogously for a car trip, a high revving engine, decreasing fuel, and rising temperature does not mean that one is any closer to the destination; as, one could be bogged down, spinning wheels, but not making notable forwards progress. Thus, the true measure of the trip's advancement is the distance travelled towards the destination in the time taken, when compared to the planned trip's distance and the expected arrival time.

Yet, how often does the Project Team (and even the Project Manager) think that all of the effort that they are putting in, will be acknowledged (let alone rewarded) by those above them in the management hierarchy ??

Recall back in [Section 2.8.2], **"Customer Satisfaction" is the combination of [Scope] & [Quality]**, and the project's **"Performance Measures" is the combination of [Time] & [Cost]**. Hence, when the [Scope] is simplified down to getting to the next stated delivery milestone, as the [Time] goes by and the [Cost] of the Level Of Effort continues to increase, then subsequently there ain't going to be much "satisfaction" with the progress being made, irrespective of the [Quality] of the output, ... and, the efforts put in.

Therefore, when selecting Key Performance Indicators, remember whose perspective that these KPIs need to inform. That is, **choose progress markers & indicators that consider the primary stakeholders concerns for [Scope], [Time], and [Cost] with the <Risks> to the [Quality], [People], and [Resources] involved with the project**. See [Section 2.7].

Integrated View Of Implementation, Management, And Governance

Back in [Section 2.3.4] the project management hierarchy & governance structure was illustrated, with at the top the **Project Steering Committee** concerned with the "big picture" **strategizing**, in the middle the **Project Working Group** (including the Project Manager) concerned with the **tactical directions**, and on the bottom the **Project Implementation Team** concerned with the **hands-on implementation** of the project.

Also recall, back in [Section 4.3.2], the depiction of the relationship between the agile Project Implementation Team's daily stand-up meeting forming a Recognize – Reassess – Revise – Reapply cycle meshing with the iterative Plan – Do – Check – Act cycle.

So, **how to integrate the running of the project with those business objectives** of the performing organization; that is, **with the governance function** of the Project Steering Committee ??

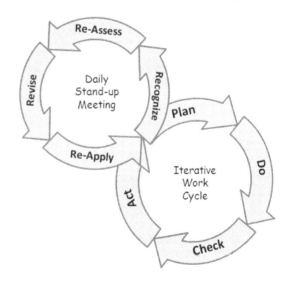

These cycles could be meshed together as cogs with Risk Management [Section 3.6.8], Stakeholder Management [Section 3.6.9], and Change Management [Section 3.6.7], as modelled in [Section 2.6]. ... Realizing that, one of the primary stakeholders who will be deciding on whether to deem the project as a Success | Failure is the performing organization's own senior management (aka the Project Steering Committee *and the Program Manager*); hence, they being some of the stakeholders who's needs & wants, concerns, expectations, perceptions, and opinions that have to be managed (in addition to those of the Customer's Representatives). Therefore, the need to incorporate Stakeholder Management into the working model of the project's undertaking. Though, as detailed in [Section 2.6], there are Risks & Issues associated with these Stakeholders as they will necessitate that Changes be made to [Baselines] and such Changes will induce Risks to the project, so on & so forth, ... and, all the while affecting how the project is managed and subsequently implemented.

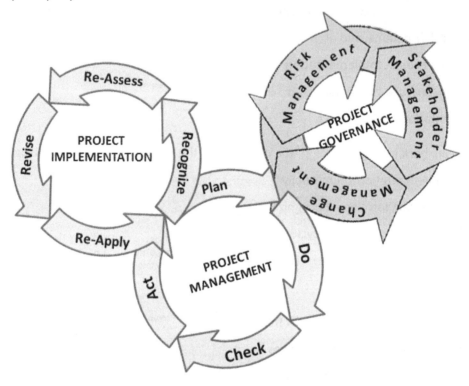

Figure 162: Relationship of Project Implementation, Management, and Governance.

Which brings us to an addendum to "Adaptive & Proactive" SDLC Project Management.

RULE 56^C: Project management must be accompanied by governance; because, doing something on time and on budget is pointless without it providing strategic benefit.

COFFEE BREAK DISCUSSION ... *Can't see the future for all of the past.*

An unrecognized problem that many an organization (especially a large organization) has, is that they don't really know the up-to-date true state of their projects (and more specifically their Programs Of Work) due to the use of performance monitoring KPIs that are all based on historical data, with very few if any "real time" forward looking triggers.

While the originally collected project performing information may have only been a few days behind the current state of the project (when the weekly Project Status Report was put together by the Project Manager), by the time that this information is incorporated into the monthly Program Management Report and subsequently percolated up through the organization's management hierarchy, then this information could have become rather outdated. Consequently, at the time when Executive Management hold their routine leadership meeting to evaluate the project | program statuses, this information is now relatively ancient history of a couple to a few weeks old. Thus, if it takes almost a full month for Executive Management's oversight review & questioning to realize that the project's progress is not compatible with the organization's greater business objectives, then there could be a sizable amount of "collateral" financial expenses involved with redirecting this wayward project (or program of projects).

> *Consider a Project Team of say 10 persons, with a "bums on seats" cost of say $125 per hour, working a 40 hour per person week, is about a ... $125 x 10 x 40 = $50,000 per week, with approximately 4.33 weeks per month over a 3 months spread, is about a $217,000 monthly burn rate, ... plus other expenses & overhead, say $250K per month to change direction of that 10 person team. ... Now scale this ten person cost up or down based on the actual number of persons involved with the affected project | program.*

Hence, it is very-very important that the Project Manager keep a diligent eye on the true state of their assigned projects and warn of changing project statuses; because in essence they function as the ship's crow's-nest lookout, vigilant for any icebergs ahead.

8. PM you ... Field Test 4

8.1. Overview

This field test examines the following topics with respect to real-world project management:

- R.I.S.C. Management
    - Risks & Issues Management,
    - Stakeholder Management, and
    - Change Management.
- The Project Manager's actions & reactions influencing the project being deemed as a success.
    - Handling stakeholder requests,
    - Delegating downwards, managing upwards, and decorum.

8.2. Introduction

That last field test concentrated on the intricacies of project monitoring & control; whereas, this current field test will focus on those higher-level project management topics that can be blurred from view due to the Project Manager and the Project Team being preoccupied with the "hamster's wheel" of getting the project done.

8.2.1. The Scenario

```
This project is your current real-life project as either
the Project Manager or Project Implementation Team member.
```

8.3. R.I.S.C. Management [Section 2.6]

As contained in the KPIs list, in [Section 7.3.6], one of the areas of **concern for** the project's primary stakeholders (i.e. the **Project Steering Committee**) **is the Risks & Issues** involved with the project. Also, as mentioned in [Section 3.4.6], the **Project Steering Committee** is additionally concerned with the project's **approved [Baselines]**, and **any such changes to these approved [Baselines]** will **require** a **Baseline Change**, see [Section 3.6.7], **via** the **Change Control Board** which happens to **involve members of that Steering Committee**.

> Recalling the addendum to "Adaptive & Proactive" SDLC Project Management.
>
> **RULE 28[A]: Any Change can inadvertently introduce undesirable side effects that place the project at Risk.**

And, such "side effects" aren't just limited to this particular project, as **the failure to successfully manage those project R.I.S.C.s will result in some amount of impact on the containing Program Of Work**, and possibly impact upon the greater business objectives.

R.I.S.C.   [R]isk [I]ssues [S]takeholders [C]hange

The purpose of R.I.S.C. Management is to ensure that the project achieves its business objectives; while delivering the agreed [Scope], by the **agreed [Time] dates**, within the **agreed [Cost] budget**, to an **agreed level of customer satisfaction [Quality]**, and the **effective & efficient utilization of** those assigned **[People] and** the allocated **[Resources]**.

R.I.S.C. Management involves the following operations, See [Figure 163]:

1.. **Identification** – identifying the Risks (& Issues) | Stakeholders | Changes, associated with the project.

2.. **Analysis** – qualitatively & quantitatively analysing these identified Risks (& Issues) | Stakeholders | Changes, for the effects that these have or will have on the project.

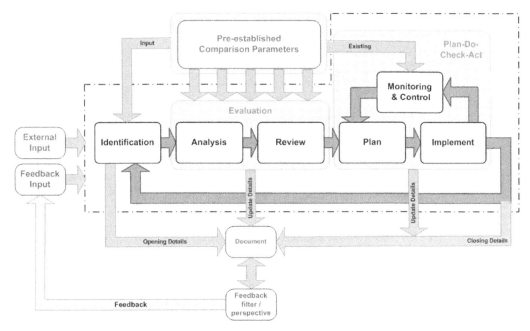

Figure 163: R.I.S.C. Management Process.

3.. **Review** – judging each of these identified Risks (& Issues) | Stakeholders | Changes, in comparison to the other Risks (& Issues) | Stakeholders | Changes, and determining the overall impact & affect that these have on the project & the program as a whole.

4.. **Plan** – the responses & strategies to be applied to each of these identified Risks (& Issues) | Stakeholders | Changes.

5.. **Implement** – the Plan-Do-Check-Act execution of those chosen response strategies.

6.. **Monitor & Control** – managing the interactions of those chosen response implementations with those identified Risks (& Issues) | Stakeholders | Changes.

7.. **Document** – for each Risk (Issue) | Stakeholder | Change, record the findings, the decisions made, the strategies to be implemented, those actions that were taken, and the outcomes of the current pass through the management process.

8.. **Feedback** – take what is "now known" about the identified risk (& Issue) | Stakeholder | Change, then feed this back to the beginning of the R.I.S.C. Management Process to be combined with any new inputs, then iteratively pass all of this through the process again & again until the matter is resolved and the associated effects are under control.

8.3.1. Risk Monitoring & Control [Section 3.6.8]

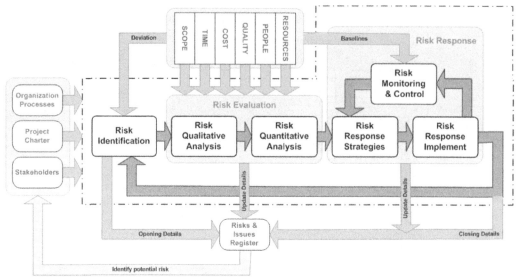

Figure 67: Risk Management Process.

One of the most common ways of tracking risks & issues (and the associated actions), ... *"other than not tracking these at all, or trawling through email trails"*, is via the use of a Risks Register and an Issues Log, or an amalgamated Risks & Issues Register (with separate worksheets in the spreadsheet for both types).

Risks Register & Issues Log

The **Risks Register contains the details of those identified risks**; and, typically consists of a spreadsheet or database that includes such information as:

- A **unique identifying number** for each identified risk; *e.g. Risk Item | Risk Number*.

- A semi-detailed **description** (or **short summation**) of the identified risk.

- The **date** when the risk was **identified** and (optionally) **who identified** the risk.

- An identifier as to **who has been assigned responsibility** for the particular risk.

- The **current status** of the risk; i.e. OPEN, CLOSED, (now an) ISSUE.

- The **date** when the risk **last changed status**.

- The perceived **impact | severity** of the risk; i.e. negligible, minor | minimal, moderate, major | critical, severe | catastrophic.

    o   The possible **[Cost] impact** if this risk was to become an issue.

    o   The possible **[Time] impact** if this risk was to become an issue.

- The perceived **probability | likelihood** of the risk becoming an issue; i.e. not likely | rare, unlikely but plausible, may happen | moderately likely, likely, almost certain.

- The **response** to be implemented; i.e. to **accept, avoid, mitigate, transfer,** or **share**.

- Any relevant **comments** about the particulars of the risk.

The **Issues Log contains information about risks when these become issues** (and the actions taken); i.e. when a risk became a reality of fact then its associated risk entry information in the Risk Register would be copied over to a corresponding entry in the Issues Log.

NOTE: Don't delete a risk or an issue entry from the Risks & Issues Register, irrespective of whether that risk / issue has been resolved or is no longer appropriate, **instead just mark its status as CLOSED or Not Applicable.**

And, this **Risks & Issues Register would be continually updated with the latest information and the decisions that have been made relating to each risk / issue.**

Risk Breakdown Structure

A **Risk Breakdown Structure (RBS)** may also be used **to group & stack similar | related risks & issues into meaningful categories**; *e.g. internal : external, hardware : software, technical : managerial, physical subsystems, componentry, applications, etc*. Where each box would represent a particular risk / issue; and, these boxes can be **colour-coded based on the given Risk Priority**, see [Figure 68]. Thereby, possibly **unearthing "clusters" of risks & issues that have common cause(s)**, and depicting the **build-up of high priority "RED" boxes in certain categories**; subsequently, the fixing of certain common cause(s) could potentially eliminate a concentration of risks & issues.

Conducting a Risk Analysis ... Extract from "A Down-To-Earth Guide To SDLC Project Management".

1.. **What risks** are perceived to be associated with this project?

- Having checked that this risk has not been covered before, give the identified risk a **unique name** and take the **next Risk Number** in the Risks Register.
- In the Risks Register **describe** each identified risk, what would cause this risk to happen, and provide **additional comments** about this risk.
- **Who knows** most about this risk, and **who** will be **responsible for resolving** this risk? Record this information in the Risks Register.

2.. **What is the "perceived impact"** to the project if this risk was to occur; i.e. **how severe** would be **the consequences to the project**?

- **Score** this risk's **impact on a scale of 1-5** where <u>1 is least impact</u> and <u>5 is most impact</u>; i.e. 1 = negligible, 2 = minor, 3 = moderate, 4 = major, 5 = severe.
- What is the possible **monetary impact** due to this risk becoming an issue?
- What is the possible **duration impact** due to this risk becoming an issue?

3.. **What is the "perceived probability"** of the risk occurring; i.e. **how likely** is it for this risk to become an issue?

- **Score** this risk's **likelihood on a scale of 1-5** where <u>1 is least likely</u> and <u>5 is most likely</u>; i.e. 1 = rare (very unlikely), 2 = unlikely, 3 = may happen, 4 = likely to happen, 5 = almost certain (imminent).

4.. Map each identified risk onto the **Risk Matrix**, see [Figure 68].

Or alternatively, **calculate the Risk Priority | Risk Ranking | Risk Index** where,

$$\text{Risk Priority} = \text{Impact} \times \text{Probability}$$

Probability / Likelihood	(1) Negligible	(2) Minor	(3) Moderate	(4) Major	(5) Severe
(5) Almost Certain	5	10	15	20	25
(4) Likely	4	8	12	16	20
(3) May Happen	3	6	9	12	15
(2) Unlikely	2	4	6	8	10
(1) Rare	1	2	3	4	5

Impact / Severity

IDENTIFIED RISK	IMPACT (RATE 1-5)	X	PROBABILITY (RATE 1-5)	=	RISK PRIORITY (1-25)
[Risk 1]	[1-5]	X	[1-5]	=	[1-25]
[Risk 2]	[1-5]	X	[1-5]	=	[1-25]
[Risk 3]	[1-5]	X	[1-5]	=	[1-25]

Some performing organizations will have **predefined numerical descriptors for grading both the Impact and the Probability**; where these numerical values will clearly differentiate between each level of impact & probability when compared to the approved [Baselines]. Therefore, rating risks based on quantifiable ranges, rather than exclusively on the subjective opinions of the risk analysts.

For example; an impact of a budget overrun when compared to the [Cost Baseline] of 0-5 % is 1 = negligible, 5-10 % is 2 = minor, 10-20 % is 3 = moderate, 20-30 % is 4 = major, and 30 %-greater is 5 = severe. Similarly, for late running schedules [Time Baseline], the number of high-medium-low priority / severity defects reported [Quality] ... etc.

For example; a likelihood of 5% chance is 1 = rare, 15% chance is 2 = unlikely, 35% chance is 3 = may happen, 70% chance is 4 = likely, and 95% chance is 5 = almost certain.

5.. **Record** the calculated **Risk Priority** in the **Risks Register**.

6.. Add a box for each identified risk to the **Risk Breakdown Structure (RBS)** and position it in the appropriate category. Possibly colour code each risk box based on its Risk Priority; i.e. use the same colours as for the Risk Matrix.

7.. Is there a **concentration of risks** starting to form in the Risk Breakdown Structure; i.e. a disproportionate growth in the columns **for certain categories of risks**, or is there a disproportionate number of high priority colours in certain categories of risks? Are there apparent common factors to the build-up of groupings of risks?

8.. What is the **Time Criticality** of responding to each identified risk?

Are there **restrictions imposed by the timing of events**, such as a pending delivery milestone, and project deliverables feeding into other work-stream projects?

9.. What is the **Perceived Urgency** of establishing a response to each identified risk?

Where the perceptions of those affected stakeholders and that of the primary stakeholders will influence the urgency of dealing with each identified risk; i.e. **'political priority' versus the calculated Risk Priority.**

10.. With the risk having been prioritized & categorized, determine what is the **"Actual Urgency" of establishing Response Strategies** for each of these identified risks.

*For example; **the sooner that the risk is likely to become an issue then the more immediate will be the urgency** of finding a response to that risk.*

11.. During this analysis, **did any Risk Response Strategies become apparent**?

Risk Response Strategies, Implementation, and Fix Verification

What are the Risk Response Strategies for each of these identified risks?

Is the chosen option to; **"Accept"**, **"Avoid"**, **"Mitigate"**, **"Transfer"**, or **"Share"**.

1.. **Define** the **[Scope] boundaries** for the implementation & fix verification,

2.. **Allocate** the **[Time] & [Cost] budget** for the implementation & fix verification,

3.. **Assign** the **[People] & [Resources]** for the implementation & fix verification,

4.. **How will it be verified** that "The Plan'" was **implemented correctly with [Quality]**,

5.. **Assign** someone to be **responsible for resolving each risk**,

6.. Get the **Risk Response Plan authorized & approved** for implementation.

To maximize the effectiveness of the Risk Response Plan (and the subsequent risk resolution), **it is essential to involve those [People] who have sufficient power and**

influence within the performing organization, 3rd parties, and/or within the customer organization, **to be able to do something about the risk** (whether it be personally resolving the risk or by coordinating others to resolve the risk on their behalf). Therefore, **the responsibility & authority for the risk resolution should be assigned to someone who is capable of dealing with that particular risk / issue**.

Noting that, **the Project Manager** (and the Project Implementation Team) **may not necessarily have the appropriate power, position, or influence to successfully resolve each risk / issue**; subsequently, **the Project Working Group and the Project Steering Committee will also have to be involved**, so as to assist in ensuring that the risk / issue can be successfully resolved. Thus, it is **essential that the Project Manager accept that they are not always capable of controlling every risk & issue by themself**. Therefore, the Project Manager has to **be prepared to deliver bad news** about the risk / issue **up the management hierarchy**, so as to get the right authorities involved in the resolution.

 It is a dangerous move for the Project Manager to self-burden themself (and the Project Implementation Team) **with solely dealing with all of the project's risks**; as though, they're ashamed to have these be known.

Nonetheless, **when responsibility for the risk has been assigned to someone else, the Project Manager should still presume partial ownership of that risk**. ... And thereby, **keep a watchful eye on the risk**; because, if it all goes horribly wrong then (more often than not) the primary stakeholders will still hold the Project Manager as the person **who is ultimately held accountable for the project's risk** (and not necessarily the persons who were assigned to deal with that particular risk). Thus, **choose wisely to whom responsibility for each risk is transferred**.

"I tend to keep a record of actions against the risk register and review daily or weekly – action dates are recorded – this means I am not blindly trusting the nominated assignee."

Now that the risk analysis is complete and the response strategies & plans have been created for the project's identified risks & issues, and these details have been updated in the Risks & Issues Register, then:

- **Regularly monitor** these risks & issues, *e.g. during the Daily Standup meeting, the weekly Project Update meetings, and the monthly Program Status Reporting.*
- **Activating** the appropriate **Risk Response Plan** when the situation requires.
- **Evaluating the effectiveness** of the engaged **Risk Response Plan**.
- When & where necessary making appropriate **adjustments to** the **Response Plans**.
- **Identifying** when a **risk / issue has changed** importance, is no longer relevant, a new risk / issue has appeared and an old one has been resolved / disappeared.
- Also, the statuses of risks & issues will have to be **periodically reported** to the relevant project stakeholders, *e.g. during the Project Steering Committee meeting.*

"IMHO, Risk Monitoring & Control is when the Project Manager can prove their beneficial contribution by functioning proactively (rather than reactionary) by dealing with risks prior to these becoming issues, and managing the elimination & minimization of the impact of such issues."

Agreed, … Risks & Issues Management is much more than producing coloured boxes to be translated into RAG Statuses in the periodic Program | Project Status Report, as **Risks & Issues Management means doing something about it before it morphs from being a potential risk into a full-blown issue.** Hence, Risks & Issues Management involves:

1.. **Recognizing** … **identification** that something could go / has gone wrong,
2.. **Reassessing** … **analysis & review** of those identified risks & issues,
3.. **Revising** … **planning** the Response Strategies to such risks & issues, and
4.. **Reapplying** … **implementing** the authorized & approved Response Strategies.

Risks & Issues Management Integrated with an Agile Implementation

It is no coincidence that this **Risk & Issues Management process** can be viewed as **having the same Recognize – Reassess – Revise – Reapply cycle** (as utilized by the Project Team during development & testing), **meshing with the iterative project's Plan – Do – Check – Act** cycle; because, the daily agile stand-up meeting is the perfect forum for the identification of those risks & issues (aka "blocking issues") confronting the project.

Though, **I wouldn't be considering each & every blocking issue as something that has to be recorded in the Risks & Issues Register**; rather, only include those blocking issues that can't be resolved self-sufficiently by the Project Team in the coming day(s), those blocking issues that will require some noticeable amount of involvement by persons | groups | 3rd parties outside of the Project Team, and especially include those blocking issues that have an affect / impact on any of the project constraint [Baselines]. That is, **only record in the Risks & Issues Register, those blocking issues that will involve the resolution via (or acknowledgement by) the Project Working Group and/or the Project Steering Committee**.

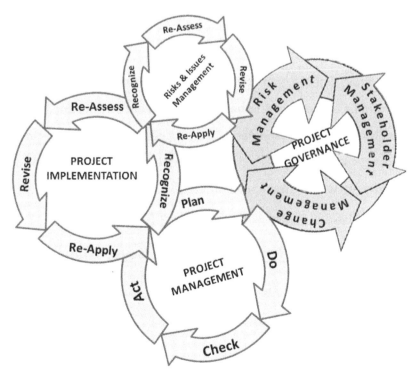

Risks & Issues Management Via Other Means

In the real-world of small sized projects and small sized performing organizations, it is common to find that the **Risks & Issues Register** is not utilized (other than as a **formalized traceability record**, whenever *"there's a Project Audit coming"*); instead, **other means are used to manage the project's currently active risks & issues**.

Don't be surprised to find that the project's risks & issues are being **tracked as action points in the minutes of the regular project meetings**; where, each minuted action has someone or some group assigned responsibility for dealing with each particular risk / issue, then at the next project meeting these minuted actions will be recalled so as to report on the progress made towards resolution. This is a reasonable enough technique, given that **the daily stand-up / weekly progress meetings often contains the quorum of those people necessary to identify, evaluate, and propose viable resolution strategies**; also, **these people will probably have the appropriate level of power & authority to get something done to resolve the risk / issue**.

"My easy way to manage this is to review the Risks & Issues register at team meetings, thereby merging a formal record with the need to monitor progress."

In addition to these regular project meetings, **a significant amount of the information pertaining to the risks & issues will be contained within email exchanges between the involved project stakeholders**. As such, these emails will be used to communicate to all of those persons (directly & indirectly) affected by the risk / issue; likewise, emails & verbal discussions will be used to coordinate with those other persons who can aid in analysing, strategizing, and resolving of the particular risk / issue.

Practical Things To Do For Risks & Issues Management

- ☑ An identified **risk / issue should not be ignored**, but rather its existence has to be **acknowledged** to the appropriate persons of concern & interest.

- ☑ Each identified risk / issue should be **entered into some form of project record** (ideally in a Risks & Issues Register).

- ☑ The **responsibility & authority** for dealing with each identified risk / issue (and actions) should be **assigned to a specific person or to the leader of a specific group**.

- ☑ The details of each identified risk / issue and **the status of the resolution** of that risk / issue should be **communicated to those parties affected by and involved with** the resolution of that particular risk / issue.

- ☑ Irrespective of whether or not the Project Manager has been assigned responsibility & authority for dealing with a specific risk / issue, **the Project Manager should keep a watchful eye on all of the project's risks & issues**, and monitor the progress being made towards the mutually acceptable resolutions of these identified risks & issues.

8.3.2. Stakeholder Monitoring & Control

If you have ever had a house built, or purchased an apartment off-the-plan, or renovated your home then you would most probably have gotten a bit annoyed (to downright bad tempered) if the person hired to do the interfacing between yourself and the builder had not been keeping you "in the loop" as to the goings on with the construction, … as it pertains to your particular information needs & wants, concerns, and expectations.

Unless you're micro-managing the build, you probably don't want (or need) to be updated on every aspect of the construction, such as when the bricks & timber will arrive on site, the days of the week when the bricklayers & carpenters started & finished their jobs, nor the logistics of the tiling & painting. However, what would probably be of particular concern to you is when the foundations were actually put down (compared to the planned schedule), when the wall frames were all erected, when the roof is to go on, when the building is to be lockable, when building inspection is to occur, and when you can expect to be handed the keys to your new home.

Yet, how often during SDLC projects has the Customer's Representatives (and even the Project Steering Committee) been seen by the Project Team (and the Project Manager) as a big hassle to be dealt with when it comes to providing the equivalent information ??

Though, herein lies one of **the primary keys to the project being deemed a "success"**; the **main arsenal weapon of Stakeholder Management is**, ... "**communications**".

Stakeholder Management Process [Section 3.6.9]

For Stakeholder Management to be successful, it is essential that each project stakeholder's needs & wants, concerns, expectations, perceptions, and opinions are (reasonably) satisfied.

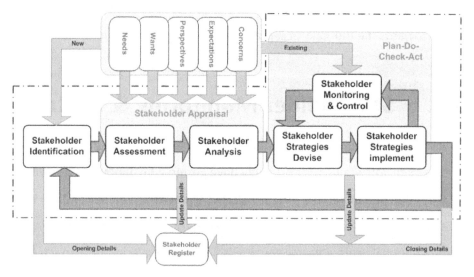

Figure 69: Stakeholder Management Process.

Thus, Stakeholder Management involves the following steps, shown in [Figure 69]:

1.. **Identify** those project's stakeholders who are **affected by** and have an **effect on the project**, and determine their roles & responsibilities with respect to the project.

2.. **Assess** these stakeholders **expectations, perspectives, concerns, "needs & wants", and categorize** these stakeholders into generic groupings.

3.. **Analyse** these stakeholders by:

- **Mapping** the **relationship** between the project and the stakeholders based on their **Power, Interest, Position, Uncertainty, Influence, and Impact**.

- **Examine** the stakeholder **characteristics** that will influence how they should be managed.

4.. **Strategize** how best **to engage** & manage these stakeholders so that they will **collaboratively participate** in the project.

5.. **Monitor** each of these identified stakeholders, and
control the responses to and interactions with each of these identified stakeholders.

NOTE: These steps would be **continually undertaken in a cyclic fashion during the entire life of the project.**

Additionally, it is best to:

1) know what these **stakeholders expect**,

2) keep these **stakeholders informed**,

3) provide these stakeholders with **what they need to know** (so that they can make their own informed decisions),

4) **when** these stakeholders **need to know it**, and

5) to make sure **things are done by when these** stakeholders **need them to be done**.

Stakeholder Register

Once these stakeholders have been identified and the information about them has been collected, it would be beneficial to include this information in a **Stakeholder Register**. Also in this register should be included; their contact details, their interests, their involvement with the project, their potential influence on the project, whether they are positively / negatively impacted by the project, their roles, their level of responsibility & authority, and eventually a list of strategies on how they should be managed.

Analysing Stakeholders ... Extract from "A Down-To-Earth Guide To SDLC Project Management".

1.. **Who are the project stakeholders**, both directly & indirectly?

- Include these stakeholders in the Stakeholder Register, along with their; contact details, roles, responsibilities, and level of authority.

2.. **What is this stakeholder's "power"** to advance or hinder the project's outcomes?

- Score this stakeholder's power on a scale of 1-5, where
  1 is least powerful and 5 is most powerful.

3.. **What is their "interest" / concern in** the project and its outcomes?

- Score this stakeholder's interest on a scale of 1-5, where
  1 is least interested and 5 is greatly concerned.

4.. **What is their "position" for or against** the project?

- Score this stakeholder's position on a scale of 1-5, where
  1 is project champion and 5 is project nemesis; i.e. 1 = project sponsor,
  2 = supporter, 3 = neutral, 4 = opponent, 5 = strives for its demise.

5.. **What is their level of "uncertainty"** or what is your level of **"predictability"** about how they will react to the project, its outcomes, to changes, and to risks?

- Score this stakeholder's uncertainty on a scale of 1-5, where
  1 is very certain and 5 are highly unpredictable.

6.. **What is their "influence" / effect on** the project's successful outcome?

- Score this stakeholder's influence on a scale of 1-5, where
  1 is least influence and 5 is most influence.

7.. **What is the level of "impact" on them** by the project and its outcomes?

- Score this stakeholder's impact on a scale of 1-5, where
  1 is little good / bad and 5 is major good / bad for them.

8.. **Tabulate** these values per stakeholder.

Stakeholder	Power	Interest	Position	Uncertainty	Influence	Impact
Prj Sponsor	5	4	1	1	3	3
Worker	4	5	2	2	5	2
Supplier	2	1	3	3	4	1
End user	1	2	4	3	2	5

For this example; while senior management (i.e. the project sponsor) is very enthusiastic about the project's new CRM system that will replace multiple proprietary databases & applications, there is the occasional grumbling from some end-user staff about the pending changes, but generally, these users don't seem too interested, also the supplier says the host-servers should turn up on time.

9.. **Draw up the relevant 5x5 grids**, preferably with the worst characteristic (and hence, the greatest risk) in the top right corner and the best characteristics (least risk) in the bottom left corner. Thereby, having similar arrangement to the Risk Matrix.

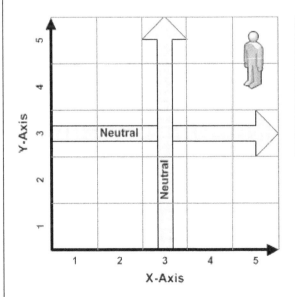

Figure 164: Outline of stakeholder grid.

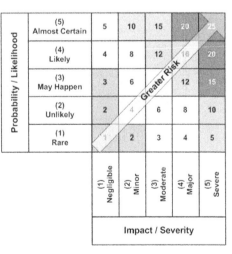

Risk Matrix

10.. Based on the numerical data that has been collected, **plot the stakeholders onto each of the Stakeholder Matrices.**

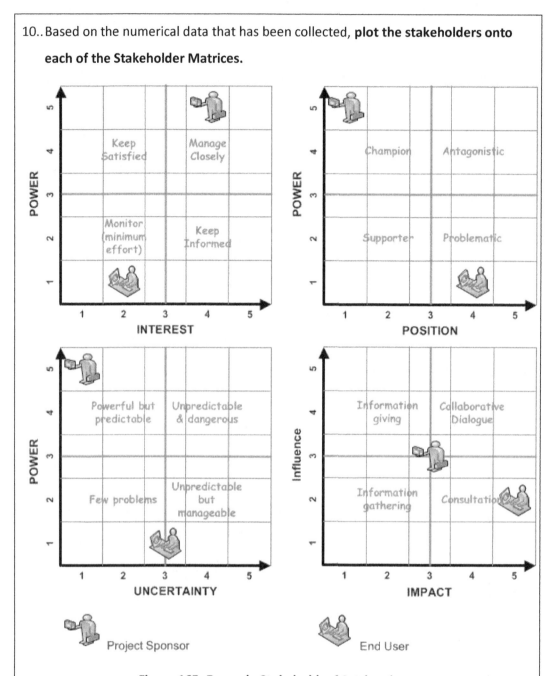

Figure 165: Example Stakeholder Matrices in use.

11.. Include the collected information into the **Stakeholder Register**; i.e. interests, involvement, potential impact, potential influence, and eventually a list of strategies on how each identified stakeholder should be managed.

Stakeholder Characteristics
... Extract from *"A Down-To-Earth Guide To SDLC Project Management"*.

Given that each of the identified stakeholders have now been mapped onto the Stakeholder Matrices, some conclusions can be drawn about each of these stakeholder's characteristics and hence what management strategies should be used on whom. This is where the grey comments on the template Stakeholder Matrices come into play.

Stakeholder	Power Vs. Interest				Power Vs. Position			
	Monitor (minimum effort)	Keep Informed	Keep Satisfied	Manage Closely	Supporter	Problematic	Champion	Antagonistic
Project Sponsor				X			X	
Worker				X			X	
Supplier	X				X	X		
End User	X					X		

Stakeholder	Power Vs. Uncertainty				Influence Vs. Impact			
	Few Problems	Unpredictable But Manageable	Powerful But Predictable	Unpredictable & Dangerous	Information Gathering	Consultation	Information Giving	Collaborative Dialog
Project Sponsor			X		X	X	X	X
Worker			X				X	
Supplier	X	X					X	
End User	X	X					X	

Relationship of Stakeholder Matrices With Stakeholder Characteristic

Matrix	Quadrant	Description Of The Stakeholder
Power Vs. Interest	Monitor	• They don't care much about the project (until such a time, as they perceive that it is affecting them). • They have limited means to affect the project.
	Keep Informed	• They care about the project, but they have limited means to affect the project. • They can help with the finer details of the project as they most probably understand the inner-workings of those bits that matter to them.
	Keep Satisfied	• They don't care much about the project, but they do have the means to affect the project if they perceive that it is adversely affecting them.
	Manage Closely	• They care about the project & its outcomes, and they have the means to affect the project if & when they choose. • Their views & opinions must be given the highest level of consideration because their opposition could make or break the project.

Matrix	Quadrant	Description Of The Stakeholder
Power Vs. Position	Supporter	• They endorse the project, but they are not really in a position of Power to be able to do much that will be advantageous to the project's success. • Unless these stakeholders can be "mobilized" to act as one then their support will not be of significant use to the project.
	Problematic	• They oppose the project, but they are not in a position of Power to do much about it. • Individually they are not a significant risk to the project's success, but if they became mobilized then they could influence the attitudes of more important stakeholders, and hence they could become a "political force" to be reckoned with.
	Champion	• They support the project, and they are in a position of Power to affect the project's successful outcome. • They are a significant ally to neutralize those potential antagonistic opponents of the project.
	Antagonistic	• They oppose the project, and they are in a position of Power to be able to do something about the project if they so desire. • They present a significant risk to the project's success, especially if they have the groundswell of support from the problematic stakeholders.

Matrix	Quadrant	Description Of The Stakeholder
Power Vs. Uncertainty	Few Problems	- They are relatively predictable, but they have limited means to affect the project. - They most probably will not change their current views of the project, so they should be managed based on their other matrices' characteristics.
	Unpredictable But Manageable	- They are relatively unpredictable, but they have limited means to affect the project. - They will probably change their current views of the project if they are encouraged to do so.
	Powerful But Predictable	- They are relatively predictable, and they have the means to affect the project. - They most probably will not change their current view of the project, so their views & opinions must be given a high level of consideration because once they are "won-over" they will probably continue to support the project.
	Unpredictable & Dangerous	- They are relatively unpredictable and they do have the means to affect the project. - They probably will change their current view of the project, so try to encourage them to support the project rather than letting them drift into opposing it.

Matrix	Quadrant	Description Of The Stakeholder
Influence Vs. Impact	Information Gathering	• They are relatively unaffected by the project and its outcome, though they have limited means of influencing the project's outcomes. • They are a source of information on what probably will and will not affect them.
	Consultation	• They are affected by the project and its outcome, but they have limited means of influencing the project's outcomes. • They most probably understand the inner-workings of those issues that affect them, and hence they should be consulted before they become problematic or antagonistic.
	Information Giving	• They are relatively unaffected by the project and its outcome, but they do have the means to influence the project if they choose to do so. • They are a source of information as to the possible objectives of the project.
	Collaborative Dialog	• They are affected by the project and its outcome, and they do have the means to influence the project. • Their views and opinions must be given the highest level of consideration because their opposition can make or break the project.

By using an arrangement of Stakeholder Matrices that ties into the Risk Matrix model (as utilized in this book), then any stakeholders residing in the top right of the grids are potentially high sources of risk for the project; see [Figure 166].

Using each Stakeholder Matrix in [Figure 165], and based on the numerical values assigned to each quadrant's squares in [Figure 166], then a weighting of the "riskiness" priority for each stakeholder can be accumulated, as demonstrated below. ... *"Though, many a project does stakeholder management by gut-feel, not matrix analysis."*

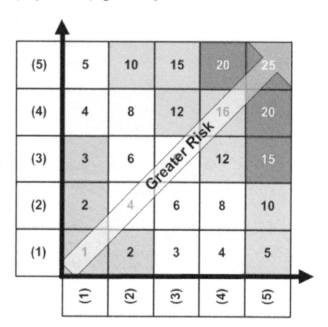

Figure 166: Stakeholder Matrix as a risk grid.

Stakeholder	Power Vs. Interest	Power Vs. Position	Power Vs. Uncertainty	Influence Vs. Impact	(Priority) Total
Project Sponsor	20	5	5	9	39
Worker	20	8	8	10	46
Supplier	2	6	6	4	18
End User	2	4	3	10	19

NOTE: An individual stakeholder's location on each grid can dynamically change over the life of the project; hence, their Priority can change equivalently.

Relationship of Stakeholder Matrices With Stakeholder Strategies

Matrix	Quadrant	Stakeholder Strategy
Power Vs. Interest	Monitor	• Don't invest too much effort into this group of stakeholders, because generally they are not really interested in the project or what you have to say. • Provide generalized "educational information" outlining the benefits & outcomes of the project so that they are aware that something is going on. • Monitor these stakeholders, and only react to them if they start drifting towards opposing the project or when the project needs defending.
	Keep Informed	• Keep this group adequately informed about the issues that matter to them. • Balance their interests against those of high-powered interested groups, while making them feel as though they are contributing to the project. • Monitor these stakeholders to be aware that no major issues are arising with them.
	Keep Satisfied	• Provide fully to these stakeholders' informational needs, but don't provide excessive amounts of communications that could bore them or worry them, because they may react burdensomely. • Respond to these stakeholders' requests in a timely fashion, but don't be confrontational as you want to maintain their neutral stance. • Involve these stakeholders according to the issues that matter to them.
	Manage Closely	• These are the key stakeholders of concern, so deal with them both proactively and reactively. • Provide fully to these stakeholders' informational needs, making all efforts to satisfy their needs so as to allay their concerns. • If these stakeholders so desire, then get these stakeholders fully engaged with the project.

Matrix	Quadrant	Stakeholder Strategy
Power Vs. Position	Supporter	• Provide educational information that encourages their continued support and involvement. • Monitor these stakeholders to be aware that no major issues are arising, and react to them if they start drifting away from supporting the project.
	Problematic	• Provide educational information that encourages these stakeholders' positive involvement, potentially moving them towards supporting the project. • Monitor these stakeholders, and react if they start mobilizing their opposition to the project or if they start drifting towards supporting the project. • Be prepared to "negotiate & modify" the project's plans / objectives if these stakeholders start to organize opposition to the project.
	Champion	• Provide educational information that reinforces their support, and encourage their involvement. • Be prepared to put work into these stakeholders' relationships so as to keep them aligned with the project's objectives, as they're the project's allies. • Ask these stakeholders to highlight the benefits and importance of the project to those problematic & antagonistic opponents of the project.
	Antagonistic	• Need to "identify & understand" the nature of and the source of their opposition to the project. • Need to construct good working relationships with these particular stakeholders so as to mitigate against an "army of hostility" to the project. • Need to carefully monitor, and then counter any opposition they raise against the project. • Need to "negotiate & compromise" the project's plans / objectives where appropriate so as to "appease" these stakeholders.

Matrix	Quadrant	Stakeholder Strategy
Power Vs. Uncertainty	Few Problems	• Don't invest too much effort into this group of stakeholders, because generally they are not going to change from their current stance. • Provide educational information as to the benefits of the project's outcomes, and encourage their involvement when it is practical to do so.
	Unpredictable But Manageable	• Provide educational information as to the benefits of the project's outcomes, and encourage their involvement when it is practical to do so. • Monitor these stakeholders, and react if they start to either mobilize their opposition or support for the project.
	Powerful But Predictable	• Need to "identify & understand" these stakeholders' desires for the project's outcomes. • Need to "win over" these stakeholders early on in the project's life. • Need to give a high level of consideration to their "views & opinions".
	Unpredictable & Dangerous	• Need to "identify & understand" these stakeholders' desires for the project's outcomes. • Need to construct good working relationships with these stakeholders so as to ensure a "coalition of support" for the project instead of inducing an "army of hostility" against the project. • Need to "negotiate & modify" the project plans / objectives where appropriate so as to "appease" these stakeholders. • Need to carefully monitor, and then counter any opposition that these stakeholders may raise against the project. *"Keep your friends close, but keep your enemies even closer".*

Matrix	Quadrant	Stakeholder Strategy
Influence Vs. Impact	Information Gathering	• Provide educational information as to the benefits of the project's outcomes. • Monitor these stakeholders, and react if they start feeling they're being impacted on by the project.
	Consultation	• Keep these stakeholders adequately informed about the issues that affect them, and encourage their involvement so as to be aware that no major opposition is mobilizing. • Try to limit the negative impact on them, but not at the expense of high priority objectives.
	Information Giving	• Note these stakeholders' concerns, and respond in a reasonably timely fashion. • Look for these stakeholders' input into the issues that are of concern to them.
	Collaborative Dialog	• Need to "identify & understand" these stakeholders' concerns. • Need to give a high level of consideration to these stakeholders' views, opinions, and objections.

Stakeholder Strategies & Implementation

As stated previously in [Section 3.6.9], there are three **major facets to Stakeholder Management strategies & implementation**:

1) **Continuous bi-directional communications** with the stakeholders.

2) **Continuous collaborative effort to reconcile** the stakeholders' expectations, perspectives, concerns, needs & wants with the project's objectives.

3) **Continuous implementation of strategies** that are based on the principle of maximizing the positive influences on the project while, minimizing those negative impacts on the project's **stakeholders**.

By actively managing these stakeholders, then one is better able to **balance the providing of benefit to the stakeholders while still achieving the project's objectives**; and hence, prevent / mitigate / limit those stakeholder related risks & issues confronting the project.

Continuous Bi-Directional Communications

Don't under estimate the part that effective & efficient bi-directional communications has to play in managing these stakeholders' expectations & perspectives.

- ☑ **Communicate clearly & openly** about the project; what are its objectives, its desired outcomes, its benefits (how will these exceed its short-term detriments), who will most probably be affected by it, and how will they be affected.

- ☑ **Listen to the stakeholders'** expectations, perspectives, concerns, needs & wants ... but especially to **their concerns**, as these concerns will **highlight** what these stakeholders **perceive as the risks / threats to them** because of the project.

- ☑ **Acknowledge that these stakeholders have been heard**, that their concerns have been noted, and not simply pay "lip-service" to them.   ... *"provide feedback."*

- ☑ **Communications** should be **tailored specifically for the targeted stakeholders** so that it **contains predominately the information that is relevant to them**, in a form that they can understand and utilize. ... *Not mumbo-jumbo gobbledygook techno babble.*

- ☑ **Establish relevant & reliable communications channels** such that stakeholders are able to; voice their concerns, provide feedback, ask questions, seek additional information, and access the information of relevance to them when they require.

As part of the communications process, progress updates would be provided to the various project stakeholders; *e.g. the Project Status Report, and the Program Management Report.* **For these progress updates**, see [Section 7.3.6], **to be of any real benefit** and to satisfy these stakeholder's expectations then **it is essential that these updates**:

1) **Be provided by when** the information **needs to be known**, *e.g. weekly, monthly.*

2) **Contains** the relevant details of **what needs to be known**, *e.g. known risks & issues, schedule progress (SPI), cost progress (CPI), expected milestone dates.*

3) **Presented in a form** that can be **readily understood & utilized** by the targeted audience, *e.g. a Key Performance Indicator Dashboard that is standardized across all of the performing organization's projects.*

4) **Presented from a consistent frame of reference for a consistent reporting duration** from one reporting period to the next.

Though, watch out for how much information is provided to these stakeholders.

 DO NOT provide stakeholders (especially the Customer's Representative) **with more information than was agreed to be provided**, and **within the boundaries of what they really need to know** about the project.

Because, the over-providing of information can result in these stakeholders manifesting expectations that such detailed information will be supplied henceforth; consequently, this can **unnecessarily over-burden the Project Team** with continually producing this additional information.

Hence why, the Customer's Representatives (and even the Project Steering Committee) could subsequently be considered by the Project Team (and the Project Manager) as a *"big hassle to be dealt with"*.

The over-providing of information also has the undesired side effect of stimulating the Customer's Representative to try to micromanage the project, and more specifically the activities of the Project Team. Subsequently, this type of micromanagement would greatly limit the flexibility of the Project Team and the Project Manager to adjust the project's plans & execution to respond adaptively to the project's ever-changing situation & prevailing circumstances.

Thus, the **over-providing of information to stakeholders can be just as detrimental** to the project (and to the Project Manager) **as** providing **too little information**.

Collaborative Effort To Reconcile

A collaborative effort to reconcile the project's objectives with the stakeholders' expectations, perspectives, concerns, needs & wants involves **those strategies employed to "build the relationships" with the project stakeholders, by involving their participation in the project, and** (where & when appropriate & practical) by also **involving these stakeholders in the decision-making process** (or at least the consideration of their interests). Thereby, **resulting in resolutions that will be mutually beneficial to both the project stakeholders and the performing organization**.

NOTE: **The more powerful the stakeholders that can be engaged to support the project then the more likely it is that the project will be successful.** This is because; these stakeholders will often have access to the [People] & [Resources] that are needed for the project to be able to succeed.

NOTE: Irrespective of whichever stakeholder strategies are chosen to be implemented, try to **fairly distribute who shall receive the benefits and who shall bear the burden.** As an unfair distribution (i.e. favouritism) could result in those disadvantaged stakeholders being transformed into unpredictable problematic & antagonistic sources of risks to the project being a success.

The following table highlights some of the strategies that should be considered for each quadrant of the Stakeholder Matrices depicted previously in [Figure 165].

Continuous Implementation Of Strategies

The **continuous implementation of the chosen Stakeholder Managements Strategies needs to be based on the principle of maximizing the positive influences on the project, while minimizing those negative impacts on the project's stakeholders; where, priority is given to the potential "riskiness" that is associated with each particular stakeholder.**

If during the implementation of the Stakeholder Management Strategies (and/or the planned project schedule, and/or the implementation of the risks & issues response strategies), opposition was to arise with the more powerful & influential stakeholders then one must be prepared to renegotiate, resolve conflicts, and **accordingly adapt the Stakeholder Management Strategies.** ... And when necessary, **adjust the project's plans, which could induce risk to the project, which could reciprocate that changes be made to the project's [Baselines]**, which will require the progress through the formalized Change Management Process. Subsequently, as these adjustments & changes progress then it is **important to keep the relevant stakeholders informed of what is & is not going on**; thereby, **keeping them "in-the-loop"**. As **the lack of frequent, "solid & well grounded" communications can be a catalyst for misunderstandings, misinterpretation, and mistrust amongst the project stakeholders.** ... *"Yep, the lack of regular updates about what's going on can be a source of much angst and unhappiness."*

Which brings us to an addendum to "Adaptive & Proactive" SDLC Project Management.

RULE 30[B]: **Stakeholders should never be taken for granted; rather, they should be managed throughout the entire life of the project, from start to finish.**

Conflicts Of Interest & Negative Impacts

If during the project's life, a **"conflict of interest"** arises where some stakeholder(s) are **obtaining undue advantage (contrary to the Terms & Conditions of the contracted agreements)**, or there is / will be **negative impacts** and/or **excessive burden** to other stakeholders (in excess of the contractual arrangements), **especially to parties outside the customer & performing organizations** then, as soon as possible, the Project Manager should **make sure that the Project Steering Committee** (and if necessary the senior management at the performing organization) **are aware of these concerns (in writing)**.

Be very wary of stakeholders who place **short-term gains ahead of the potential long-term viability**; because, serving these stakeholders particular needs & wants could turn out to be detrimental to others' success.

COFFEE BREAK DISCUSSION ... *Short-Term gains versus Long-Term pains.*

The main issue here is how to balance the view held by groups such as Sales | Business Development of short-term gains versus the long-term considerations of the Project Team & Operations (and subsequently the performing organization); when, the sales persons have their incentives & remunerations based on the amount of sales that they make for the month / quarter / year, instead of the actual cost & duration of the implementation & delivery, and the ongoing profitability of maintenance & support.

This difference in perspective can lead to members of the sales group selling as much as they possibly can for whatever (acceptable sounding) price & duration, without worrying about the practicality & profitability of the implementation & delivery; potentially this can result in their "pushing of vaporware".

One solution to this problem is for the Executive Management of the performing organization to modify the **incentive & remuneration schemes** so as to **tie** these **to the profitability & timeliness of the implementation & delivery, and the ongoing viability of maintenance & support contracts**.

However, if the Chief Executive Officer (CEO) happens to have a salesperson's slant, they might see sales as the be all and end all measure of business success; hence, such a change in perspective could be difficult to achieve. Therefore, it may be highly constructive for those who bear a significant amount of the pain from such short-term thinking, i.e. the Project Teams & Operations, that they exert upwards pressure on the management hierarchy to change their ways. Possibly towards the Chief Financial Officer (CFO) to demonstrate the lack of profitability of such short-term gain thinking.

The Project Manager (and the Project Team) should also **document in the Post Implementation Review (PIR), what has been the impact of unclear requirements, the misalignment of the allocated Level Of Effort and time duration to the realities of the actual implementation**. Then, provide the Executives with a summary of the PIR(s), so as to demonstrate the commonality of the problems with short-term target views.

8.3.3. Change Monitoring & Control

NOTE: See [Section 9.4.1] for business change management.

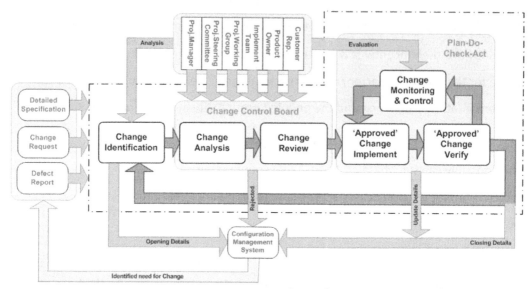

Figure 64: Change Management Process.

Back in [Section 3.6.7] a simplified [Baseline] Change Request process workflow and a Defect Reporting & Defect Repair process workflow were respectively depicted in [Figure 65] and [Figure 66] which are reproduced a few pages over from here.

When No Formalized Change Management Is In Place

However, in the real-world of small sized performing organizations and with small projects there may not be such **formalized Change Management in place** for either Baseline Changes and/or Defect Repairs. As a result, it may be necessary for the Project Manager to coordinate a "pseudo" Change Management Process, with the Project Manager taking on the role as one-half of the **Change Control Board (CCB)** and **Defect Review Board (DRB)** opposite the representative of the customer organization.

NOTE: **Formalized Change Management** may also be referred to as **Perform Integrated Change Control** (as per the PMBOK ® Guide).

As an aside, **when no formalized Change Management process is in place then don't be surprise to find that "coincidently" the project's [Scope Baseline] was never formally agreed to**, because the Detailed Specifications had not been signed off. ... And, to complicate things further, **there will be no specific point in time when some form of [Scope Baseline] was declared**. These oversights will subsequently enable **[Scope] changes** to be **made ad-hoc** in response to the project's ever-changing situation & prevailing circumstances. ... And, to make matters worse, these [Scope] changes may be expressed verbally with the limited written summation of the change requested.

As a **consequence** of the above, **[Scope] creep & shrinkage will occur virtually unchecked**; but, **later on during the [Quality] acceptance testing** of the project's deliverables there will be **some dissension** between the performing organization and the Customer's Representatives **over whether the provided functionality is correct or not**. Where, the interpretation of the "correctness" will be based on the individual's understanding of what was supposed to have been included in the project's [Scope]. These "disagreements" over [Scope] could eventually turn into a "political" and financial wrangle over whether such-n-such **constitutes a contract deviation or the clarification of pre-existing** ~~requirements~~ **specifications**. ... And, what exasperates the situation is when there is a long time between delivery releases (i.e. as with a waterfall life cycle); in such a situation, [Scope] creep & shrinkage could occur without being detected until a significant way into the project's Implementation Phase, at which point, the necessary corrective actions for these scope deviations could result in a significant amount of rework. Such **[Scope] rework will consequently have a catastrophic ripple effect on the project's [Time] & [Cost]**; and hence, on the project being deemed a Success | Failure.

 The lack of some kind of formalized Change Management Process can be a catalyst for problems during the later phases of the project's life cycle. Combined with an unsigned Detailed Specifications, and a non "frozen" [Scope Baseline] then the project is likely to culminate in an argument over what was & was not agreed to, and who is right & who is obviously wrong.

NOTE: The project cannot be run without some form of "responsible" Change Management, as **an adhoc "make it up as you go" arrangement could easily result in the project haemorrhaging both financially and duration wise**.

So, how can the project be delivered successfully when there is no formalized Change Management Process in place ??

Change Control Board

If the proposed change affects either the [Scope Baseline], the [Time Baseline], the [Cost Baseline], and/or the [Quality Baseline] then some form of "due diligence" Change Management with review & approval is necessary.

Ideally this review & approval would be via way of an established **Change Control Board (CCB)** that would be **composed of representatives from** the **Project Steering Committee**, relevant members of the **Project Working Group**, appropriate **Subject Matter Experts** from within the performing organization, members of the **Project Implementation Team**, … and, **representation** for the **customer organization** (i.e. the "Product Owner" in agile terminology) so as **to ensure that the customer's perspective is being taken into consideration** and that decisions are not being "rubber stamped". … And, recording the Change Request and the details of the subsequent investigation, decisions, and outcomes in a **Change Management System (CMS)**.

"IMHO, an effective Change Control Board would have members with different perspectives & expert judgment, so that the benefits & disadvantages (pros & cons) of the change can be openly explored, analysed, and discussed before being approved / rejected."

So, whenever there is a change proposed or necessitated by the realities of the current situation & prevailing circumstances, then pull together representatives from; senior management, other relevant managerial ranks, and knowledgeable senior implementers to decide on the merits & ramifications of the proposed change(s). … And, **engagement with the Customer's Representatives.**

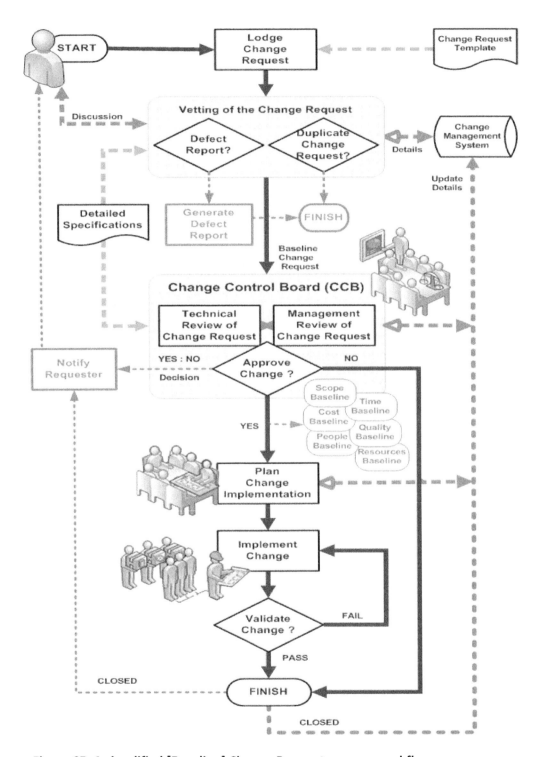

Figure 65: A simplified [Baseline] Change Request process workflow.

Only then, after such diligent consideration, would **the [Baseline] Change Request be authorized by the Change Control Board** (including obtaining signed off approval from both the customer & performing organizations' representatives). Consequently, **the relevant project constraints and affected plans would be "re-baselined"**.

Depending on the implementation methodology being used, these changes would entail:

- For **agile**; changes would be translated into new – updated – removed **story points**, and these story points would be placed **into the Product Backlog**, for the Project Implementation Team to choose to implement later on when The Team deem it most appropriate.

- For **iterative**; changes would be planned into **future releases**, and hence the project parameters adjusted accordingly for those affected future releases.

- For **waterfall**; changes would be integrated into appropriate points in the **re-baselined** project plans.

- **NOTE:** **It is essential that all involved parties (especially the implementers) are communicated to about the change and its progress** towards completion. Where necessary, individual parties may have to be personally updated, as to how the change will affect them specifically.

    "That being stakeholder communications management."

What you **don't want** is for **changes** (most probably [Scope] changes) **to be incorporated into the project without the project's constraints** (of [Time] and [Cost]) **being accordingly compensated for**.

Where compensated could mean (de) increase in the [Time] duration, (de) increase in the allocated [Cost] budget, the (re) affirmation of [Quality] standards, the (re) assignment of [People], the (re) allocation of [Resources], and actual financial compensation to the affected parties.

Defect Repairs

For the following reasons, **all Defect Reports** (whether generated internally or externally) **will need to go through some form of Change Management Process** via way of a **Defect Review Board** (similar to a Change Control Board) to decide what does & does not constitute a defect.

- ☒ Sometimes **[Scope] changes are inadvertently snuck through as Defect Reports**. This often happens when the Customer's Representative finds functionality that doesn't comply with their "needs", "wants", "perceptions", or "expectations". That is, it is not as they "intended it to be", or it is not as they now "think it should be"; contrary to what was agreed to in the approved Detailed Specifications.

    "Oh, but" there ain't **no agreed** & signed off version of the **Detailed Specifications**, just the Customer Requirements to go by. Hence, there may have been some **misunderstanding & misinterpretation** between the Customer's Representatives and the Project Implementation Team **as to what has actually been specified**.

- ☒ There should be **several severity categories for defects with differing priorities**. Without such a grading scheme and Change Control Process logically & consistently sequencing the order that defect repairs are to be implemented then, alternatively such repairs would be done in a haphazard manner without real consideration being given to the possible ramifications to the project's [Scope], [Time], [Cost], [Quality], [People], [Resources], and <Risks>.

Hence, **it is essential that defect reporting & defect resolution be undertaken methodically, analytically, and supervised** so as to reduce the number of spurious defects being handled. Do this by answering the following questions:

1) **What should the deliverable's features & functionality be compared against?**
    ... the approved Detailed Specifications.

2) **How & where should Defect Reports be recorded?**
    ... in a Defect Tracking System.

3) **How should Defect Reports be severity categorized and prioritized?**

... consistently across all defect reports and across defect reporting input streams.

4) **How should defects be resolved?**

... in an orchestrated manner, in accordance with the approved Detailed Specifications.

5) **How should defects be verified as resolved?**

... in an orchestrated manner, in accordance with the approved Acceptance Criteria.

Defect Tracking System

With respect to Stakeholder Management, [Section 8.3.2], it is highly advantageous from a perception of quality, that defect reporting / tracking be segregated into two distinct areas; one area for those defects found "internally" by the performing organization, and the other area for those defects that were found "externally" by the customer. ... Because:

- Those **defects found externally** will **relate directly to the customer's perception of the deliverable's "fitness for use"**; i.e. the **effectiveness of the project's Quality Control verifying conformance to the specifications**. Whereas, those **defects found internally** while detailing non-conformances with the specifications, these could also relate to non-functional issues such as non-compliance with coding standards, deviations from policies & procedures; i.e. the **application of Quality Assurance** to the project.

- **From the primary stakeholder customer's perspective** (i.e. their senior management), **the measure of the [Quality] of the deliverables is proportional to the tally of the defects that have been found & remain open | unresolved**. Hence, it would be detrimental to pollute the visible quantifiable metrics for [Quality] by inflating the numbers with internally reported (non) defects.

> Which brings us to the next rule of "Adaptive & Proactive" SDLC Project Management.
>
> **RULE 57:   It is better to find the defects behind closed doors; rather than, have these exposed on the showroom floor.**

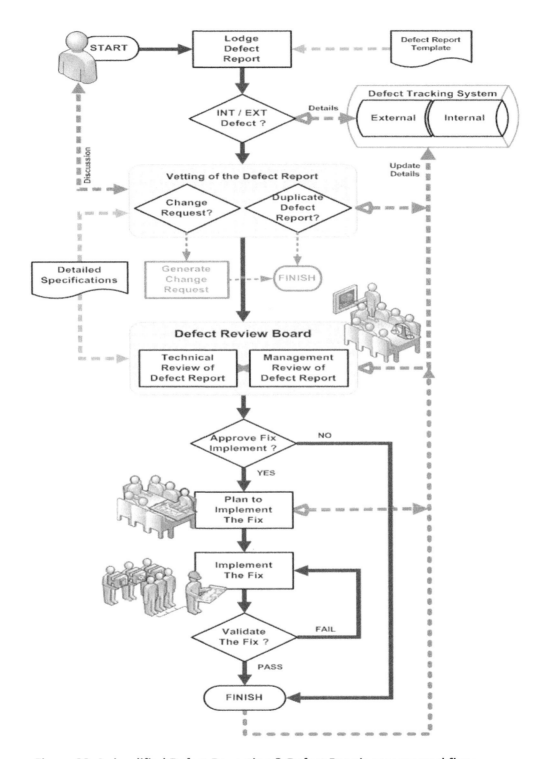

Figure 66: A simplified Defect Reporting & Defect Repair process workflow.

Defect Repair Prioritization & Severity

While the Defect Review Board (possibly consisting only of the Project Manager and the Customer's Representative) sort through these reported defects, it would also be highly advantageous that these defects be differentiated based on their severity & priority:

- **Severity** ... is **how much impact** the defect has on the deliverables & objectives.
- **Priority** ... is **how soon this particular defect should be resolved when compared to the other defects**.

All open | unresolved defects should be periodically (re) graded based on the level of impact | severity that each defect has on the operability of the project's deliverables; using a severity grading scheme as follows:

1) **Critical** – is where the defect renders a section of the **deliverable** (or the entire deliverable) as either **completely inoperable** or majorly **unusable**, or where **data is lost or corrupted**, and there is **no reasonable workaround**. *For example; a crash bug.*

2) **High** – is where the defect renders a section of the **deliverable as unusable or not fit for use**; but, there is **a reasonable workaround available as an interim solution**.

3) **Medium** – is where the defect is for **some particular feature / functionality** of the deliverable that is **not operating as specified** and this **deficiency is causing difficulties**; but, the **rest of the deliverable is usable**, and there is **no loss or corruption of data**.

4) **Low** – is where the defect is for a **minor non-conformance** with a part of the specifications; but, this **does not impede that deliverable's practical usability**.

Just because a defect has a low severity grading, this does not mean that the defect can't have a high priority. There may be some low severity defects (or cosmetic changes) found early in the project's life, where for reasons of stakeholder perception, it would be better to fix these sooner rather than later; as **to let these linger on would be a constant reminder** to the Customer's Representatives **of the issues with the deliverables**, and also raises concerns as to the priority of satisfying the stakeholder's concerns.

The Politics of Defect Repair

While, thinking about Stakeholder Management, there are some "politics" involved with defect resolution:

1.. Fix ~~as soon as possible~~ immediately those defects | issues that **embarrass or humiliate the customer organization** (i.e. cause them to "loose face"). Such as, those defects which are publically viewable by the customer's customers; *e.g. those easily encountered crash bugs and out-of-service issues*.

2.. Fix as soon as possible those defects with the **highest priority**, followed by those with high severity & priority, and so on down to the lowest priority, then lowest priority & lowest severity. Though, this may necessitate fixing a 'low severity but high priority' defect before a 'high severity but low priority' defect; where the stakeholders "perception" occasionally overrules the factuality of significance. ... *"Just who is the one paying for this work to be done ??"*

3.. Fix those defects | issues **causing several other defects**; i.e. fixing this one issue will resolve or partially resolve several other defects, or eliminate the apparent potential for additional defects to occur in that affected area.

4.. Fix those defects that the **Customer's Representative keeps noticing**; *"oh, that bug, is still there"*. Even if this defect has a low severity & low priority, from the perception of [Quality] and evidence that the customer is being heard, it may be advisable to, *"just fix the damn thing, and move on"*.

5.. Try to keep the rate that the 'Critical' and 'High' graded defects are being **fixed in pace with the rate that defects are being reported**; i.e. evidence that the project is making forwards progress and is still under control.

Which brings us to an addendum to "Adaptive & Proactive" SDLC Project Management.

RULE 57^A: Defect count is a pseudo indicator of the project's Quality. ... *"And the count of Change Requests indicates the quality of the Customer Requirements."*

WARNING: The failure to manage changes & defects adequately will cause the project to haemorrhage financially [Cost] and duration-wise [Time] due to the amount of rework, over-work / under-work [People] & [Resources] that could result from these changes, ... and, not forgetting the impact that changes can have on the [Quality] of the project's deliverables.

This Project Is My Small Business

If you have come up through the development ranks, then you will most probably have experienced the situation where the Project Implementation Team take ownership of the project, as though it was their personal challenge to deliver, rather than being just a job. Similarly, as the Project Manager it can be beneficial to think of the project as "your small business"; where, if you don't manage your small business in a sensible & profitable way then you will go out of business; i.e. the project's primary stakeholders / your senior management will cancel the project or cancel your project management reign.

Hence, it is your responsibility as this small business owner to ensure that:

1) The project produces the expected results and that the resultant [Scope] of **deliverables contains the agreed features & functionality**.

2) The project deliverables are handed over to the customer's representatives when it is agreed that these will be **delivered as per the [Time] milestone dates**.

3) The project is a **profitable endeavour** that **doesn't exceed the allocated [Cost] budget**.

4) The resultant **deliverables do conform with the agreed [Quality] Acceptance Criteria**.

5) The project **effectively & efficiently utilizes those assigned [People]**.

6) The project **effectively & efficiently utilizes those allocated [Resources]**.

Though, remembering that **the customer is not the only primary stakeholder of the project**; there is **also the performing organization's senior management who will be concerned with the profitability & timelines of this endeavour.**

8.4. Actions & Reactions

As the Project Manager, you may think that you have the project's R.I.S.C.s under control; but, **how the Project Manager acts & reacts to the current situation & prevailing circumstances will greatly influence the stakeholders perceptions of this control** (or lack thereof). One of these is how the Project Manager handles stakeholder "requests".

8.4.1. Handling Stakeholder Requests

Stakeholder Requests are those stakeholder's concerns, needs, wants, expectations, and perceptions that at present the Project Manager can choose to; ignore, deal with later on, or deal with immediately. As the Project Manager, the choice is yours; but, as with every choice in life there is some <Risk> associated with that choice, hence choose wisely.

Thus, it is advantageous for the Project Manager to:

- **Deal with stakeholder requests well informed and in a timely manner ...**

    That is, to give these stakeholders what they need, when they need it, in a form that they can use. Though, before the Project Manager gets the urge to act hastily on a stakeholder's request, the following should be determined:

    (1) **What are the latest facts pertaining to this stakeholder?**

    For example; what is the current business situation, are they pushed for time, are they running out of budget.

    (2) **What is their demeanour towards the performing organization** (*are they not too impressed with the effort so far*).

    (3) **What are the up-to-date details of their request** and not just hearsay, and not what where the facts about this request some time ago?

    (4) **What are the actions / solutions** that the performing organization intends to take **in order to respond to this request**, and what will this response entail?

All stakeholder requests should be replied to within a respectful amount of time; even if that response is only to say that, *"more time will be required to formulate an appropriate response"*.   ..."*Just don't leave them hanging, waiting.*"

- **Be more Proactive rather than solely Reactive …**

    When dealing with these stakeholders, the Project Manager could operate reactively or proactively. That is, the Project Manager could choose a <u>**reactive**</u> **/ "defensive"** management style of **only responding** to stakeholder requests **once** these have been **initiated**; alternatively, the Project Manager could try to be <u>**proactive**</u> **/ "pre-emptive" dealing with** those potential stakeholder requests **before these have been asked**.

    > *For example; it is coming up to the end of the financial quarter, so maybe it would be beneficial to provide the primary stakeholders with a synopsis of the project's financials compared to what has & hasn't been achieved on the project during the current financial quarter.*

    Admittedly, dealing with the project's stakeholders **reactively is** a lot **easier** than dealing with them proactively; because, being **proactive requires that the Project Manager has a better understanding of each stakeholder's needs, wants, concerns, expectations, and perspective**, … as well as having a good understanding of the relationship between each stakeholder's power – interest – position – uncertainty – influence – impact; see the stakeholder analysis in [Section 8.3.2].

    Though, it is not that hard to partially pre-empt some of these stakeholder requests, given that **for all SDLC projects there will be several stakeholder requests that will be inherently common** from one project to the next; for example:

    (1) **What are the current states of the four traditional project constraints used to judge a project's progress / Success : Failure**?

    - **How much** of the project's **[Scope Baseline] has been covered so far**, and **how much [Scope] remains** to be covered?

- **How much of the [Time Baseline] has been covered so far**, how much billable [Time] has been recorded against the project's Cost Account Code, and **how much [Time] is estimated to complete** the remainder of the current release and to complete the remaining releases?

- **How much [Cost Baseline] has been spent so far**, how much is it expected to **[Cost] to complete** the remainder of the current release and to complete the remaining releases?

- How many defects have been detected thus far, and **what is the perceived [Quality] of the project deliverables**?

(2) **What are the current states of the other project constraints / variables?**

- **What [People] have been utilized** to get the project to where it is now, and **what [People] will be required when, where, and how many to complete** the remainder of the current release and to complete the remaining releases?

- **What [Resources] have been used** to get the project to where it is now, and **what [Resources] will be required when, where, and how many to complete** the remainder of the current release and to complete the remaining releases?

(3) **What are the R.I.S.C.s involved with the project?**

- **What risks & issues currently confront the project**, and how will these be mitigated, responded to, and possibly resolved?

- **Which stakeholders** have **requests on** this project and/or have changed in their power – interest – position – uncertainty – influence – impact that will affect this project?

- **What baseline changes** to the project does this stakeholder need to be made aware of and/or will be required to act upon?

Thus, **if you were that stakeholder then what would you need, want, and expect to be told**? Hence, as the Project Manager try to answer these questions before the stakeholders have to ask these (and before they have to chase up the answers to these questions). Thereby, you will be perceived as a highly effective & efficient manager.

- **Prioritize the order in which stakeholder requests are dealt with ...**

Because, not all of the stakeholder requests can be dealt with simultaneously; and, some of these stakeholder requests should not be addressed at all (let alone immediately). ... *"Let's discuss that tomorrow, or lets discuss on Friday."*

There are not enough hours during the work day to **satisfy every stakeholder request**; so **DO NOT try to**, else the Project Manager (and/or the Project Team) could suffer from overworked burnout, as well as unnecessarily impacting on the project's [Time] & [Cost].

"And, doing this as unclaimed overtime will eventually impact on those home-life stakeholders; let alone, distorting the recorded Level Of Effort which will subsequently neutralize the true Earned Value Performance Measures that are used to track this project's progress. ... And, such LOE distortions could mean that the next similar project is under allocated required work hours, because that is how many were recorded last time."

The Project Manager has to **prioritize** these stakeholder requests based on:

(1) Those requests that have an actual **imminent timing factor as to the necessitated speed of the response**.

(2) Those requests from **stakeholders who possess great power – interests – position – uncertainty – influence – impact over the project's successful outcome**. That is, requests from those stakeholders that appear in the top right hand corners of the Stakeholder Matrices.

(3) Those requests from primary stakeholders **who make the direct decision over whether the project is considered a Success | Failure**.

(4) Those requests from **stakeholders who have been waiting** sometime for some form of response.

- **Be proficient at Time Management ...**

The only truly effective multi-tasking that the Project Manager can hope to achieve is the pseudo multi-tasking of doing one activity (or part of an activity) at any one time – for a reasonable (but limited) duration of time – before switching to the next highest priority activity. That is, the Project Manager should deal with one activity (i.e. a stakeholder request) and work on that activity until it has been taken to a point where:

(1) that activity **can't effectively be worked on anymore** because there is something that is hindering its effective progress, or

(2) that activity is **dependent on some other activity** that has to be completed or partially completed first (*e.g. a quotation received from a third party*), or

(3) that activity has **received enough attention** for today (*e.g. a couple of hours effort*), or

(4) that activity is of a **lower priority** than a higher priority stakeholder request that has just been received.

Unfortunately, what can result when the project manager tries to deal with multiple activities all at the same time is the jumping from one activity to the next without having effectively completed anything to a stakeholder satisfying level. Thus, **attempting to deal with multiple activities all at the same time can result in** project management inefficiencies and a lack of productivity because time is **wasted churning between tasks** to eventually end-up **"firefighting" these activities as problems arise**.

This task churning can also happen to the Project Implementation Team, when project management reassign team members between incomplete tasks, and "bouncing" them from one ongoing project to another project. Effectively inducing a form of project dementia where *"I know I was doing something, but I just cannot remember what it was, let alone where, I was up to when I was last here"*.

- **Effective Constraint & Restraint Management ...**

    The Project Manager has to develop that elusive art of sufficiently servicing all of those relevant stakeholder requests without losing a grip on the project's [Scope], [Time], [Cost], and [Quality]; i.e. maintaining control over those traditional determinates of whether the project is deemed a Success | Failure.

    NOTE: By the Executing Phase these four project variables will probably have become project constraints due to; the sign-off of the approved Detailed Specifications, the schedule having been accepted / milestoned, the project's budget having been allocated, and the Acceptance Criteria having been defined. ... If some of these have yet to be formally accepted, then at least there will be a solid expectation as to what these are.

    As the Project Manager (and the members of the Project Team) try to service every stakeholder request, the Project Manager may feel that much of their day (and that of the affected project team members) is taken up with dealing with these stakeholder requests instead of getting on with the project's actual implementation.

    For example; "I spend more of my day turning out reports, than I do managing the project's activities."   ... "Insert frustrated smiley face here."

    Therefore, **before** the Project Manager (and the affected members of the Project Team) **undertake any action in response to a stakeholder request**, they should stop for a moment and question; "*does acting on this stakeholder's request contribute to the project's overall progress towards the successful completion of the project?*", and "*does servicing this request clear the way to moving the project forwards?*".

    If the answer to those questions is "**NO** it does not directly contribute", **then it may be prudent to restrain from undertaking that action** (or at least delaying it to another day), and then **enquire as to its necessity**.

It is very easy to try to **satisfy every stakeholder request** without these actions effectively contributing towards the production of the project's deliverables, and thereby **potentially no longer working within the boundaries of the agreed project constraints**.

However, **depending on whom the primary stakeholders are** and what is their attitude & demeanour, the Project Manager (and the members of the Project Team) can **sometimes** find themselves having **no "political" option other than to do this non-project-advancing activity** specifically to appease this all-powerful & highly influential project stakeholder.

As stated previously, **the determination of whether a project is deemed a Success | Failure is based entirely on the expectations, perspectives, and opinions of the project's primary stakeholders**.

- **Effective Time Record Management …**

    As the Project Manager, you could find that (due to the amount of stakeholder requests being handled) the recorded 'project management' hours are in fact closer to third of the total project hours reported thus far.

    "Yes, customer handholding can consume a lot of time and effort."

    However, excessive 'project management' hours are not going to look good for the project (even if these hours were necessary for the project's successful completion), as this can be interpreted as a sign of a lack of control over the project. Thus, the next question that the Project Manager should ask is; **"was the undertaken activity really 'project management' or was it an extension of an implementation activity?"** That is, just how much of that time recorded against the 'project management' job code was in reality related to technical & semi-technical activities such as document review, architectural design, engineering support, or acceptance testing of the deliverables?

In addition, members of the Project Team may be using this 'project management' Cost Account Code as a catch all for any other tasks that were not specifically assigned to them. Consequently, when the project's (weekly / monthly) timesheet report is generated, it may come as a bit of a shock to find out that there has been a disproportionate amount of time booked to 'project management'.

"Yep, so much for that approved project plan that had the 'project management' hours as being only 10-15% of the project's total people hour allotment."

Recommend that, **those activities that were not really 'project management' be recorded against an appropriate Cost Account Code, such as 'technical authority support' and 'business analyses'**. Also, it may be politically wise to write-off some of those excessive 'project management' hours against the project's other tasks such as 'support' and even 'design' & 'development'.

Therefore, in addition to 'project management', it is advisable to have **additional job codes for the situation when the project management is servicing "technical" related stakeholder requests that were not planned for** as scheduled tasks in the project's [Time Baseline] schedule.

"For SDLC projects, I usually include a 'Business Analyst & Technical Authority' task of 10-15% of the implementers' total allotted time in addition to the 10-15% for project management activities."

The 'project management support' task would also be used for those situations when senior management get directly involved with the project.

- **Go back to the basics to maintain control, "To Do" lists ...**

For the management of day-to-day project matters, the use of a diary / notebook / calendar application can prove to be highly beneficial.

"I find that when I am stressed; if I write lists of all of those tasks that I have to do, then sort these into prioritized order - I get back into a mindset of being able to control the situation. As, too many tasks in my head and not written down causes me problems."

The simple reality of fact is that, **the Project Manager is just going to have to cope with continual interruptions and with spontaneous requests that will have to be reacted to in a timely fashion**. ... But, **the Project Manager does not have personally to do these actions all by themselves**; rather, the Project Manager need only orchestrate getting these actions done in a timely & efficient manner. ... "**Delegation**".

Which brings us to the next rule of "Adaptive & Proactive" SDLC Project Management.

RULE 58: **To do or not to do, or leave it to another day, ... or delegate it to another, ... that is the question.**

8.4.2. Delegating Downwards And Managing Upwards

If you have come from a technical implementer's background, then one of the problems that you may encounter as a Project Manager is getting used to the idea of delegating (the right amount of) work to others under your command, instead of doing too much yourself. ... And, then there is the art of "managing upwards".

Delegation

While [Section 4.4.1] covered delegation from the perspective of being a "humane" project manager; based on your own interpersonal traits, as the Project Manager, are you willing to trust the members of the Project Team to not only undertake their own assigned tasks as they deem necessary (without your continual oversight), ... And, are you also open to the possibility of allocating them some of your own activities that you have to get done; i.e. **will your management style incorporate Theory Y?** ... *"of Yes we can."*

OR

So as to ensure the success of the project, do you think that you would try to control & organize every aspect of the tasks that you've assigned to them, ... and, forget about giving them some of your own activities to do, as you know that you can do it better (let alone the possibility that they could screw it up)? Hence, **would your management style be based on Theory X where you feel that you need to micromanage everything that your assigned people do possibly down to the smallest details?** ... *"of No you can't."*

"IMHO, many a project manager doesn't really want to be Theory X, but they're afraid that if they don't manage that way then their people will only stuff-up. Alas, this manager doesn't understand that coaching, mentoring, and freedom (to occasionally fail) can help their people to establish their own problem solving & solution creation methods."

And, if someone hadn't been willing to delegate back in your younger days, then would you now be in your Project Manager's role (or would you have been given this opportunity to be a project manager) ??

Managing Upwards

At various points in your career you will encounter project & program managers who don't appear to be very technically astute (and possibly not even project management savvy), yet senior management think highly of them; how can this be so ??

Well, what they have going for them is a good understanding of how to. ... ... "manage upwards".

Contrary to popular belief, "managing upwards" is NOT the art of "*buttering up*", "*butt kissing*", "*brown nosing*", "*unashamed self-promotion*", nor "*yes men-ing*".

Managing Upwards is in factuality **a skill learnt by personal experience** of. ... ...

Think of it this way; of those persons in your Project Team, what can they do to help you succeed as a manager ??

(1) They can do what they **committed to do by when** they **committed to doing it**, and they can **do it to an acceptable | expectable | usable standard of quality**.

(2) They can **tell** you **what** you **need to know by when** you **need to know it**; so that, you still have time to make choice decisions about what to do, instead of being left with no alternatives but (n)one. ... And, they can **make** you **aware of the relevant facts that need to be known, without flooding** you **with superfluous information.**

(3) When **faced with** a problem come **decision to be made**, they **provide** you with **sensible alternative options, list the benefits** of each option, the **costs & time involved**, the **potential impacts** of doing & not doing each option, the **risks involved**, and they are prepared to **state what is their actual recommendation** (and not just what they think you want to hear).

(4) They **don't provide** you with (what they reasonably could have determined as) **incorrect facts, mistruths, nor outright lies**; being those things that if presented by you to your bosses (or to the Customer's Representatives), on being found out would tarnish your credibility. ... And, by telling the truth, this also means that they **state when things can't realistically be done by when these were requested / stipulated**.

(5) They **don't hide the truth** about (their part of) **the project's actual predicament**, and they **don't give "last minute surprises"** that it can't be done by when it was previously stated as being *"all okay"*; which is now contrary to the status that you last reported up the chain-of-command.

(6) They **are loyal to you and to the Project Team** so that when the going gets tough they **dig in to support you and The Team**; not leave you to defend for yourself. ... And, when things happen to have gone wrong, they **don't "throw"** you and/or The Team **"under the bus", nor** join the other party to **"lay in the boot"**.

(7) They **communicate** with you **via means that** you **require, requested, and understand**; so that, this **information can be readily passed up the chain-of-command**. Thus, they don't lump you with a hundred page document to be read through so as to construct a few summation slides in a presentation. ... And, they **don't speak convoluted** techno-babble so as to belittle yours and others technical astuteness.

(8) They **turn up to meetings on time, prepared** in advance **to participate**, and are **cognizant to the time** duration that has been allocated for the meeting. ... And, they **immediately do those resultant actions from the meeting** without being told to act.

(9) They **do those activities** that they are **responsible for with minimal supervision**, and they **don't try to palm** some of this **off** onto you **when they** themselves **should do it**. While, it's okay that they ask you to review their work, they understand that this doesn't mean that you're going to fix-up their mistakes & make the required changes.

(10) They can **self-manage their own work time** and they can **move straight onto the next appropriate task without you having to tell them** what to do. Thus, they only require a tap in the right direction, not driving them on step by step.

(11) They are **not clock-in clock-out workers** who arrive & depart at specific times of the day **irrespective of what is the current work situation**. If they need to deliver something or assist with the delivery for a pending deadline, then they are **prepared to put in the extra effort & time to get it across the line**; instead of doing the equivalent of, "*stuff it, time to go home, this can be done tomorrow*", when it actually can't wait.

Alas, **to successfully manage upwards then do those** above listed **things, upwards to your manager and to their managers so as to help them to succeed**.

Which brings us to an addendum to "Adaptive & Proactive" SDLC Project Management.

RULE 58[A]: Do unto those above you, as you would respect from those below your position in the Chain-Of-Command.

8.4.3. A Project Manager's Decorum

How one "carries oneself" will also affect how effective a project manager will be (or not). To be an effective project manager, one needs to be perceived by the project's stakeholders as:

- **In charge & in control of the project ...**

    The Project Manager has to be **"visibly seen" as** the person who **oversees the orchestration of the project's activities** and is **driving the project forwards** to delivering what they said would be done by when they said it would be done. Thus, from the stakeholders' perspectives, the Project Manager is the **"public face" representing the project.**   ... *"And, serving the stakeholders' interests."*

- **Calm & in self-control ...**

    The Project Manager has to be **visually perceived as unflappable**, **collected**, and **not panicked by** the project's **changing situation & prevailing circumstances**. Thus, from the project stakeholders' perspectives, the Project Manager should appear **unruffled** as he / she **methodically & logically works** their way **through project matters, the issues, and their concerns**.

- **Master of the project's domain ...**

    The Project Manager has to be **perceived as knowing what the project is about**, and **knowing what is going on** with the project; by **understanding** the relevant principles & **concepts** behind **project management** and the **SDLC methodology** being used, has **some understanding** of the **technology / architecture / processes & procedural basis** of the project, and is **aware of what is going on in & around the project**.

    Which brings us to an addendum to "Adaptive & Proactive" SDLC Project Management.

    RULE 58[B]: You cannot expect to manage those stakeholders who don't believe that you know what you're doing.

- **The 'GO TO' person for project related things ...**

    The Project Manager has to be thought of as the **first Point Of Contact for finding things out about the project**, and considered the "go to person" for that project.

- **Confident & trustworthy ...**

    Based on past performances and current appearances, the Project Manager's attitude has to **encourage confidence & trust** from the project's stakeholders; because, without confidence & trust, the project stakeholders will be second-guessing everything that the Project Manager and the Project Team does and has previously done. Hence, the more burdensome will be the dealing with stakeholder requests.

Hmm, coincidently this list is starting to head down the route of "building a good first impression, of PM you" in [Section 4.2.2].

- **Dress appropriately for the occasion ...**

- **Introduce yourself ...**

- **Confident, but not arrogant ...**

- **Be a good listener, not a know-it-all ...**

- **Show interest in each person ...**

- **Know the stuff that they would expect the Project Manager to know ...**

- **Quickly come up to speed**, about the project and the organizations involved ...

- **Be enthusiastic about the project ...**

    Which reiterates a previous rule of "Adaptive & Proactive" SDLC Project Management.

    RULE 35: You cannot hope to manage that which you don't know or reasonably understand what is going on.

9. PM you ... Field Test 5

9.1. Overview

Alas the final field test, and effectively the last major chapter of this textbook (project). This field test examines the following topics with respect to real-world project management:

- Project closure.
- Transitioning from project activities to BAU (Business As Usual) operations.
- Business Change Management and Benefits Realization.

9.2. Introduction

That last field test concentrated on the intricacies of the Monitoring & Control Phase, and the characteristic traits of an effective & efficient project manager; whereas, this current field test will focus on those particulars of the project that are often given minimal consideration, and sometimes, completely overlooked. These being, Project Closure, transitioning to BAU operations, Business Change Management, and Benefits Realization.

9.2.1. The Scenario

```
This project is an in-house development of an application /
system that will be used within the performing organization.
That is, where the implementers and the customer reside
within the same organization; hence, where it can be
perceived as convenient to "make small adjustments on the
fly" and to have "small changes pushed through" without the
rigmarole that is involved with dealing with external system
& service providers.
```

9.3. Project Closure

Now that the project is coming to an end, the following project closure questions, from [Section 3.7], need to be answered, preferably with a *"Yes"*, else reasons *"why not"*:

- ☐ Did **WE** (the implementer) **GET IT RIGHT**?

    - o Do the project's deliverables **conform to** the agreed **[Scope Baseline]**?

        From the ~~Customer Requirements~~ approved Detailed Specifications to being mapped in the Requirements Traceability Matrix (RTM) to the Pass : Fail tick-off in the Testing Coverage Matrix (TCM).

    - o Do these project deliverables & associated artefacts **conform to** the agreed **[Quality Baseline]**?

        The Acceptance Test Reports have been signed off as conforming to the agreed Acceptance Criteria and these test results & documentation have been retained as the project's Objective Quality Evidence (OQE).

    - o Have these project deliverables & associated artefacts been **provided within** the agreed **[Time Baseline]**?

        As per the signed off schedule and the check-off of the Milestone List.

    - o Have these project deliverables & associated artefacts been **produced within** the agreed **[Cost Baseline]**?

        Milestone activated progress payments have occurred and the associated project invoicing has been sent to the relevant parties (i.e. to the customer organization) for payment.

    - o Have the assigned **[People]** been **effectively & efficiently utilized**?

    - o Have the allocated **[Resources]** been **effectively & efficiently utilized**?

        "And, pray tell, what is your evidence that this is indeed fact?"

- ☐ Does **EVERYONE** (implementers & stakeholders) **AGREE it is ALL DONE**?
    - ○ Have all of the **implementation** activities been concluded | **closed out**?
    - ○ Have all of the **agreed baseline** constraints been **settled** "satisfactorily"?
    - ○ Have **Concessions** & **Waivers** been generated, reviewed, and approved **for** those **non-conformances**?
    - ○ Has the agreed **Objective Quality Evidence (OQE)** been produced, accumulated, and signed off by the relevant **Quality Authorities**?
    - ○ Have the **project deliverables** been **transferred** to the appropriate recipients with signed off acceptance on delivery?
    - ○ Have the agreed **project artefacts & documentation** been **handed over** to the appropriate recipients, and where appropriate, signed off as delivered?
    - ○ Has the project's **Intellectual Property** been accumulated, filtered, archived, and indexed for future reference (in an Intellectual Property Register – IPR) ?
    - ○ Is there **anything else** that is **specific to** this **industry / market-segment**, that has to be completed (and delivered)?
    - ○ Is there **anything else** prescribed by **national, state, or statutory authorities**, that has to be completed (and delivered)?

    "Are these deliverables of a professional & acceptable appearance? How these are presented (including delivery packaging) will directly reflect the [Quality] image of the project's deliverables, and subsequently the competency of the performing organization."

    - ○ Have the project's **externally procured & acquired resources** been **returned** to their authorized & approved recipients, and **usage fully paid for**?
    - ○ Have **people been timely & reasonably informed in advance that their services will not be required**, have they been told what they will be going onto next, and paid out on amounts owing to them?

- Have those persons departing done sufficient **knowledge handover** to their successor or to whoever remains?
- Have the associated **Timesheet & Cost Account records been updated** prior to close out, and have those required reports been produced?
- Have the relevant **Timesheet Job Codes** and **Cost Account Codes** been **disabled** from further use?
- Have all **residual claims** been **resolved** by when the **Final Invoice** is to be **submitted** for payment?
- Have all associated **financial accounts** been **finalized**, and **closed out** where necessary?
- Are there any **outstanding issues** that need **to be resolved**, and have these been noted in the **Risks & Issues Register**?
- Have all of the **contractual obligations** been **fulfilled**?
- Has this project's **Exit Criteria** been **satisfied**?
- Has a **Post Completion Review** been **held**?
- Has a **Project Closure Report** been **produced**?
- Has the **Project Closure Report** been **signed off** to mark **the formal end** of this project | milestone release?

And, if that seems like one big tick-off TO DO list, then that is correct.

Think of **the project's closure as being like preparing for a trip to Mars**; once it's on its way, **an** *"oops, we forgot that,"* situation can potentially have a noticeable detrimental effect on its "Business As Usual" BAU operational performance, and subsequently on the **mission's success**. ... And where, fixing it up later on may not be a viable nor affordable option.

Which brings us to the next rule of "Adaptive & Proactive" SDLC Project Management.

RULE 59: Think of Project Closure as those completion activities in preparation for a journey to the planet Mars.

RULE 59^A: Resolving forgotten "End Game" activities can noticeably affect the project running over [Time], and can result in inflated [Costs].

NOTE: Several of the outputs of the Closing Phase serve to complement "bookend" the outputs from the Initiating Phase and the Planning Phase.

Cross-Reference of the Closing Phase outputs to the Initiating & Planning Phases.

Initiating Phase	Planning Phase	Closing Phase
Project Charter	Project Management Plan	Project Closure Report
Customer Requirements	Detailed Specifications (Acceptance Criteria)	Acceptance Test Reports, Baseline Change Authorization
	Baseline Change Requests	Objective Quality Evidence
Detailed Specifications (features & functionality)	Requirements Traceability Matrix	Testing Coverage Matrix, Defect Reports

WARNING: The Project Manager is often the person who is held responsible & accountable for overseeing the closeout of the project, is the person who subsequently ties off those outstanding project activities, and ensures that things are correct for the signatories to mark the end of the project's contracted arrangements.

533

"Yet, how often has project closure been perceived as a rush to get to the exit doors; or alternatively, the project is abandoned to slowly fade away as person after person is requisitioned off to work on other activities."

- ☐ What did **WE** (the implementer) **LEARN** from this?

    - An **introspective self-assessment & reflection** as to the significant achievements, the positives & negatives, notable mistakes & failings, the standout performances & why, and how to do better next time.

    - An **extrospective assessment** as to the mutual benefits of this working relationship, and how was the **project's burden & risks distributed**.

    - Did this **project meet** the performing organization's **Selection Criteria** for acceptable projects, and **did undertaking this project make sense by providing "net positive" value** to the performing organization?

    - What were the **reasons for** the project being (un)**successful**; was it due to management, those restraints imposed, an understanding & relevant knowledge for the tasking, or the technical & people difficulties involved?

    - Does the project's **resultant outcomes** correspond to those **desired outcomes** (i.e. those reasons why the project was undertaken in the first place), and have the **benefits** been **realized**?

- ☐ Can **EVERYONE** (implementers & stakeholders) say **GOOD BYE** as friends?

    - Have the **participants been thanked** for their efforts & contributions?

    - Have **rewards & recognitions** been distributed to the **contributors**?

    - Have **final payments** been **made within** a **reasonable** amount of **time**?

    - Have **outstanding grievances** or **underlying conflicts** been **resolved**?

"When the current project ends, it is often a case of getting straight onto the next project. But, what about the benefits that the completed project provides to the business, has this benefit been realized?"

9.4. Business Change Management

9.4.1. The Effect Of Change On The Business

Now that the project's deliverables are coming closer to the reality of being handed over to the customer organization (*who happens to be the performing organisation in this field test's case*); **once the project deliverables are** "in situ" (i.e. **deployed**) then **there is going to be some resultant changes to the customer organization's business operations**.

<u>**Types Of Business Change**</u>

These resultant changes to the recipient organization's business will either be:

(a) **Strategic | "Transformational"** … of **affecting the business's major strategies** and/or **dramatically altering the operational characteristics of the business**, … even going as far as **affecting the makeup of the business's objectives**. See [Figure 167].

> *For example; changing from manual labour-intensive manufacture to automated machine production where there is upheaval to the infrastructure, to the processes & procedures used, to the skilling & education of the workers & support staff, and possibly to the numbers of employees engaged.* … "*an industrial revolution.*"

(b) **Tactical | "Transitional"** … replacement of an existing incumbent system / service / application, and/or reworked processes & procedures, that don't greatly affect the business's strategies nor its overall operational characteristics, but these changes do **affect how the business conducts some of its main activities to achieve its business objectives, potentially affecting the makeup of some groups within the organization**.

> *For example; the replacement of an incumbent system with a newer product, or a major upgrade to the current system, such as changing the Standard Operating Environment (SOE) from Windows XP / 7 to Windows 10. While there is some upheaval to the business, the underlying principles & purpose remains the same as the business continues on as before, but with some noticeable changes made.*

(c) **Continuous Improvement** ... **adjustments, patches, minor upgrades** to the existing system / service / application, and/or **revisions** to processes & procedures, where essentially it all remains the same, but "*a bit better than it was before*".

For example; a defect fix patch, a service pack, and "tweaking around the edges".

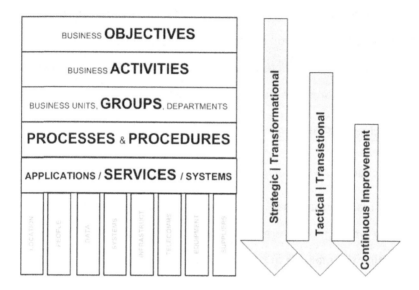

Figure 167: Top Down effect of a project's deliverables on business change.

Analysis Of Business Change

But, you being a former implementer (be that an engineer / developer / tester / analyst) ain't going to accept some qualitative declaration "opinion" of the project as Strategic | Transformational, Tactical | Transitional, or Continuous Improvement. There has to be some quantifiable way of deciding which category of business change that the project is.

How about some grading scheme as was used for Risks in [Section 8.3.1], or Stakeholders in [Section 8.3.2]. ... Could have a series of questions related to what is important to the business, and weight the answers on a 3 – 2 – 1 points scoring scale based on whether the answer is; applicable to the entire organization, business unit | department, or team; affect is high, moderate, or low; most impact, medium, or least ... then tally these scores and let the numbers denote the scale of the proposed project's effect on the business.

For example, consider a telesales company that is contemplating replacing their analogue PABX system with a digital phone system & collaboration tools.

Questions	3 – Strategic	2 – Tactical	1 – Continuous Improvement
What coverage of the organization's business strategies will be (majorly) affected?	Entire Org. / Division	**Business Unit** Customer Service, Sales & Marketing	Team
What coverage of the organization's operations will be (majorly) affected?	Entire Org. / Division	**Business Unit** Helpline Support, Telephone Sales	Team
What coverage of the organization's personnel will be effected?	**Entire Org.** Everyone uses the phone for work	Business Unit	Department / Team
What is the level of risk to the business if this change is not made?	**High** Telco switching over to all digital service after date	Moderate	Low
What proportion of the business's customer base would be affected (i.e. unavailable to be accessed)?	**Most** Given 75% sales & all support done over the phone	Some	Minimal
What proportion of the business's revenue streams would be affected if the change is not made?	**100 – 31%** Given 62% revenue is via telephone sales	30 – 16%	15 – 0%
What type of project is this change to be made?	**New** Sys, Serv, & Apps	Upgrade	Patch
TOTAL	5x3 = 15	2x2 = 4	0x1 = 0

As categorized in [Figure 167], **the project's deliverables & outcomes** can have varied effects on the customer organization, where these **effects can relate to**; the organization's **Objectives & Activities**, its systems / **Services** / applications, its **Processes & Procedures**, and possibly to the **Groups** / departmentalized structure. … And, potentially, affecting the **"behavioural"** aspects of the organization's personnel; how they interact both internally to fellow staff and externally towards the customers, clients, vendors, and suppliers.

Hence, **given the potential scale of the effects that could result from the impact of the project's deliverables & outcomes, it is due-diligent to manage the "rollout" of such changes to the business; i.e. "Business Change Management"**, as opposed to Project Change Management that was covered in [Section 3.6.7] and [Section 8.3.3].

THEORY ... *Change Management*	[Ref: Prosci® Change Management]

Well, after much nagging from your better-half, the results of your health check have come back, and it doesn't take a medical degree to know what the results are going to be; that, you have to lose weight and lower your cholesterol & triglyceride counts, else you (and your family) are "looking down the barrel" of your ailing demise.

Thus, there is an **AWARENESS of the need for change**, and the **DESIRE** of yourself to **participate in** such a **change**; and, your family's willingness to **"SUPPORT"** this change. ... But alas, there is the realities of your everyday life; the hour & half each way to & from work, the constant push-pull of work & home life commitments where meals have to be grabbed on the run, ... with excuse after excuse scrolling through your mind, as though subconsciously **"RESISTANT" to making any such changes**. Though, with the advice given, you have the **KNOWLEDGE on how to make the change**; but the question remains. ... Do you have the cap-**ABILITY to implement the required changes**; let alone the **"BEHAVIOURAL" adjustment** necessary, and "**the SKILLS & the WILL**" to carry on through with these changes. ... And, are you prepared to **REINFORCE these changes**, so that these are **"SUSTAINED"** longer than your typical New Year's resolution ??

As exemplified above, there are two **participatory components to** such **change**; the **INDIVIDUAL** being you, and the **ORGANIZATIONAL** being your family & society. Additionally, there are three **aspects to the implementation** of the change solution:

(1) **Technical** ... of the systems / services / applications being a treadmill / gym / diet.

(2) **Processes & Procedures** ... being, going to bed an hour earlier, getting up an hour earlier, then doing a daily 20-30 minute sweating walk / stride / jog, and the daily preparation of a healthy lunch & snacks for the workday ahead.

(3) **People** ... of getting the family, friends, and work colleagues on board to encourage & support your healthier lifestyle (and your efforts thereon).

And, more importantly, can you keep this healthier lifestyle up, or will you just slip back into your unhealthy ways; so, **who & how to reinforce this change over the long term** ?

9.4.2. Business Change Management Process

Due to the potential resultant changes to the business and the outcome impacts thereof, this **necessitates that these changes be prepared for** (and handled) **during the various phases of the project's life cycle**; see [Figure 168].

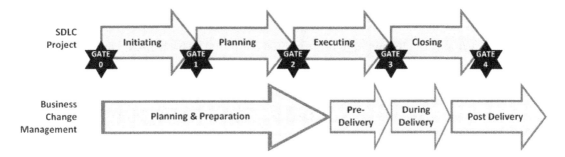

Figure 168: Relationship between the SDLC Project and Business Change Management.

In actuality, **the project's Business Change Management can be considered as a project, being a project to transition the SDLC project's deliverables & desired outcomes** (i.e. capabilities for change) **into the customer organization's Business As Usual (BAU) operations**. Where this **Business Change Management project runs parallel to the SDLC project** (as twins in a program of work); in much the same way that the customer organization (or IT Group) would have their own in-house program of projects which are preparing for the deployment of those SDLC project deliverables.

Consider that there are four **timing stages to a project's Business Change Management**:

1.. **Planning & Preparation** ... of figuring out (along time) before the change is to occur:

- o **What** potentially **will change, and what subsequently will be affected**?
- o **Who** potentially **will be impacted by these changes**?
- o **How** will these **changes affect** those identified **People**?
- o What are the **Benefits and** the **Negative Impacts** thereof?
- o What are the **Business Change Management Strategies** to be implemented?

- What are the **Communications Paths** to be utilized to detail such changes?
- Who are the **Targeted Audiences to be informed** about these changes?
- What are the **Key Messages** that have to be **imparted per audience**?
- What are the optimal **timings & frequency for this information delivery**, and who should impart these messages?
- What is the **Knowledge** (educational contents) **to be transferred**?
- What **Training** (reskilling) has to be **provided to whom** and **by who**?
- What **Support Personnel & facilities** are required to see all of this through?

As was done in [Section 3.3.4], the answers to those previously listed questions can be obtained by conducting a "Top Down Analysis", see [Figure 25] below.

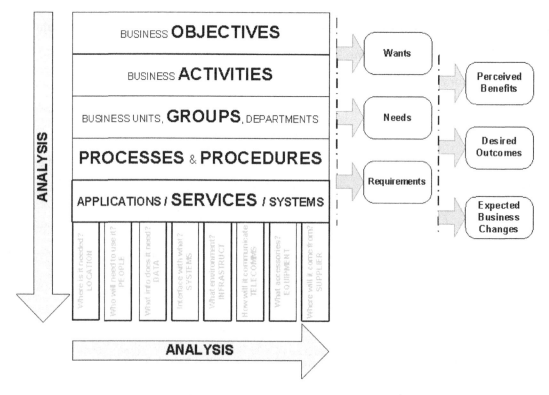

Figure 25: Top Down Analysis to determine "Wants", "Needs", and "Requirements".

Where **the "Perceived Benefits", "Desired Outcomes", and "Expected Business Changes" are the reasons why this project is being conducted**.

Thus, **what are the "Strategic Drivers" and/or the "Tactical Imperatives" for undertaking this project**, see [Section 3.3.3]; because, if this question can't be reasonably answered, then *"why the hell is this project being undertaken ??"*

As, **every project undertaken MUST provide some tangible benefit (or create a capacity for positive change) to the organization that commissioned the project**.

> *"Yet, how often at small, medium, and even at large organizations have I found that there has been a complete disconnect between the project's objectives and the realization of the benefits obtained from the project's deliverables & outcomes. That is, there are no tangible measurements that map the project's objectives with BAU operational KPIs; such as, the reduction in expenses, the reduction in the time to market, an increase in productivity, an improvement in delivered quality, etc. ... Nor, is there some orchestrated means by which to provide & verify the desired benefits to the organization."*

See Benefits Realization in [Section 9.6].

2.. **Pre-Delivery** ... prior to the project deliverables being handed over to the customer's representatives and subsequently deployed, those questions listed previously need to start being acted upon; if not by the Project Team, then at least by a Change Lead.

> *For example; awareness campaigns, status update communications, educational training (of "train the trainer"), and Go Live notification & preparation details.*

During this stage, **the Change Lead / Project Team should be on the lookout for**:

- ☒ **resistance to the pending changes**,
- ☒ **misunderstandings of what is going on and what is going to change**, and
- ☒ **concerns about what is to come**.

Hence, these project stakeholders will need to be engaged at all levels of the customer organization; i.e. "**Stakeholder Engagement**", as per [Section 8.3.2]. ... Specifically, continuous bi-directional communications, continuous collaborative effort to reconcile, and continuous implementation of strategies that maximize the positive outcomes and to minimize the negative impacts. Additionally, the Change Lead / Project Team should **seek confirmation that the business is ready for the changes ahead**; especially **ensuring that the "People Leaders" within the organization are fully on board with** participating, supporting, and encouraging **these pending changes** that are to be rolled out. ... *"As a lack of buy in is a sure way for the changes to be delivered flat, with a lack of impetus to be carried on through for the short to medium term, let alone for long term sustainability."*

Which brings us to an addendum to "Adaptive & Proactive" SDLC Project Management.

RULE 59[B]: "Change is as good as a holiday"; but, an unplanned vacation can end up being a very expensive waste of time, and not much fun for the participants.

3.. **During Delivery** ... most of the focus will naturally be on the SDLC project's delivery / "Go Live", and subsequently dealing with those stakeholders involved directly with the delivery activities; however, the other **stakeholders will need to be "kept in the loop"** as to what is going on (**as it pertains to their particular interests & concerns**).

> For example; communications about "what's happening", the stating of the Business Continuity Plan if something does go wrong, hand-holding during the switchover, and transitioning people from the previous operations to the new / revised processes & procedures for the new / revised operating platform.

And, once the project's deliverables are in-situ, the business needs to be **"smoothly" transitioned into its Business As Usual BAU operations.**

4.. **Post Delivery** ... the SDLC project is now in its **Warranty Support** stage, and is starting to **wind down to the conclusion of the project's Closure Phase**; yet, the project stakeholders will still need to be monitored & controlled so as to ensure that no unforeseen problems arise after the project has been closed out. Hence, ...

- Are the **project deliverables** being **used as** was **commissioned**?
- Do the **end-users understand** what the **deliverables can & can't do**?
- Have the end-users and/or maintenance & support staff's **educational training** "reskilling" been sufficient, or do these **need to be revised & reapplied**?
- Do the **Operational KPIs reflect** what were the **SDLC project's objectives**?

During this stage, **the Change Lead / Project Team should be on the lookout for:**

☒ a **lack of participatory support** from the affected People Leaders / end users / associated staff,

☒ has there been **lower than expected adaption** of the delivered system / service / application,

☒ is there **limited usage of** certain delivered **features & functionality**, ... and,

☒ is there apparent **disgruntlement towards the changes** to the work situation?

 WARNING: do not underestimate the impact that insufficient end-user training can have on the perception of the project deliverables as broken; hence, the customer may consider the project to be a failure, even though what was delivered was exactly what was commissioned.

Thus, **Recognize – Reassess – Revise – Reapply those stakeholder communications & education strategies**; because, now is when the project's primary stakeholders will definitively deem the project to be a Success or a Failure. ... And, their decision will be influenced by the expectations, perspectives, and opinions of those change-affected end-users (both internal & external to the customer organization).

9.5. Business As Usual Operations

Back in the theory [Section 1.1], there was a statement differentiating Project Work from Business As Usual (BAU) activities;

> "A PROJECT IS A LIMITED DURATION UNIQUE ENDEAVOUR THAT PRODUCES A ONE-OFF SET OF DELIVERABLES THAT ARE NOT BROUGHT ABOUT BY CONTINUALLY ONGOING REPETITIVE OPERATIONS; I.E. A PROJECT IS NOT A BUSINESS AS USUAL (BAU) ACTIVITY."

The Blurring of the Line Between Project Work and BAU Activities

How often has a project dragged on into what is Business As Usual operations ??

How often have former project stakeholders assumed that the finished project was still ongoing; and so, they strongly argued that their discovered non-conforming feature / functionality should be pushed through as a Defect Report, instead of as an "after the completed contractual agreement" Change Request ??

There are a **number of reasons** for this **blurring of the line between** what is **Project Work** and what is **BAU Activities**, these vary from:

(1) There is **no definitive project closure activities** conducted; such as the **Post Implementation Review (PIR)**, and then the **Post Completion Review (PCR)**, see [Section 3.7.4]. Nor is there the subsequent production of the **Project Closure Report** that would be signed off by representatives from both the customer & performing organizations to **signify THE END** of the project.

(2) The project's **Timesheet Job Codes** and the **Cost Account Codes** were **never disabled**, i.e. *"the books were never closed out"*; hence, work efforts that happen to be related to the project's deliverables continue to be booked against these codes. ... And, there

is no one left overseeing the concluded project to stringently monitor & control the usage of these codes, so these attract usage like rubbish to a vacant lot.

(3) There is **no clearly defined Warranty Support Period** for the project's operational deliverables, *e.g. 30 days from the Go Live date of the system / service / application*; hence, in-the-field non-conformances to the ~~Detailed Specifications~~ Customer Requirements are reported as outstanding Defect Repairs. ... And so, the project drags on fixing defect after defect, as the line between what is a Defect Report and what constitutes an "after contracted agreement" Change Request blurs into a *"fix it"* quest.

> **NOTE:** A non-conformance found during the warranty period is a defect; whereas, a non-conformance found after the contracted agreement has ended is a Change Request (unless it would be legally considered a Latent Defect).

(4) There is **an "all powerful" customer organization (representative) who refuses to sign off on the project's deliverables** *"due to those outstanding defects"*, and/or due to *"the continuation of field trials / evaluation testing"*; thereby, obtaining an extended free usage without having to pay up. Thus, **why the insistence on progress payments is so important for the short-term viability of the project and the long-term financial survivability of the performing organization**, see [Section 6.3.3].

Separation of Project Work and BAU Activities

At this point, the warranty period has expired and the project has been closed out, with the **project deliverables** having been **transitioned to BAU operations via** those implemented **Business Change Management strategies**. ... And, evidently the project has been deemed a success, with the customer & performing organizations' management now engaged in negotiations for another major release; i.e. a Phase Two project per se.

But, ..., ..., ..., here in lies a potential pitfall (especially for a performing organization who happens to also be the customer organization), in that there could be serious consideration given to a strategy of using the same system / service platform for both

Project Work and BAU operations. While this strategy may seem like a cost-effective plan, the reality of the implementation is that this is an expensive disaster in the making.

It should be a tactical imperative to have a clear separation between Project Work and BAU operations; as the simplest project tasking or the littlest misstep which puts the development / test platform out of commission, this could | would also render the BAU platform as inoperable. Consequently, **would the Lost Potential Cost** (of Lost Opportunity Costs plus Real Costs), see [Section 3.3.3], **exceed those "measly cost saving" of skimping on having independent project and BAU platforms** ??

Similarly for **project development and testing**, there **needs to be a clear separation of these environments**; because, for the sake of valid & reproducible **Acceptance Testing there has to be relative stability & isolated test environment**, which *"ain't gonna happen"* given the dynamic nature of development. ... And, while separating these platforms / environments for development, test, and BAU; it is also **beneficial to have a separate environment dedicated to training**, so that end-users can learn without the potential of screwing up BAU operations, nor being confused by future features & functionality that happen to be in the development | test environment but aren't yet available in daily operations.

Some risk adverse organizations will take this separation of platforms & environments even further and have two distinct BAU platforms; a primary in daily use, and a secondary ready to swing into action if there is an unavailability problem with the primary; i.e. **Business Continuity**.

Which brings us to the next rule of "Adaptive & Proactive" SDLC Project Management.

RULE 60:   BAU operations and Project Work don't always make good bedfellows; but, when given their own rooms, get on like a house on fire.

9.6. Benefits Realization

So, …, …, …, **what was the point of conducting that project, if there are no tangible benefits** (i.e. positive outcomes) **for the organization that commissioned the project** ??

How can these benefits be ~~realized~~ surmised; before all of that [Time], [Cost], [People], and [Resources] were expended on that project; hence, **what are the expected benefits to the commissioning organization** (and **to its "beneficiaries"** such as its end-users and customers), prior to committing to the project's undertaking, as well as, during the project's Initiation – Planning – Execution – Closure Phases ??   … *"A good question."*

Thus, **how to ensure that the project's deliverables, outcomes, and business changes are going to (and do) result in real benefits for the commissioning organization**; such as:

- ☐ **"Performance"** …
    [Time] & [Cost]
    - decrease in expenditures,
    - increase in revenue growth, increase in sales,
    - decrease in unproductive work time,
    - increase in productivity & production output,

- ☐ **"Satisfaction"** …
    [Scope] & [Quality]
    - improvement in quality control results, reduction in the number of dead-on-arrival manufactured units, decrease in warranty claims, decrease in customer support loads,
    - increase in customer satisfaction as per a decrease in customer complaints,
    - increase in customer base numbers, increase in customer retention, increase in ongoing customer utilization of the product / service / system,
    - increase in market share, increase in market ranking,
    - increase in brand recognition & social media penetration.
    - increase in staff retention,   … *"all tangible."*
    - increase in staff morale & participation.   … *"intangible."*

That is, **benefits that are measurable in terms of increases in [XXX] or decreases in [YYY]**; such that, there is **an increase in the GOOD "advantages"** and/or **a decrease in the BAD "disadvantages"**. ... *"These benefits should have been identified as part of the Business Case justifying the initial project approval."*

Relationship of Benefits Realization to R.I.S.C. Management

Though, what is advantageous for one project stakeholder may be disadvantageous to another project stakeholder; hence, **(ethically) there needs to be a weighing up & balancing of the GOOD versus BAD**. ... Which coincidently, brings this topic back around to **Stakeholder Management** in [Section 3.6.9] and [Section 8.3.2], of maximizing the positive outcomes and minimizing the negative impacts; where, these advantages | benefits come about by **Project Change Management** in [Section 3.6.7] and [Section 8.3.3], as well as by **Business Change Management** in [Section 9.4], and **Risk Management** in [Section 3.6.8] and [Section 8.3.1].

And, going by [Figure 169], **Benefits Realization isn't something that is done exclusively after the project finishes, nor prior to** the project's start; rather, **Benefit Realization needs to occur before, during, and after the project**. ... Else, there could be an uncoupling of why you're going to do it, while doing it, and after the fact of did it make a difference.

Figure 169: Relationship of R.I.S.C. Management to Business Change Management, and Benefits Realization.

"Yet, how often at small, medium, and even large sized organizations, have I witnessed a complete disconnect between the project and benefits realization; as there is no tracking of the difference that the project's outcomes made when compared to those benefits realized."

But, here in lies **the dilemma** for many a project commissioning organization, of … **how to define what are the expected benefits of conducting a particular project when there could be some amount of time between the conclusion of the project** and **when those supposed benefits are to be felt** ??

Thus, Benefits Realization has a time relevance aspect of **short-term, medium-term,** and **long-term**; which is, why a **Business Case should always be produced prior to committing to the commencement of any project**, so as to **quantify the benefits that could be realized versus the [Time] & [Cost] of the undertaking**. … Given that, this expenditure could be better utilized for some other projects and/or BAU activities that would provide greater beneficial returns to the organization, i.e. a better **Return-On-Investment**.

"Hence why, when proposing a project there will be some questioning by Executive Management as to the quick wins, of putting runs on the board, as well as the long-term gains. Remembering that, Executive Management are often judged by the Board Of Directors and the Company's Shareholders based primarily on what the Executive have achieved during the reporting period of the past month, quarter, & year. Thus, a project that doesn't provide short-term to medium-term tangible or evidential intangible benefits, won't be looked upon favourably."

Which reiterates a previous rule of "Adaptive & Proactive" SDLC Project Management.

RULE 15: **What is the point of doing something, if there is nothing positive to be gained (tangibly and/or intangibly).**

"And, early benefits realization also encourages ongoing support for the project."

Defining The Project's Benefits To Be Realized

How to determine the benefits of a project (especially for an enhancement / extension to a previous project), so as **to ensure that it truly does provide real benefit to the business;** other than, "*we think it will*" ??

Hence, **the project's stated objectives need to result in the production of Deliverables & Outcomes that <u>enable</u> Expected Business Changes that <u>result in</u> particular Desired business Outcomes which will help <u>to achieve</u> the realization of certain benefits to the organization**, see [Figure 170].

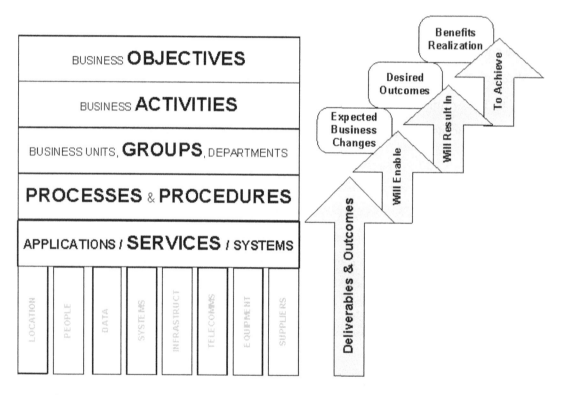

Figure 170: Relationship of Benefits Realization to Bottom Up Implementation.

QUESTION: by conducting **THIS PROJECT**, its deliverables of systems / services / applications, and processes & procedures, ... **WILL RESULT IN [AAA]** expected changes to the business, ... that **WILL ENABLE [BBB]** desired outcomes, so as **TO ACHIEVE [CCC]** benefits to be realized **as per the Business Objectives**.

Mapping Benefits Realization to BAU Key Performance Indicators

Based on a Bottom Up implementation of the project, see [Figure 170], then **how to define tangible measures (KPIs) for that question of "will enable", "will result in", "to achieve"** as indicators of the "change capability created" by the project being conducted ??
Because, **without** such **numerical descriptors, then how to judge the effectiveness of the benefits due to that particular project** ??

"At which point, the answer NOT spoken by many a project manager is, ... I don't know, I just manage the implementation of the project, ... that's for the program manager and senior management to decide."

And, while benefits realization (or rather, **Benefits Management**) **falls into the realm of Program Management**; so as to be a successful **Project Manager** by managing upwards, see [Section 8.4.2], it would be self-beneficial to **suggest relevant BAU operational KPIs that could be used to measure the resultant benefits of the conducted project.**

Whatever are the chosen BAU operational KPIs for Benefits Realization, these should:

- ☐ **Map the project's objectives to the greater Business Objectives**; which could entail **rolling the project's objectives up into the program | portfolio's objectives**, so that **these can be translated into** the business's **objectives**.

- ☐ Have these particular **BAU operational KPIs** been **mapped to** those **Key Initiatives** (KIs) that were **set out for the business to achieve as per the strategic Objectives**?

- ☐ **Will / do the project's deliverables result in outcomes & business changes that** (will) **advance the organization towards** achieving its **Business Objectives**?

NOTE: That the **emphasis is on BAU operational KPIs and not Project KPIs** (like Earned Value Performance Measurements), the reason being that the **benefits to the business's operations are independent of whether the project is/was delivered on [Time] within [Budget]**; rather, was the project delivered with [Quality] deliverables that conform to the [Scope] of features & functionality.

COFFEE BREAK DISCUSSION ...*Why benefits realization often becomes an afterthought.*

Surprisingly, or is that unsurprisingly, a project's Benefit Realization is often given very little thought by the Project Team, and especially none by the Project Implementation Team. There are several reasons for this lack of interest in Benefits Realization:

- ☒ "***It ain't my problem***; that's a Program Management / Senior Management issue, which they should've figured out before handing us this project to be implemented."

- ☒ "***How do I define what the benefits are***; that'll require a detailed analysis of the customer organization's business objectives, activities, and hierarchy; let alone, knowledge of their processes & procedures, and their systems | services | applications," ... then append the point above.

- ☒ "***Who and how will we collect the operational data, who's responsible*** for doing that, and ***who'll make the judgement*** of the operational benefits; because, that's BAU activities, and not Project Work that we're responsible for doing."

- ☒ "As far as I am concerned, those **stated business objectives** are just pie-in-the-sky, feel-good, public-relations jargon that has **nothing to do with the daily running of the business**."

- ☒ "**There is no statement of the business objectives**; that we staff know of", as these are just assumed to be understood, given the nature of the industry that the business is operating in.

- ☒ **The interpretation of** the project's **potential benefits** is ~~changed~~ **mutated as these transfer from project to program to portfolio** perspective where these are then transitioned to BAU operational capabilities and KPI measures.

- ☒ There is **inadequate or a disconnection of Business Change Management** in the organization; where, it is a hit & miss affair that the Executive Management who are responsible for projects, happens to be communicating effectively & efficiently with the Executive Management who are responsible for BAU operations.

10. The End

10.1. Adapt and Improvise, Else

Well, here I sit at the back of this early morning bus, heading off to work in a sea of stationary red taillights, as the rain comes tumbling on down. ... I come to the realization that, *"this is THE END"*, the final chapter to this textbook on SDLC Project Management.

Personally, I feel that this book is an improvement on my previous textbook, **"A Down To Earth Guide To SDLC Project Management"** (excluding all of its cartoons); just as, that 2nd Edition book (with the "spaceman" cover) was an improvement on the 1st Edition, which in turn greatly improved on the unpublished first draft print.

In my opinion, this current book that you are reading, *"is more EVOLVED"*.

Like taking the project management methodologies & techniques (and past experiences) from those earlier books, taking these out the back into a darkened alley-way, and giving them a good thrashing with fists, feet, elbows, knees, and a head-butt to a bleeding nose and a busted lip; thereby, better preparing these methodologies & techniques to handle the real-world where projects live or die based on how well the Project Team and the Project Manager are able to adapt & improvise, to the ever changing situation & prevailing circumstances.

And I suppose, that is the point of this book's theory and those field test lessons learnt.

To survive, let alone to thrive, your personal project management techniques ("PM you") **will need to evolve** as an ongoing "work in progress"; **continually adapting & modifying to better suit the current situation & prevailing circumstances** that you, your Project Team, and your employer (being the performing organization) find yourselves in for that particular project.

Noting that, **given the current situation & prevailing circumstances for each project, there will be different intermixing of [Scope] & [Quality], [Time] & [Cost], [People] & [Resources], as well as the R.I.S.C. involved with Risks & Issues, Stakeholders, and those project & business Changes to be managed.** Consequently, those project management techniques that were highly effective for one project, these same techniques may not result in optimal outcomes for another project.

> Which brings us to the next rule of "Adaptive & Proactive" SDLC Project Management.

RULE 61: Adapt and Improvise to overcome; else, expect to come undone.

And yes, you may have heard the beginning of that expression before, being the adapted mantra of the United States Marine Corps. Though, **the reality of fact is that, during your career as a project manager, your work is definitely going to change over time**; and so, **to survive in a world of potential & probable uncertainty, you will have to adapt** (preferably proactively adapting) to those ever-changing situations & prevailing circumstances.

> *For example; my project management career did not start out in the industry that I am currently working in. I started out in telecommunications R&D, where a Waterfall System Development Life Cycle was the predominate norm, with Iterative releases for the implementation, Object Oriented coding was establishing a beachhead, and Agile, well that was just a twinkling of something yet to come.*
>
> *THEN, the "dot Com crash" brought that telco-boom tumbling on down with devastating consequences; and so, along with some two thirds plus of my fellow work colleagues, it was off to join other industries where things were done slightly differently, with the emphasis placed on other constituent parts of the SDLC. By now, Object Oriented had gained real traction, and Agile was evolving from smaller & smaller duration Iterative release cycles to something of a true methodology.*

As with life, projects come and projects end; and all the while, the SDLC in use was shaded slightly differently from one organization to the next, with Burndown Charts & Task Boards being (and not being) used alongside of daily stand-up scrum meetings at various organizations, and Earned Value Performance Measures being effectively utilized at another organization but not a clue of its existence at the next organization where they didn't even capture timesheet data.

THEN, the Global Financial Crisis (GFC) came along to turn industry after industry upside-down, with the "Great Recession" slumping many surviving organizations' prosperity for the following years to come. And all the while, the SDLC used at the next organization was slightly different to those before; and so, my project management techniques needed to be tailored for the makeup of each organization's projects, for the managerial style of the organization's leadership, for that industry's practices, and for the composition of the Project Team.

Thus, it is an odds on certainty that the industry I will finish my career in, is not the industry which I am currently employed. Though, without all of these changes in industries, the ups & the downs, the good project management mentors and the not so good ones; my own personalized project management techniques would not have been forced to evolve, nor would I have gained such varied experiences.

Which brings us to the next rule of "Adaptive & Proactive" SDLC Project Management.

Rule 62: **Life is a wheel; sometimes you're on the up, and other times you're being grind down into the ground.**

Rule 62^A: **A project is a wheel; sometimes it's on the rise, and other times it's spiralling into the ground.**
The real challenge is to pick it up, and start it moving again.

10.2. PM you Don'ts

Don't Get Hung Up On The Rules; Adapt Instead

Those rules that have been seeded throughout this book are not actual rules that have been decreed from on high by some "all powerful" Project Management Body; these rules are simply reminders & thought provoking points for you, the reader. When in practicality, the Project Management Rules that will be utilized on each project | program of work, these rules will be established by; the performing organization's processes & procedures, the performing & customer organizations' business rules, the project stakeholders' opinions perceptions & expectations, the industry in which the project is being conducted, the makeup of the Project Team, and by "you" as the project's manager.

So, while it is useful to have the PMBOK® Guide, an Agile handbook, and other such project management methodologies & philosophies; the simple reality is that, **it will be "you" as the Project Manager who has to adapt & improvise on your own "PM you" techniques so as to better suit the current situation & prevailing circumstances that confront you, the Project Team, your employer, your customer, and your project's stakeholders. ... Else, expect for the project to come undone.**

"You do realize that, as author, you have mentioned 'the current situation and prevailing circumstances' three to four times, in just these last few pages. ... I think we get the point."

Which brings us to the next rule of "Adaptive & Proactive" SDLC Project Management.

Rule 63: Rules ain't rules, per se; rather, these are highly recommended advisories, that may need to be adapted to better suit the current situation and prevailing circumstances. ☺

Don't Get Hung Up On Job Titles; Learn Instead

As you transition through your project management career (and experience a PM's life) at various organizations, you will discover that what is titled as a "Project Manager" role at one organization (with a certain amount of responsibility for [People], [Resources], [Cost] budgets, [Time] durations of deliveries, and the business criticality of the [Scope] & [Quality] of deliverables); that very same Project Manager's role at another organization could be titled "Senior Project Manager / Program Manager / Program Director".

While such job titles may look impressive on your resume / curriculum vitae, the simple reality is that, no matter what your job title; **if you don't know & understand the basics of project management and the underlying principles & processes of the system development life cycle** (and the sequence of the constituent parts), **then you are going to look "a bit of a fool" to those who do know & understand these things**.

"Where your subordinates and rank-equals, behind hand covered mouths, will joke and mock at your lack of knowledge and understanding, of those things that they consider, in your job role, you should really know by now. Thus, the expression, "promoted to the level of their incompetence."

Yes," *how embarrassing, nay, pathetic*" is it for a Senior Manager of an SDLC organization, not to understand the relatively simplest of project / program schedules; let alone, being unable to interpret Earned Value Performance Measures and S-curve plots as it relates to the Level Of Effort, Management Reserve, and Profit Margin (as demonstrated in [Section 5.4.2] and [Section 5.4.3]). Because, irrespective of how well that manager can "talk-the-talk"; you can be assured that, from the program manager on down in the organization, there is behind-covering-hands mocking & ridiculing of that manager having *"never passed Project Management 101"*.

> **RULE 64:** When it all goes wrong, people will quickly turn away from who has the title, to who can fix the situation.

Don't Get Hung Up On Paper Qualifications; Understand Instead

In the pursuit of respectability, a manager may collect certificates, degrees, masters, and doctorates until their business card's name-ending acronym list of qualifications looks like a convoluted chemical equation; yet, when it comes to the practical application of this learning, they evidently didn't take in the real purpose of all of that study. ... **It is not the collection of acronyms; but rather, the collection of usable knowledge, understanding, and experiences that are of the utmost importance to project success.**

Hence, don't be surprised when someone with umpteen (boasted) qualifications to their name is behind-covering-hands mocked, when all of those acronyms can't be sufficiently utilized to deal with the realities of the current situation & prevailing circumstances that are failing the project / program / organization.

Thus, I would advise that you **base your "PM you" techniques around the principle of continual learning, and a drive to understand**; so as to be, a more effective & efficient Project Manager. ... And coincidently, many of the best lessons learnt in your PM career won't come from textbooks ☺, classrooms, one-week intensive training courses, nor studying for exams; rather, **knowledge & understanding comes from actual experience, and with consultation & mentoring from veterans in their field of expertise**.

"In much the same way that a graduate straight out of university, while having studied so much, has to be hands-on retrained and mentored to become useful for the actual work to be done."

And, **at some point in your career, it will come time for you to repay all those who taught you a thing or two; therefore, mentoring those who come after you.**

Which brings us to the next rule of "Adaptive & Proactive" SDLC Project Management.

RULE 65:   What will you learn today ??
What will you teach tomorrow ??

"And, at some point in your PM career, possibly when you have been at one particular organization for a while, you may come to the realization that you're not really learning anything new, as the work that you are doing is just a variation of the taskings before that and before that. Consequently, you may start pondering whether a stable position, a nice job title, and a reliable pay packet at your current employer is lacking in the impetus that you need to get up in the morning. Such impetus could possibly only come about with new challenges at another organization."

Don't Make Yourself A Target For Behind-Covering-Hands Mockery

Let's be truly honest, ... people can be cruelled (with their words), even to their fellow kin, be that; to their co-workers, to their rank equals, to the management hierarchy above, and to those they consider as subordinate to their own position. While these **behind-covering-hands jabs** of *"just joking"* **may not be intended to cause harm**, as it was *"only in jest"*; it is **human nature to isolate the different, the weak, those perceived as potential threats, and to bound together as a collective common-interest group** (possibly at the expense of others).

The previous subsections mentioned mockery over a lack of knowledge & understanding; this is not the only reasons that people, especially fellow staff, will mock a manager for. Think about your current work situation, your previous work, and **what you know was said (and what you said) behind-covering-hands**; now, consider that **those same things could & would be said about you**, given the particulars of the situation & circumstances.

- ☐ The manager, who is sending out emails (and work instructions) in the dead-of-night; for how long will they think of you as dedicated to your job? ... OR, will they think, *"are you @#$% serious"*, when that message notification pings them at 1 AM?

- ☐ The manager, who reads umpteen reports over the weekend, and spends their evenings writing work documents; for how long will they think of you as a hard worker?... OR, will they consider you as someone who doesn't have a life outside of work?

- ☐ The manager, who is the first one at their desk in the morning, and is often seen as the last one to go home at night; for how long will they think of you? ... OR, not, as they're thinking about their own family, their own social life outside of work.

While, these examples maybe seen by hierarchical management as dedication & promotion worthy; too many of the staff at the organization, such a PM's behaviour is (subconsciously) perceived as a direct threat to their work-home life balance, because it could set unrealistic expectations on work norms, when they do have a life outside of work that they have to contend with.

And, let's not start on the behind-covered-hands mockery worthiness of your personality quirks, the peculiarities of your dress sense, the way that you move, the way that you speak (your accent), and your favourite catchline phrases.

Which brings us to an addendum to "Adaptive & Proactive" SDLC Project Management.

RULE 65[A]: What will you do and say today ??
What will others say tomorrow ??
What will you pay for, later on ??

Don't Climb Snakes, And Fall Down Ladders

See Office Politics.

10.3. Office Politics

While the following is not what you would expect to find in a textbook on SDLC Project Management; **"Office Politics" is a reality of fact when working in an organization**. … And thus, **Office Politics is something that you, as the Project Manager, will have to come to grips with, even utilize yourself (in a good way)**; else, you and your Project Team will be pushovers, and possibly be used as cannon-fodder for others' self-interested gains.

The term **"Office Politics"** often has negative connotations of meaning, **the act of manipulating others so as to advanced one's own vested interests, potentially at the expense of others**. Where, possibly these self-interests could be contrary to the best-interests of the organization with whom "the Manipulator" has been employed.

> Which brings us to the next rule of "Adaptive & Proactive" SDLC Project Management.
>
> **RULE 66[6]: Office Politics is a fact of life; so, learn to deal with it.**

As an implementer, you were relatively insulated from the political intrigue, scheming, and self-advancement manoeuvring that occurs behind the scenes in organizations; especially occurring within medium to large sized corporations, where the Manipulator's tactics can be concealed by the happenings & goings on of the many people within the organization.

However, by moving across into project management, you will now be exposed to the political machinations of those within the management hierarchy; potentially becoming prey for the empire builders, the narcissists, and the corporate psychopaths. Where answering such an innocuous question as, *"what's your honest opinion on"* something (or of someone within the organization) could inadvertently place you on the wrong side of the border in the Manipulator's political games; … OR, you could be unknowingly adding to the Manipulator's information arsenal that will be utilized for your pending detriment, your Project Team's detriment, and/or to someone else's detriment.

Information, The Key To Office Politics

For "Office Politics", **the primary key to success is the acquisition & control of information; and, the subsequent "management of the flow of information" to whom**. Hence, **a critical part of the Manipulator's strategy is to obtain access to the information that they may not already know; and, to moderate (and manipulate) the information that they wish others to know & not know**.

Thus, what you as an implementer would have previously considered as *"just friendly conversation"*, this is **one of the means by which the Manipulator gains access to the information which could prove useful** to them. Hence, the Manipulator will deploy such conversational techniques as; lowering themselves to your level with the pseudo mateship tone of their words, with false platitudes & empty complements, and literally, with a pat on your back. ... And, in the worst cases, a touch or a look which treads a fine line on the verge of sexual manipulation (or an artificial "bro-mance").

With the information extracted, the **Manipulator is then able to coerce / influence / manipulate you and/or others, so as to advance their own self-interested agenda**; such as, **strategically positioning themselves and/or their allies to some self-beneficial end**.

OR, the Manipulator may use this information to **pacify a person who may potentially interfere with their plans**, and/or **realigning someone to their side of a pending engagement** with someone else, and/or to **disadvantage & neutralize** you and others **who happen to be restricting their progress towards their long-term goals**.

 I would recommend that, you **don't engage in work related gossip**; because, this will fuel the potential for negative office politics. ... Also, **don't bitch & moan, and emotionally complain** about the situation and/or circumstances of your work, or worse, **don't tell of your personal weaknesses**; as this, will simply add to the stockpile of information that the Manipulator can utilize against you, and against others who you happen to mention or refer to.

I would recommend that, you **keep the topic of such conversations focused on the betterment of the organization and the improvement of the business at large.** ... And, **only state those things that you are prepared to have fully exposed to the general populous of the organisation** (with you as the known source of such information).

Which brings us to an addendum to "Adaptive & Proactive" SDLC Project Management.

RULE 66^A: Always expect that at some time, what is "confidential information" will eventually be revealed.

Information, Has Time Limited Powers

As the saying goes, *"knowledge is power"* / *"INFORMATION IS POWER"*; or rather, **the acquisition & control, and the subsequent manipulation, misrepresentation, misdirection, and suppression of the information flow is the basis for power**.

However, *"INFORMATION HAS A TIME LIMITED USEFULNESS"*; where, today's supersensitive & confidential information, is tomorrow's old news which no one cares about anymore. Therefore, **the information that the Manipulator seeks to acquire**, this information **usually has to be utilized in a relatively short amount of time so as to maximize its effectiveness for the Manipulator's short-term gains**. ... Unless, that information is being stockpiled for future underhanded blackmailing purposes; and, this type of blackmailing strategy is utilized by a different kind of "Office Predator", whom you really don't want to be entangled with for longer than is necessary to you.

Which brings us to an addendum to "Adaptive & Proactive" SDLC Project Management.

RULE 66^B: To a Machiavellian, *"the ends justifies the means"*, as *"all's fair in love and war"*, that is Office Politics.

What Motivates the Office Politician, To Be The Manipulator

It's fair to say that **what motivates the Office Politician**, is the same things that motivate local / state / federal politicians; that being, their **underlying premise that they are seeking to improve the situation & circumstances for those constituents who they represent, in accordance with, what they personally believe is best (for all of those involved)**.

Unfortunately, other aspects of **human nature often kick in** with the injection of personal **wants, needs, and greed**; hence, the **desire for more power & influence over others** to get things done as they believe these things should be, the **desire for more resources to control** so as to aid in / enforce their power & influence, the **desire to rise up in social standing** via promotion in title & rank, the **desire to be better remunerated** for all of that responsibility which they have amassed, and the **desire for a better standard of living**.

> Which brings us to an addendum to "Adaptive & Proactive" SDLC Project Management.
>
> **RULE 66^C:** An organization is a mini nation, or a confederation of nations; hence, the Office Politician has the same desires and goals as does the nation's politicians.

In much the same way as with a nation, **an organization has limited resources, has limited available capital budget, and has limited personnel on hand**; with certain things to be accomplished in a limited amount of time (i.e. within the project / program's duration or by the next performance review & promotion cycle, compared to the next election cycle). **Hence, the Office Politician has to compete with other department heads, group leaders, and managers, so as to obtain access to the [People], [Resources], [Cost] budgets, and [Time] that they believe necessary to be able to successfully deliver the [Quality] of [Scope] that they are responsible for delivering**. ... Else, they fear that they will fail to achieve their assigned objectives; let alone, failing to achieve their own personal goals.

Office Politics, Just Another Form Of Stakeholder Management

Well, this topic on **Office Politics** has been a bit of a depressing read; just as, it **can be depressing to deal with in the real world of working in some organizations**. ... And, Office Politics is especially **frustrating for the Project Manager who simply wants to get on with successfully completing the project**; yet, feels that they are being hamstrung by certain people in the management hierarchy who are *"more interested in playing political games"* for their own self-promotional & self-advancement purposes.

> Which reiterates a previous rule of "Adaptive & Proactive" SDLC Project Management.

RULE 66: Office Politics is a fact of life; so, learn to deal with it.

Hence, as the implementer that you once were, if this Office Politics was a technical problem to be overcome, then what would you do ?? ... **Identify**, **Assess**, **Analyse**, **Devise** strategies & plans, **Implement** those strategies & plans which were deem viable & practical, **Monitor & Control** the outcomes, and **Repeat** the process until the situation was resolved satisfactorily.

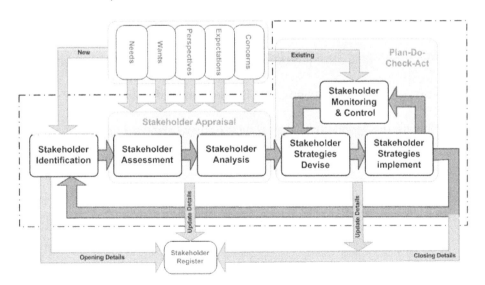

Figure 69: Stakeholder Management Process.

Thus, those Office Politicians are just another breed of project stakeholders who you will have to manage; as per Stakeholder Management, in [Section 3.6.9] & [Section 8.3.2].

1.. **Identification** … "*shut-up, watch, and listen*" to how people within the organization (publically & privately) act & react to each other, what is said & not said when given the current situation & prevailing circumstances.   … *"And, you can't do that, if you're the one doing all of the talking."*

2.. **Assessment** … (privately away from view of anyone else in the organization), take the published Organization Chart, and lay the "players" out on a table (at home); and, like in those detective TV shows, pin each one of their "mugshots" up on the wall. … *"Maybe not literally pinning them to the wall, even if in some cases this would definitely feel satisfying and be highly deserving."*

Based on your observation of the comings & goings on in the organization, reposition each of them into "groupings of power", then move them about with lines forming the hierarchical relationships; while, answering the following questions:

- ☐ (Irrespective of what the published Organization Chart depicts), who is actually reporting to whom; who is subservient to whom?
- ☐ Who is respected and why; who is rejected / ignored and why?
- ☐ Who is routinely communicating with whom; who is not talking with whom?
- ☐ Who gets along with whom; who is in known conflict with whom?
- ☐ Who are the "good guys" that appear to be working for the benefit of the organization; who are the "baddies" that appear to be only in it for themselves?
- ☐ Who is on the rise; who is on the fall?   … *"You can judge this by how many people are they responsible for, or what portion of the business resides within their domain."*
- ☐ Who has the real power of influence in the organization?

3.. **Analysis** ... based on your identification & assessment of those "players" within the management hierarchy; who will you need to make use of as allies & associates to help ensure that the project and the Project Team are successful (for the beneficial gains of the organization)? ... And, who will you have to interact with, even though to you, they appear to be operating primarily with a self-motivated agenda?   ... *"And, if so, what are the drivers for this Manipulator? Because, if what you're trying to achieve for the benefit of the business can coincidently aid or at least not hinder the advancement of their self-interested agenda, then they could become a willing contributor to helping you achieve what is right for the organization."*

4.. **Devise** ... what is the plan & strategy to **build up your own "social network"** (given those existing relationships) with those people that you have identified & assessed as being advantageous and necessary for the success of the project (and for the organization to benefit).

5.. **Implement** ... engage with these ~~"players"~~ "Organizational Stakeholders" as per the techniques suggested for Stakeholder Management, in [Section 8.3.2].

I would recommend that, you **follow up all verbal interactions with the project's "Organization Stakeholders" with a summation email (possibly CCing your immediate supervisor) to reiterate the key points of the discussion and to outline what are the actions requested of you and/or your Project Team; thereby, establishing a formal record of what was requested.**

As it's an odds on certainty that, if there are dubious schemes at hand by a Manipulator, then they will be asking you not to mention such discussions in emails. ... And, if it can't be recorded in an email just between you, them, and your supervisor, then this is a request that you definitely should reconsider doing, as evidently the desired outcome ain't primarily in the best interests of the organization.

COFFEE BREAK DISCUSSION … *The down side of the "screw-em-over" rise to the top.*

As with any quick rich scheme, there is the possible consequential flipside to playing Office Politics, in that; if the person or group who were "screwed over" by the Manipulator, do happen to realize what has happened to them, then the victim may:

(a) go inform someone of authority in the organization's hierarchy, as to who has been using unethical business practices, and/or

(b) the next time that the Manipulator makes a request of them, they could be a lot less forthcoming with providing the requested information, and/or

(c) the activity that the Manipulator has asked them to do, they could totally ignore or do it with as little effort as they can get away with.

Hence, this type of deliberate "self-advancement at others' expense" strategy has diminishing returns; and so, the Manipulator has to maximize the use of such techniques, then quickly move on, else they risk being exposed or boxed in.

Thus why, the "Master Manipulator" will strive to be quickly promoted up through the management ranks; until, they hold a position that is as far as they can go at that particular organization (or until they get the inkling that *"the writing is on the wall"* for them). Whence, they will suddenly depart for another organization (or to another business unit within the corporation); where, they can resume their political maneuverings, so as to further advance their resume credentials & pay packet.

And, what happens when the Manipulator has boxed themself in, or they have limited upward prospects remaining; they will be prepared to turn on those around them who could potentially threaten their power base and/or the retention of their position in the hierarchy. Subsequently, the business unit / department / group that the Manipulator is responsible for, could end up having a relatively high turnover of staff (and/or a high proportion of "drone workers" without opinion or individual thoughts); where, those with real potential who could actually benefit the organization, these people are either pushed out or leave of their own accord due to limited opportunities afforded to them.

10.4. Is Being A PM Right For You?

Before you decide to throw your hat into the ring and take up the challenge of project management, there are a few questions that you personally should give serious consideration to, (especially given your work & home life situations & circumstances); else, you could risk noticeable failure (at work or at home) and/or being unhappy with the project management career that you have chosen.

So, before you make the jump from an implementer to a project manager, have an introspective look at yourself, and honestly answer these following questions:

(1) **Why ??**

(2) **Why have you decided that project management is the next step for you?**

Is this something that you really want to do for yourself, because being an implementer *"no longer feels right"* for you; OR, is this something that you're expected to do, so as to advance your career at your current employer's organization (or at least to advance in terms of your salary)?

(3) **Are you willing to accept that some of what you have already learnt & done as an implementer will not be transferable (let alone applicable) to a new career in project management?**

Sadly but true, your copious amount of technical knowledge, skills, and experiences may not necessarily mean that you will be a competent project manager.

(4) **Are you going to be comfortable saying *"we made this"*, instead of *"I made that"*?**

As the Project Manager, you will no longer be the one who works through those technical problems; rather, other people will work on these types of things, while you strive for a beneficial business resolution. That is, **your role as the Project Manager will no longer be *'TO DO'*; rather, the PM's role is to make sure *'IT GETS DONE'*.**

(5) **Are you really going to be comfortable with giving others the implementation reigns, while you remain hands-off?**

Are you sure that you can accept this demarcation of roles; OR, are you the type who on seeing that they are not doing it as you would (as an experienced implementer, that you were), are you likely to push them aside with a, *"get out of the way"*, as you steamroll on through with an, *"if you want something done right, you have to do it yourself"* attitude.

(6) **Are you really-really going to be comfortable with entrusting your project's assigned implementers with the freedom to do their work in a manner that they deem as appropriate & necessary to get their tasking done?**

Hence, will your management style be based on Theory Y (of *"YES we can"*); OR, would you be more Theory X (of *"NO you can't"*) where you would feel the need to micromanage everything that your implementers do & did?

(7) **Are you prepared to go from** (what was) **the top of the technical pile to** (possibly) **being at the bottom of the management heap?**

As an implementer you were respected for your skills, your local knowledge of the work practices, your years of technical experience & expertise; however, once you move across into project management (especially at a new organization), the GAME RESETS, and you will be starting at the bottom rung of the ladder, as a mere junior project manager. So, can you swallow your pride, and start learning all over again?

(8) **Are you really-really prepared for all of that good technical work that you have previously done, to be put aside and to start the long climb back up the "earn respect mountain" again?**

As that technical field could, in retrospect, start looking very appealing once you're in the toing & throwing of Stakeholder Management, and having to rework the proposed project's plans, schedule, and the possible budget costings again, ... and again, ... and.

(9) Are you prepared to be first in line for retrenchment, if things do go financially wrong at the organization?

By moving across into project management, you have repositioned yourself as a junior. And as a junior, you are no longer considered as essential personnel; hence, a financial crisis could mean that you are one of the first to be let go.

(10) Are you prepared to possibly carry the destiny of your employer's organization on your shoulders, and are you willing to potentially hold the fate of your colleagues' livelihood in your hands?

As Project Manager, you will be responsible for people, resources, and budgets of delivering on time; how will you handle the situation where it all goes wrong and your employer's survival necessitates that some of your colleagues and/or your Project Team who has to be let go. ... Especially when, it is you who has to advice on who should be retrenched, or it is you who (colloquially) "has to pull the trigger"?

(11) Just how far up the management hierarchy, do you really want to go?

As the higher up the management hierarchy that you progress, then the more of a "Project Accountant" and "Office Politician" you will have to become; and, the less & less involved you will be with understanding the technical aspects of the [Scope] & [Quality] (resolving those intriguing technical problems). ... As more & more, it will become the routine of balancing [Time Baseline] milestones versus budgeted [Costs Baselines] versus the RISCs, while shuffling the interplay of available [People] & [Resources]. ... And, ever so politely dealing with those persons who have their self-interested agendas and self-advancement manoeuvrings in play.

(12) Are you prepared for your morals to potentially be at odds with the business ethics of those above you in the management hierarchy?

As all it takes is a financial crisis or sustained economic downturn for someone above you in the management hierarchy to push past the boundary of your sense of business ethics and your personal morals? ... So, how will you be able to look at yourself in the bathroom mirror, when you know that what you are being asked to do is, questionable at best, let alone, possibly outright fraudulent?   ... *"There have been occasions when I have stood my moral ground; and, faced the consequences. Then much later on, I heard via the grapevine, that they eventually ended up facing the consequences of their unethical actions."*

(13)   Just how much of your personal time are you prepared to commit?

You do realize that, your decision to become a project manager (and eventually a higher management rank) could mean that in addition to your day job, you will also have a night job of catching up on that work which you didn't get done during the day?

(14)   Just how much of your spouse's | partner's time, and your family's "quality time" are you prepared to commit?

As your decision to move into project management doesn't just affect yourself, it will also affect your spouse | partner, your family, and your friends due to possible limitations placed on your available time.

Noting that, these home-life stakeholders are not going to be prepared to put their lives on perpetual hold while you attend to the latest project. ... So, **how will you cope with the situation & circumstances of when the time available to you has competing commitments of work and home life; what will you miss, and what will you gain by choosing the other?**

> Which brings us to the next rule of "Adaptive & Proactive" SDLC Project Management.

RULE 67:   Everyone has two parts to their life, work and home; when one of these imposes too much on the other, then eventually there will be negative consequences that someone will have to pay. ... Most likely, it'll be you.

11. References & Index

References

- Boyde, Joshua (2014), "A Down-To-Earth Guide To SDLC Project Management" 2nd Edition.
- PMI (2008/13), "A Guide to the Project Management Body of Knowledge (PMBOK® Guide) 4th/5th Editions", Project Management Institute, PA, USA.
- www.Wikipedia.org, the Free Encyclopaedia. Wikimedia Foundation, Inc.
- Belbin, Meredith. (1981), "Management Teams", London, UK.
- Clarkson, M. (1999), "Principle of Stakeholder Management", Toronto, Canada.
- Cleland, D. (1998), "Stakeholder Management", New York, USA.
- Deming, W. Edwards (1986), "Out of the Crisis", Cambridge.
- Herzberg, F., Mausner, B. & Snyderman, B.B. (1959), "The Motivation to Work", New York, USA.
- Kniberg, Henrik (2007), "Scrum and XP from the Trenches", USA.
- Kniberg, Henrik & Skarin, Mattias (2010), "Kanban and Scrum, making the most of both", USA.
- Lewin, K.; Lippitt, R.; White, R.K. (1939). "Patterns of aggressive behavior in experimentally created social climates", USA.
- Maslow, Abraham (1954), "Motivation and Personality", New York, USA.
- McGregor, Douglas (1960), "The Human Side of Enterprise", USA.
- www.prosci.com, Prosci Change Management. Prosci Inc.
- Kilmann, Ralph; Kenneth W. Thomas (1977). "Developing a Forced-Choice Measure of Conflict-Handling Behavior: The "MODE" Instrument". Educational and Psychological Measurement
- Tuckman, B. W. & Jensen, M. A. (1977), "Stages of small-group development revisited", USA.

From my past experiences & lessons learnt at those various employers; as well as, from those dark crevasses of my mind, wherever I recalled this stuff, while being imprisoned on another motionless bus, stuck in another meaningless traffic jam, as I head off to work ... ARRRRRRR !!!

Index – Grid

Key Word	Theory	Project Phases	PM You	Field Tests
Acceptance Testing		3.5.5, 3.7.3,		6.3.3,
Agile - SDLC	2.3.3,	3.5.3, 3.5.4, 3.5.6,		7.3.4, 8.3.1,
Baseline, (Project Constraint)		3.4.7, 3.4.10, 3.6.7, 3.7.3,		5.6.2, 9.3,
BAU Business As Usual Operations				9.5,
Benefits Realization				9.6,
Budgets		3.4.5, 3.6.5, 3.7.3,		5.4.2, 5.4.3,
Burn Rate	3.6.5,			
Burndown Chart	3.5.6,	3.5.6, 3.6.10,		7.3.4,
Cash Flow	3.6.5,	3.3.7,		
Change Control Board		3.6.7,		8.3.3,
Change Management (Business)				9.4,
Change Management (Project)	2.6.3,	3.4.9, 3.6.7,		8.3.3,

Key Word	Theory	Project Phases	PM You	Field Tests
Change Requests		3.6.6,		5.6.2,
Closing Phase		3.7, 3.7.7,		9.3,
Communications Management		3.4.8,		8.3.2,
Completion Review		3.4.10,		
Conflict Resolution	4.4.1,		4.4.1,	
Contracts	3.3.7,			6.3.2, 6.3.3, 6.3.4,
Cost - Project Variable		3.4.5, 3.6.5, 3.7.3,		5.4.2, 5.4.3, 6.3.3, 6.3.4, 7.2.3,
Cost Of Quality		3.6.6,		
Customer Requirements		3.4.3,		5.5.3,
Customer Satisfaction	2.8.2,	3.7.3,		7.2.3,
Defect Reports & Repair		3.6.6,		8.3.3,
Delegation				8.4.2,
Detailed Specifications		3.4.3,		5.5.3,

Key Word	Theory	Project Phases	PM You	Field Tests
Earned Value Performance Management (EVPM)	3.6.5, 5.4.3,			7.3.5,
Executing Phase		3.5,		7.2.2,
Failure - Dealing With				5.6, 5.6.3, 5.7.1,
First Impressions			4.2.2,	5.2.2,
Gating	3.2.2,	3.3.9,		
Governance	3.2.2,			7.3.6,
Hybrid - SDLC	2.3.4,			
Initiating Phase		3.3,		6.3.2, 7.2.2,
Intellectual Property				5.3.2,
Interviewing			4.2.3,	
Iterative - SDLC	2.3.2,			5.3.1,
Key Performance Indicators (KPI)	3.6.5,			7.3.5, 7.3.6, 9.6,

Key Word	Theory	Project Phases	PM You	Field Tests
Leadership			4.4.2,	
Lessons Learnt		3.7.5,		6.2.1,
Management Reserve	3.6.5,	3.3.6,		5.4.2,
Management Style	4.4.2,		4.4.2,	
Managing Upwards				8.4.2,
Maslow's Hierarchy Of Needs	4.4.1,			
Monitoring & Control Phase		3.6,		
Motivating	4.4.1,		4.4.1,	
Office Politics				10.3,
Payback Period		3.3.3,		
Payment Schedule, Progress Payment		3.3.7,		
People - Project Variable		3.4.4, 3.6.10, 3.7.6,		
Performance Index				5.4.3, 7.3.5,

Key Word	Theory	Project Phases	PM You	Field Tests
Performance Measures	2.8.2,	3.7.3,		7.2.3, 7.3.3, 7.3.4,
Plan Do Check Act	2.4,	3.5.1, 3.5.2, 3.5.5, 3.6.6,		
Planning Phase		3.4,		6.3.2, 7.2.2,
PMBOK	2.8.1, 3,			
Post Completion Review (PCR)		3.7.4,		
Post Implementation Review		3.5.8,		
Profit Margin		3.3.6,		5.4.2,
Progress Payments		3.3.7,	5.3.2,	6.3.3,
Project	1.1,			
Project Autopsy		3.7.5,		5.7.2,
Project Charter		3.3.6,		5.3.2,
Project Constraints, (Baselines)	1.2,			5.3.2,
Project Cost				5.4.2,

Key Word	Theory	Project Phases	PM You	Field Tests
Project Documentation	3.2.1,			
Project Implementation Team	2.3.4,			
Project Management Process	2.8, 3.2,			
Project Manager - Characteristics			4.4.1,	8.4.3,
Project Recovery	2.5.2,			5.5.2,
Project Reporting				7.3.6,
Project Rescue	2.5.1,			5.5.1,
Project Steering Committee	2.3.4,			
Project Variables	1.2, 2.7,	3.4.7,		
Project Working Group	2.3.4,			
Quality - Project Variable		3.4.8, 3.6.6, 3.7.3,		7.2.3,
Quality Assurance	3.5.4,	3.5.5,		
Quality Control	3.5.4,	3.5.5,		

Key Word	Theory	Project Phases	PM You	Field Tests
R.I.S.C.	2.6,	3.4.9,		8.3, 9.6,
Recognize Reassess Revise Reapply	2.5.1,		4.3.2,	8.3.1,
Recruitment			4.2.3,	
Reporting				7.3.6,
Requirements Traceability Matrix (RTM)		3.7.3,		5.5.3, 6.3.3,
Resources - Project Variable		3.4.4, 3.6.11,		
Risk Management	2.6.1,	3.4.9, 3.6.8,		5.3.2, 8.3.1,
Risk Matrix		3.6.8,		8.3.1,
Risks & Issues Register				8.3.1,
Rolling Wave				5.3.1, 5.6.1, 5.6.2,
Safety Margins		3.3.6,		5.4.2,
Schedule		3.4.4,		5.4.1, 5.5.1, 6.3.3, 7.3.1, 7.3.3, 7.3.4

Key Word	Theory	Project Phases	PM You	Field Tests
Scope - Project Variable		3.4.3, 3.4.4, 3.6.3, 3.7.3,		5.5.3, 7.2.3,
S-Curve	3.4.10, 3.5.6, 3.6.5,			5.4.3, 6.3.3, 7.3.2, 7.3.5,
SDLC - Life Cycle	2.3,			
Sprint Review Meeting		3.5.8,		
Stakeholder Analysis				8.3.2,
Stakeholder Management	2.6.2,	3.4.9, 3.6.9,		8.3.2, 8.4.1, 10.3,
Stakeholder Matrix		3.6.9,		8.3.2,
Stakeholder Strategies				8.3.2,
Stand-up Meeting	4.3.2,	3.5.6,	4.3.2,	
Stress - Work Related Stress				5.6.3,
Task Board		3.5.4, 3.6.10,		7.3.4,
Team Building	4.3.2,		4.3, 4.4.1,	
Testing Coverage Matrix (TCM)		3.7.3,		6.3.3,

Key Word	Theory	Project Phases	PM You	Field Tests
Time - Project Variable		3.4.4, 6.3.4, 3.7.3,		7.2.3,
Topic Down Analysis		3.3.4,		6.3.1, 9.4,
Waterfall – SDLC	2.3.1,			5.3.1, 5.6.2,
Work Breakdown Structure		3.4.3, 3.4.5,		5.5.3,

Made in United States
North Haven, CT
08 April 2023